Mathematical Research

Multivariable Control

Kurt Reinschke

Band 41

AKADEMIE-VERLAG BERLIN

In this series original contributions of mathematical research in all fields are contained, such as

— research monographs
— collections of papers to a single topic
— reports on congresses of exceptional interest for mathematical research.

This series is aimed at promoting quick information and communication between mathematicians of the various special branches.

In diese Reihe werden Originalbeiträge zu allen Gebieten der mathematischen Forschung aufgenommen wie

— Forschungsmonographien
— Sammlungen von Arbeiten zu einem speziellen Thema
— Berichte von Tagungen, die für die mathematische Forschung besonders aktuell sind.

Die Reihe soll die schnelle Information und gute Kommunikation zwischen den Mathematikern der verschiedenen Fachgebiete fördern.

Manuscripts
in English and German comprising at least 100 pages and not more than 500 pages can be admitted to this series. With respect to a quick publication the manuscripts are reproduced photomechanically. Authors who are interested in this series please turn directly to the 'Akademie-Verlag'. Here you will get more detailed information about the form of the manuscripts and the modalities of publication.

Manuskripte
in englischer und deutscher Sprache, die mindestens 100 Seiten und nicht mehr als 500 Seiten umfassen, können in diese Reihe aufgenommen werden. Im Interesse einer schnellen Publikation werden die Manuskripte auf fotomechanischem Weg reproduziert. Autoren, die an der Veröffentlichung entsprechender Arbeiten in dieser Reihe interessiert sind, wenden sich bitte direkt an den Akademie-Verlag. Sie erhalten dort genauere Informationen über die Gestaltung der Manuskripte und die Modalitäten der Veröffentlichung.

K. J. Reinschke

Multivariable Control

Mathematical Research · Mathematische Forschung

Wissenschaftliche Beiträge
herausgegeben von der
Akademie der Wissenschaften der DDR
Karl-Weierstraß-Institut für Mathematik

Band 41

Multivariable Control

by

K. J. Reinschke

Multivariable Control

A Graph-theoretic Approach

by

Kurt J. Reinschke

Akademie-Verlag Berlin 1988

Autor:

Prof. Dr. Ing. Dr. rer. nat. Kurt J. Reinschke
Ingenieurhochschule Cottbus
Wissenschaftsbereich Informatik und Prozeßsteuerung
Cottbus
7500

Die Titel dieser Schriftenreihe werden vom Originalmanuskript der Autoren reproduziert

ISBN 3-05-500452-3
ISSN 0138-3019

Erschienen im Akademie-Verlag Berlin, DDR-1086 Berlin, Leipziger Str.3-4

Lizenznummer: 202 · 100/411/88
Printed in the German Democratic Republic
Gesamtherstellung: VEB Kongreß- und Werbedruck, 9273 Oberlungwitz
Lektor: Dr. Reinhard Höppner
LSV 1085
Bestellnummer: 763 791 4 (2182/41)

03800

Preface

This monograph is addressed to engineers engaged in control systems research and development, to graduate students specializing in control theory and to applied mathematicians interested in control problems. The author's objective is to present a graph-theoretic approach to the analysis and synthesis of linear time-invariant multivariable control systems.

After 1950, the demands of control practice led to the development of control methods for plants having more than one input and/or output. These multivariable control systems have attracted much attention within the so-called "modern control theory", which is primarily based on state-space methods. The state-space theory has provided the control engineers' community with profund new concepts such as controllability and observability and has clarified essential system theoretic questions which were poorly understood in the "classical control theory".

Nevertheless, seen from the point of view of practising engineers, the state-space theory has also serious disadvantages:

- The plant is modelled by ordinary matrix differential equations where the entries of the occurring matrices are regarded as numerical values which are exactly known. Experience shows that the results of the controller design may be largely sensitive to small variations of the chosen numerical values for the matrix entries in the description of the plant. In practice, however, the engineer has to cope with more or less uncertain and varying plant parameters.
- The procedures for the plant analysis and controller synthesis are based on cumbersome matrix manipulations. Control engineers lose desirable "feeling" and visual insight.
- In the case of large-scale systems the order of matrices to be investigated increases rapidly. Sparsity, as a typical feature of large-scale systems, must be taken into consideration. This requires specialized advanced matrix techniques.

After 1970, the "geometric approach" to linear multivariable control arose as an attack "against the orgy of matrix manipulation" (see Wonham, 1974). This geometric approach is based on an abstract coordinate-free representation of linear vector spaces. The comparatively simple language of matrix arithmetic is translated into the more abstract language of high-dimensional vector spaces. Unfortunately, this level of abstraction does not correspond to the traditional kind of reasoning of control engineers.

The "graph-theoretic approach" is another attempt to overcome the disadvantages of the state-space theory: The given control system is modelled by a suitably chosen graph representation. The system properties under investigation may be expressed by properties of the graph. That is why the investigator obtains a better insight into the structural nature of interesting properties. Provided he succeeds in showing that a desired property holds generically (i.e. independently of numerical parameter values), the indeterminate parameters may be considered to be degrees of freedom during further steps of design or optimization. If the graph-theoretic characterization shows that a desirable property does not hold, the investigator is able to suggest system modifications with the aid of which the desired property could be fulfilled. As for large-scale systems, it should be realized that the graph-theoretic system representation reflects exactly the non-vanishing couplings. Zero matrix entries do not at all appear in the graph representation. Consequently, the graph-theoretic approach has proved to be especially suitable for sparse large-scale systems.

The reader of this book is expected to have some previous acquaintance with control theory. He should have taken, or be taking concurrently, introductory courses in frequency-responce methods and in state-space methods. Some working knowledge in graph-theory would be an advantage, although it is not necessary. The Appendices 1 and 2 provide the reader with all the graph-theoretic tools needed for our purposes. In Appendix 1, the basic concepts are explained and the notations used in this book are introduced. In Appendix 2 several possibilities for graph-theoretic interpretations of determinants are dealt with. A crucial role plays an interpretation of determinants by cycle families published by A.L. Cauchy as early as in 1815 and re-invented by C.L. Coates in 1959.

The main part of this book consists of four chapters.
Chapter 1 starts with an appropriate graph-theoretic representation of large-scale dynamical systems. Based on this representation, important system properties such as decomposability, structural controllability and observability can be checked easily. Besides, the concept of "generic validity of a system property" is discussed in some detail.
Chapter 2 - 4 cover material on controller synthesis.
In Chapter 2, static state feedback is assumed. The problem of pole placement is reconsidered, seen from the graph-theoretic point of view. The graph-theoretic approach to two classical problems - disturbance rejection and decoupling by static state feedback - supplies nice new results.
In Chapter 3, static output feedback is assumed. Based on the graph-theoretic interpretation of the open-loop and the closed-loop characteristic polynomial coefficients, poles and zeros of multivariable systems - including zeros at infinity and their multiplicities - are characterized both algebraically and graph-theoretically. Then, the well-known problem of arbitrary pole placement by static output feedback is attacked. A necessary and sufficient criterion for local pole assignability and a new sufficient condition for global pole assignability are derived.
In Chapter 4, an outline is given for further exploitation of the graph-theoretic approach to controller synthesis. Apart from static output feedback under structural constraints, the remaining problems - dynamic controllers, implicit system descriptions, non-linear plants, mixed system consisting of logical and dynamical components - are just briefly mentioned. This reflects the author's deliberate intent to leave the book "open" at the far end. That is, only the direction is shown in which further results by means of graph-theoretic tools may be found. There seems to be much beyond the confines of ths monograph, that should be tackled graph-theoretically.

A few words about references are in order. Each chapter is divided into sections. Thus, Section 23 refers to the third section within the second chapter. For purposes of reference, formulas, theorems, lemmas, corollaries, examples and figures are numbered consecutively within each section. A reference such as Fig. 23.11 refers to the eleventh figure of Section 23. In referring to the bibliography the name of the author and the year of publication of the source contained in the bibliography are written down. To indicate the end of a proof a full triangle, ▲ , is used.

Finally, I would like to acknowledge the help and support I have had from other people. I am grateful to the Director of the Zentralinstitut für Kybernetik und Informationsprozesse der Akademie der Wissenschaften der DDR, Prof. V. Kempe, for the possibility to work in this field over a period of some years. A debt of gratitude is owed to Prof. M. Thoma, President of the IFAC, and to Prof. H. Töpfer (Dresden). They encouraged me to prepare such a monograph. I have profited by (mainly epistolary) discussions with many experts in other countries, in particular with Prof. O.I. Franksen and Prof. P.M. Larsen of the Danish University of Technology, Prof. F.J. Evans (London), Prof. D.D. Siljak (Santa Clara, Calif.), Dr. N. Andrei (Bukarest), Prof.D. Hinrichsen and Dr. A. Linnemann (Bremen), Prof. H. Schwarz and his co-workers (Duisburg), Dr. K. Tchon (Wroclaw), Dr. A.J.J. van der Weiden (Delft), Dr. L. Bakule (Prague).
In this country, we have had stimulating debates concerning the topics of this book in the Control Theory Group headed by Prof. K. Reinisch (Ilmenau).
A careful study of parts of the manuscript was undertaken by my colleagues Dr. J. Lunze and Dr. P. Schwarz. They helped to improve the presentation by their criticism. This support is most gratefully acknowledged. Another debt is acknowledged to Mrs. Schöpke who turned the author's rough pencil drawings into neat and workmanlike figures.
Thanks are dure to the staff of the Akademie-Verlag, especially Dr. R. Höppner, for friendly and effective cooperation.
Last but not least, I greatly appreciate all the various forms of help of my wife while the manuscript was in preparation.

Kurt Reinschke

Contents

Chapter 1. Digraph modelling of large-scale systems

11 Mapping of state-space models into digraphs

There are several possibilities to describe dynamic feedback systems mathematically. In this monograph, we base on the state-space description. This means, we assume the plant under investigation to be modelled by equations of the form

$$\dot{x}_i(t) = f_i(x_1(t),x_2(t),\dots,x_n(t);u_1(t),\dots,u_m(t);t) \qquad \text{for } i = 1,\dots,n \tag{11.1}$$

$$y_j(t) = g_j(x_1(t),x_2(t),\dots,x_n(t);u_1(t),\dots,u_m(t);t) \qquad \text{for } j = 1,\dots,r \tag{11.2}$$

The plant can be controlled by different control strategies. In the case of static state feedback the control law is given by

$$u_k(t) = h_k^s(x_1(t),x_2(t),\dots,x_n(t);t) \qquad \text{for } k = 1,\dots,m \tag{11.3}$$

In the case of static output feedback we have

$$u_k(t) = h_k^o(y_1(t),\dots,y_r(t);t) \qquad \text{for } k = 1,\dots,m \tag{11.4}$$

Usually, the column vectors

$$x(t) = (x_1(t),x_2(t),\dots,x_n(t))' \tag{11.5}$$

$$u(t) = (u_1(t),u_2(t),\dots,u_m(t))' \tag{11.6}$$

and $$y(t) = (y_1(t),y_2(t),\dots,y_r(t))' \tag{11.7}$$

are called state vector, input vector, and output vector, respectively. The symbol "'" means transposition.

With a given dynamic feedback system (11.1), (11.2), (11.3) we associate a directed graph (digraph) G^s defined by a vertex-set and an edge-set as follows:

The vertex-set is given by m input vertices denoted by I1, I2,..., Im, by n state vertices denoted by 1, 2,..., n, and by r output vertices denoted by O1, O2,..., Or.

The edge-set results from the following rules:

If the state-variable x_j really occurs in $f_i(x,u,t)$, i.e. $\partial f_i/\partial x_j \neq 0$, then there exists an edge from vertex j to vertex i.

If the input-variable u_k really occurs in $f_i(x,u,t)$, i.e. $\partial f_i/\partial u_k \neq 0$, then there exists an edge from input vertex Ik to state vertex i.
If the state-variable x_i really occurs in $g_j(x,u,t)$, i.e. $\partial g_j/\partial x_i \neq 0$, then there exists an edge from vertex i to vertex Oj.
Finally, if the state-variable x_i occurs in $h_k^s(x,t)$, what is generally assumed in the case of state feedback, then there exists an edge from vertex i to vertex Ik.
For illustration, a characteristic part of a digraph G^s has been sketched in Fig. 11.1.

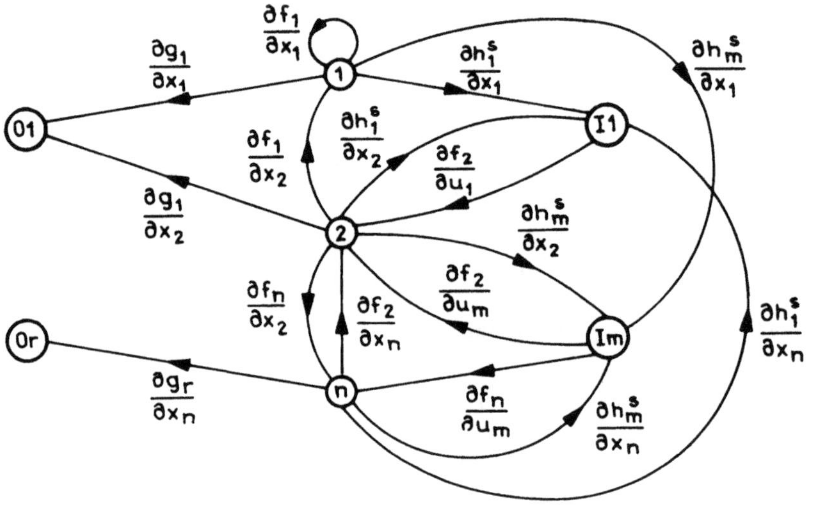

Fig. 11.1

Similarly, with a given dynamic system (11.1), (11.2), (11.4) we associate a digraph G^o.
It has the same vertex-set as G^s.
The state edges indicating $\partial f_i/\partial x_j \neq 0$, the input edges indicating $\partial f_i/\partial u_k \neq 0$, and the output edges indicating $\partial g_j/\partial x_i \neq 0$ result from the same rules as in case of G^s.
Instead of feedback edges leading from state vertices to input vertices we have now feedback edges from output vertex Oj to input vertex Ik if and only if y_j occurs in $h_k^o(y\ ,t)$.
Fig. 11.2 shows a characteristic part of a digraph G^o.

Obviously, the digraphs G^s and G^o contain less information than the equations (11.1, 11.2, 11.3) and (11.1, 11.2, 11.4), respectively.
We shall say that the digraphs G^s and G^o reflect the structure of a closed-loop system with state feedback and with output feedback, re-

spectively.
As far as small-scale systems are concerned, for example, systems with n = 2 or n = 5 state variables, it seems to be unnecessary to investigate the system structure separately. In case of large-scale systems, however, we should start with a structural investigation in any case. The author hopes to convince the reader of this book that the digraph approach is extremely useful, in particular for large-scale systems.

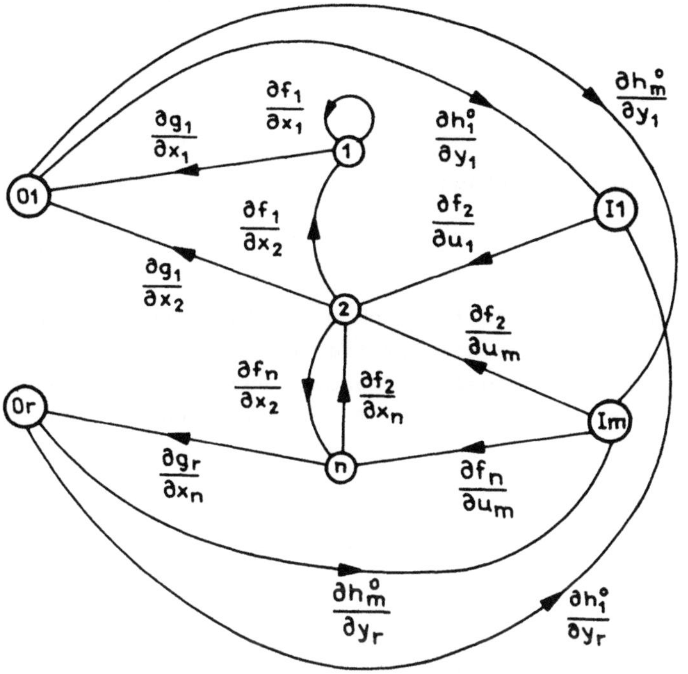

Fig. 11.2

A striking feature of large-scale systems is their <u>sparsity</u>. This structural property becomes evident in the digraphs G^s or G^o. These digraphs reflect a priori only the non-vanishing couplings of the system. So, instead of n^2 state edges we have really to take into account only a small percentage of this number in most applications. Moreover, the digraphs G^s and G^o give us an immediate impression of the information flow within the closed-loop systems. The considerations of the following chapters will demonstrate that the investigator receives more insight into the structural nature of properties under investigation than with the aid of the conventional numerical treatment. He should exploit the recognized structural properties for design purposes before he performs numerical computations.

In this monograph, we deal with the digraph approach for linear multivariable control systems. This means the plant is modelled by matrix equations of the form

$$\dot{x} = Ax + Bu \tag{11.8}$$

$$y = Cx \tag{11.9}$$

where $x(t) \in \mathbb{R}^n$, $u(t) \in \mathbb{R}^m$, and $y(t) \in \mathbb{R}^r$. The matrices A, B, C have real elements and are of dimension $n \times n$, $n \times m$, $r \times n$, respectively.

In case of static state feedback we have

$$u = Fx \tag{11.10}$$

and in case of static output feedback

$$u = Fy \tag{11.11}$$

As a rule in control practice, the feedback matrices F are subjected to structural constraints because some of the feedback gains may not be chosen freely. Assume the admissible feedback pattern to be characterized by the freely changeable entries of F. If all the entries of F are presumed to be freely assignable, then we shall sometimes use the symbol E instead of F.

In order to investigate multivariable control systems we shall consider suitably chosen square compound matrices.
It will be seen to be useful that the total information about the open-loop system is summarized in a compound square matrix of order r+n+m,

$$Q_0 = \begin{pmatrix} 0 & C & 0 \\ 0 & A & B \\ 0 & 0 & 0 \end{pmatrix} \tag{11.12}$$

In context of controllability investigations the following square matrix will prove to be most appropriate,

$$Q_1 = \begin{pmatrix} A & B \\ E & 0 \end{pmatrix} \quad \text{of order } n+m \tag{11.13}$$

in context of observability investigations,

$$Q_2 = \begin{pmatrix} 0 & C \\ E & A \end{pmatrix} \quad \text{of order } r+n \tag{11.14}$$

in context of both controllability and observability,

$$Q_3 = \begin{pmatrix} 0 & C & 0 \\ 0 & A & B \\ E & 0 & 0 \end{pmatrix} \quad \text{of order} \quad r+n+m \tag{11.15}$$

in context of output feedback, possibly structurally constrained,

$$Q_4 = \begin{pmatrix} 0 & C & 0 \\ 0 & A & B \\ F & 0 & 0 \end{pmatrix} \quad \text{of order} \quad r+n+m \tag{11.16}$$

In Appendix A1.3, it has been discussed that there are several possibilities of constructing an associated digraph having a one-to-one correspondence with a given square matrix. In the sequel, we shall base on the second graph-theoretic matrix characterization introduced in A1.3.

Definition 11.1

Let Q be a given square matrix of order q.
Q may be represented by a digraph G(Q) with q different vertices $v_1, v_2, \ldots, v_q$. There exists an edge (v_i, v_j) from vertex v_i to vertex v_j if and only if the entry q_{ji} of Q does not vanish. The edge weight is equal to the numerical value of q_{ji}.

Example 11.1: Let be $n = 3$, $m = r = 2$, and the plant equations (11.8, 11.9) be given by

$$\begin{pmatrix} \dot{x}_1 \\ \dot{x}_2 \\ \dot{x}_3 \end{pmatrix} = \begin{pmatrix} a_{11} & a_{12} & 0 \\ 0 & 0 & 0 \\ a_{31} & a_{32} & 0 \end{pmatrix} \begin{pmatrix} x_1 \\ x_2 \\ x_3 \end{pmatrix} + \begin{pmatrix} b_{11} & 0 \\ 0 & b_{22} \\ 0 & 0 \end{pmatrix} \begin{pmatrix} u_1 \\ u_2 \end{pmatrix} \tag{11.17}$$

$$\begin{pmatrix} y_1 \\ y_2 \end{pmatrix} = \begin{pmatrix} c_{11} & 0 & 0 \\ 0 & 0 & c_{23} \end{pmatrix} \begin{pmatrix} x_1 \\ x_2 \\ x_3 \end{pmatrix} \tag{11.18}$$

For this example system, the digraph $G(Q_3)$ has been drawn in Fig. 11.3.

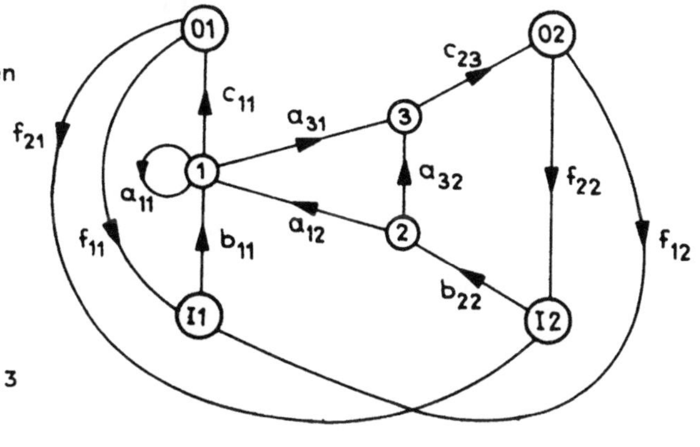

Fig. 11.3

Generalized symbolic represent tions of the digraphs $G(Q_o)$, $G(Q_1)$, $G(Q_2)$, $G(Q_3)$, and $G(Q_4)$ are shown in the Figures 11.4.1, b, c, d, and e, respectively.
The hyper-edge symbols ⇛ correspond to matrices which may be regarded as generalized edge weights. The hyper-vertices u, x, and y of the generalized digraphs of Fig. 11.4 are associated with the input vector, the state vector, and the output vector, respectively.

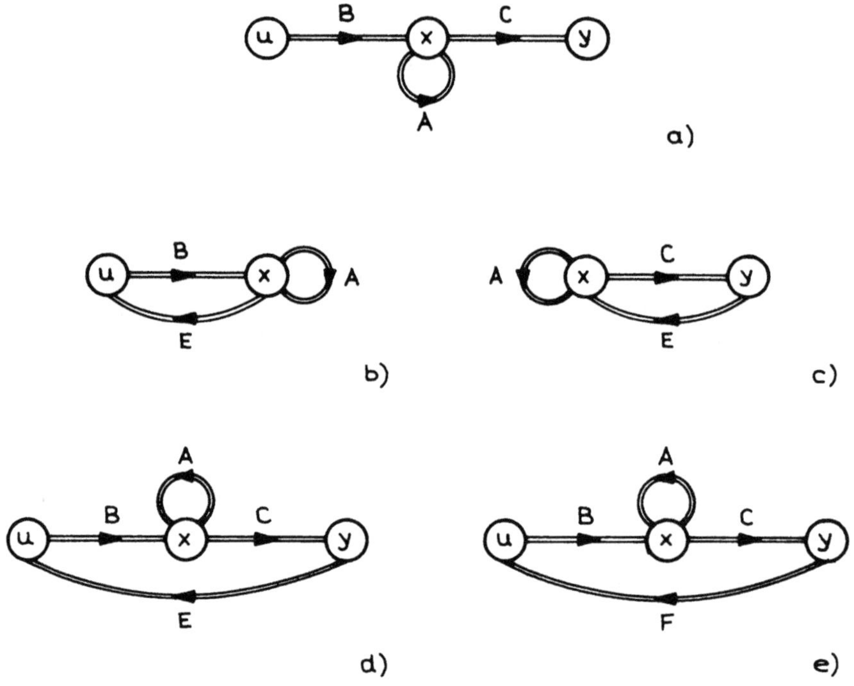

Fig. 11.4

12 Structure matrices and their associated digraphs

In the framework of the traditional control theory, the entries of the matrices A, B, C, F are regarded as numerical data given with 100 percent precision. For physical reasons, however, the parameters involved in the entries of A, B,... are only approximately known. Consequently, it seems to be more adequate to regard the most entries of A, B, ... as indeterminate. Only some entries which are often precisely zero have exact numerical values.

In context of "structural controllability" introduced by C.T. Lin in 1974 and of related "structural investigations" one has been used to take into account only the "structure" of the matrices A, B,.... This means, instead of numerically given matrices A, B,... the corresponding structure matrices [A], [B], ... of the same dimensions are considered.

Definition 12.1

The elements of a structure matrix [Q] are either fixed at zero or indeterminate values which are assumed to be independent of one another.
A numerically given matrix Q is called an admissible numerical realization (with respect to [Q]) if it can be obtained by fixing all indeterminate entries of [Q] at some particular values.
Two matrices Q' and Q" are said to be structurally equivalent if both Q' and Q" are admissible numerical realizations of the same structure matrix [Q].

We shall denote the indeterminate entries of a structure matrix by "L" and the entries fixed at zero by "0" or, often more conveniently, by an empty place.

Example 12.1: Consider a plant with $n = 6$ state-variables, $m = 2$ inputs and $r = 3$ outputs that is described mathematically by the following matrices introduced by Eq. (11.8) and Eq. (11.9):

$$A = \begin{pmatrix} 0 & 0 & a_{13} & 0 & 0 & 0 \\ a_{21} & 0 & 0 & a_{24} & a_{25} & 0 \\ 0 & 0 & a_{33} & 0 & 0 & a_{36} \\ 0 & a_{42} & 0 & 0 & 0 & 0 \\ 0 & 0 & 0 & a_{54} & a_{55} & 0 \\ 0 & 0 & 0 & 0 & 0 & 0 \end{pmatrix}, \quad B = \begin{pmatrix} 0 & 0 \\ 0 & 0 \\ 0 & 0 \\ 0 & 0 \\ b_{51} & 0 \\ 0 & b_{62} \end{pmatrix}$$

$$C = \begin{pmatrix} 0 & c_{12} & 0 & 0 & c_{15} & 0 \\ c_{21} & 0 & 0 & 0 & 0 & 0 \\ c_{31} & 0 & c_{33} & 0 & 0 & 0 \end{pmatrix} \qquad (12.1)$$

Assume the vanishing elements to be fixed at zero while the other elements have unknown real values. Then the structure matrix $[Q_3]$, see (11.15), is given by

$$[Q_3] = \begin{pmatrix} 0 & [C] & 0 \\ 0 & [A] & [B] \\ [E] & 0 & 0 \end{pmatrix} = \left(\begin{array}{ccc|cccccc|cc} 0 & 0 & 0 & 0 & L & 0 & 0 & L & 0 & 0 & 0 \\ 0 & 0 & 0 & L & 0 & 0 & 0 & 0 & 0 & 0 & 0 \\ 0 & 0 & 0 & L & 0 & L & 0 & 0 & 0 & 0 & 0 \\ \hline 0 & 0 & 0 & 0 & 0 & L & 0 & 0 & 0 & 0 & 0 \\ 0 & 0 & 0 & L & 0 & 0 & L & L & 0 & 0 & 0 \\ 0 & 0 & 0 & 0 & 0 & L & 0 & 0 & L & 0 & 0 \\ 0 & 0 & 0 & 0 & L & 0 & 0 & 0 & 0 & 0 & 0 \\ 0 & 0 & 0 & 0 & 0 & 0 & L & L & 0 & L & 0 \\ 0 & 0 & 0 & 0 & 0 & 0 & 0 & 0 & 0 & 0 & L \\ \hline L & L & L & 0 & 0 & 0 & 0 & 0 & 0 & 0 & 0 \\ L & L & L & 0 & 0 & 0 & 0 & 0 & 0 & 0 & 0 \end{array}\right) \qquad (12.2)$$

In analogy to Definition 11.1 we have a one-to-one correspondence between square structure matrices $[Q]$ and digraphs $G([Q])$ with unweighted edges.
For the example just discussed, the digraph $G([Q_3])$ can be seen in Fig. 12.1.

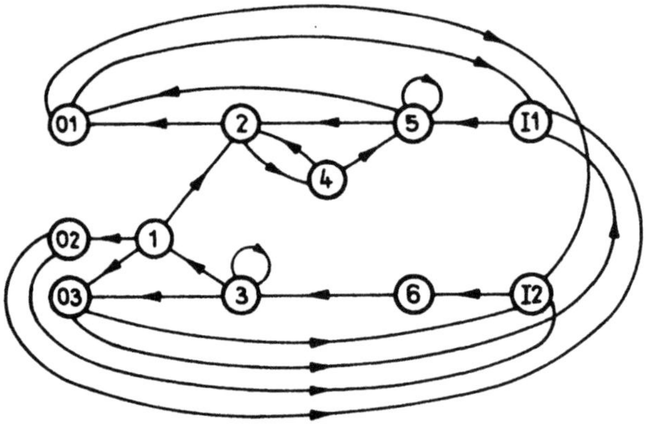

Fig. 12.1

The so-called "structural control theory" deals with classes of systems whose system matrices are structurally equivalent. Instead of numerical system properties the corresponding "structural properties" are investigated.

Definition 12.2

A <u>property holds structurally</u> within a class of structurally equivalent systems if the property under investigation holds numerically for "almost all" admissible numerical realizations.

This Definition may be explained in detail as follows:
Interpret the class of structurally equivalent systems as an Euclidean d-dimensional space $\mathbb{R}^d$. Then the set of actual systems having the property under investigation in the usual numerical sense forms a subset $S \subset \mathbb{R}^d$. For the class under consideration the property holds structurally if and only if S is dense in $\mathbb{R}^d$. The notion "dense" is taken from Topology. (A subset $S \subset \mathbb{R}^d$ is said to be dense in $\mathbb{R}^d$ if, for each $r \in \mathbb{R}^d$ and every $\varepsilon > 0$, there is an $s \in S$ such that the Euclidean distance $\varrho(s,r) \leq \varepsilon$.)

To give an example of a structural property let us introduce the notion of "structural rank of a rectangular structure matrix" [Q].

Definition 12.3

A <u>set of independent entries</u> of [Q] is defined as a set of indeterminate entries, no two of which lie on the same line (row or column).
The <u>structural rank</u> (for short, s-rank) of [Q] is defined as the maximal number of elements contained in at least one set of independent entries.

It should be noted that the s-rank of [Q] is equal to the maximal rank (in the usual numerical sense) of all admissible numerical matrices Q,

$$\text{s-rank } [Q] = \max_{Q \in [Q]} \text{rank } Q \qquad (12.3)$$

In the literature, the notations "generic rank" and "term rank" are used with the same meaning as "structural rank" (see, for example, Johnston et al. 1984 or Andrei 1985 and the numerous references cited there).

13 Appropriate state enumeration

13.1 Decomposition based on connectability properties

For large systems there exists a high degree of sparsity in the system matrix A. The associated digraph G([A]) reflects this sparsity in a most evident manner. Moreover, the digraph representation of [A] has an important invariance property: The enumeration of the vertices does not play any role. In other words, the digraph G([A]) is invariant with respect to permutation transformations of [A]. An appropriate reordering of the vertices, however, has proved to have many useful implications, especially in case of large systems. For this purpose, we try to decompose the digraph G([A]) into subgraphs based on connectability properties between its vertices. Such a decomposition of the digraph G([A]) should always be made as basis for the investigation of large-scale systems.

In Appendix A1.2, notions and notations suited for the decomposition of digraphs are explained. We have to look for subgraphs of G([A]) whose vertices are strongly connected. Here let us remind the reader of the fact that two vertices j and i are said to be strongly connected if a path exists from vertex j to vertex i as well as a path from vertex i to vertex j.
The subset of vertices strongly connected to a given vertex i forms an equivalence class K(i) within the set of all the n vertices of G([A]). Each equivalence class of strongly connected vertices, together with all the edges incident only with these vertices, constitutes a subgraph G([Q]) belonging to a square submatrix [Q] of [A]. In terms of matrix theory, the property of strong connectability of G([Q]) is called irreducibility of [Q] (see, for example, Gantmacher 1966).

If G([A]) does not contain a cycle that touches a given vertex i then we shall say that the vertex i constitutes an "acyclic" equivalence class. The corresponding subgraph G([Q]) is the isolated vertex i, and the corresponding square submatrix [Q] is a zero element placed on the main diagonal of [A].

The set of equivalence classes can now be enumerated in such a way that transitions from equivalence classes of lower indices to equivalence classes of higher indices are impossible.

The reordered structure matrix $[\tilde{A}]$ results from [A] by a permutation transformation,

$$[\tilde{A}] = P'[A]P \qquad (13.1)$$

[A] =	1	2	3	4	5	6	7	8	9	10	11	12	13	14	15	16
1	L				L											
2				L			L	L				L			L	
3		L		L		L	L			L				L		
4																
5									L		L					
6													L	L	L	
7				L			L						L			
8		L					L						L			
9									L					L		
10					L									L		
11	L															
12		L														
13				L												
14	L				L									L		
15		L					L									
16				L									L			

a)

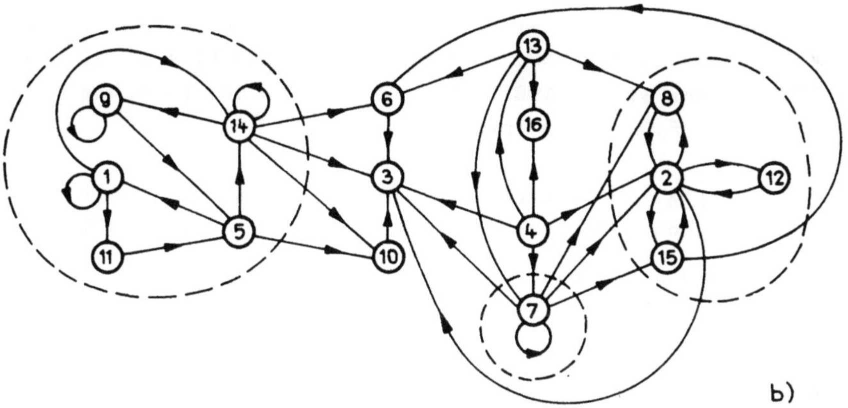

b)

[Ã] =	3	16	6	10	14	9	1	11	5	15	8	12	2	7	13	4	
3			L	L	L								L	L		L	K_1
16															L	L	K_2
6					L					L					L		K_3
10					L				L								K_4
14					L		L		L								K_5
9					L	L											K_5
1							L		L								K_5
11							L										K_5
5						L		L									K_5
15													L	L			K_6
8													L	L	L		K_6
12													L				K_6
2										L	L	L		L		L	K_6
7														L	L	L	K_7
13																L	K_8
4																	K_9

c)

Fig. 13.1

The transformed matrix $[\tilde{A}]$ has the following properties:

- $[\tilde{A}]$ is an upper quasi-triangular matrix, more exactly, all the parts of hyper-columns below the diagonal blocks belonging to strongly connected subgraphs consist of zeros.

- The non-vanishing main diagonal blocks are irreducible structure matrices.

Example 13.1: Fig.13.1 shows a 16 × 16 structure matrix [A], the associated digraph G([A]), and the reordered structure matrix [A].

13.2 An algorithm for reordering the states

Reordering procedures for large-scale matrices have been described and practically applied in the most varied fields for many years (see, for example, Kemeny and Snell 1960, Kaufmann 1968, Kevorkian 1975, Bunch and Rose 1976, Evans et al. 1981). Let us outline one possible method closely related to F.J. Evans' proposal.

The adjacency matrix (or occurrency matrix) N of G([A]) results from [A] by replacing each indeterminate entry "L" by the number "1", i.e.

$$n_{ij} = \begin{cases} 1 & \text{if an edge leads from vertex } j \text{ to vertex } i \\ 0 & \text{else} \end{cases} \qquad (13.2)$$

for $i,j = 1,\ldots,n$.

The elements $n_{ij}^{(l)}$ of the powers N^l obtained by multiplying N by itself l times can also be interpreted graphically,

$$n_{ij}^{(l)} = \begin{cases} \text{number of paths of length } l \text{ with} \\ \text{initial vertex } j \text{ and final vertex } i \end{cases} \qquad (13.3)$$

for $i,j = 1,\ldots,n$.

The connectability matrix (or reachability matrix) [R] is an $n \times n$ structure matrix defined by

$$r_{ij} = \begin{cases} L & \text{if a path leads from vertex } j \text{ to vertex } i \\ 0 & \text{else} \end{cases} \qquad (13.4)$$

for $i,j = 1,\ldots,n$.

It can easily be proved that

$$[R] = [\sum_{l=1}^{n-1} N^l] \qquad (13.5)$$

A digraph G([A]) is not decomposable if the entire digraph is strongly connected. This is valid if and only if no entry of the connectability matrix [R] vanishes. Then the proposed reordering process would be without any success. In control practice, however, such a situation may generally be excluded at least for large structure matrices [A].

The strongly connected subgraphs of G([A]) can be obtained from the symmetrical structure matrix

$$[S] = [R] \wedge [R'] \tag{13.6}$$

where "$\wedge$" means elementwise conjunction, i.e.

$$s_{ij} = r_{ij} \wedge r_{ji} \qquad \text{for} \quad i,j = 1,\dots,n$$

as follows:
Provided that the main diagonal element $s_{ii} = L$, then the non-vanishing elements of the ith row (or the ith column) of [S] correspond to the equivalence class K(i).
If $s_{ii} = 0$ then the isolated vertex i forms an "acyclic" equivalence class.

To find algorithmically a suited ordering between the equivalence classes, a further structure matrix should be derived from [R], namely

$$[T] = [R] \wedge [\overline{R'}] \tag{13.7}$$

i.e.

$$t_{ij} = r_{ij} \wedge \overline{r}_{ji} \qquad \text{for} \quad i,j = 1,\dots,n$$

where $\overline{L} = 0$ and $\overline{0} = L$.

While $s_{ij} = L$ means that the vertices i and j belong to the same equivalence class, says $t_{ij} = L$ that there is a path from j to i but no path from i to j.

Now we count the L-entries within the jth column of T and notice the sum as element w_j of a row vector w. In other words,

$$w_j = \begin{cases} \text{number of vertices to which a path leads from} \\ \text{vertex } j \text{, but not vice versa} \end{cases} \tag{13.8}$$

$$\text{for} \quad j = 1,\dots,n.$$

It is easily verified that two vector elements w_i and w_j are equal if the corresponding vertices i and j belong to the same equivalence class. Reversely, $w_i = w_j$ does not imply that i and j belong to the same equivalence class.

The permutation matrix P can be derived from the integer vector w as follows:

First, order the elements of w according to their magnitude.

Second, order the elements with the same values in such a way that all those elements whose corresponding equivalence classes agree appear one after another without interruptions.

Assume that both reordering steps result in a vector

$$(w_{i_1}, w_{i_2}, w_{i_3}, \ldots, w_{i_n}) \qquad (13.9)$$

Then one can use

$$P = (e_{i_1}, e_{i_2}, e_{i_3}, \ldots, e_{i_n}) \qquad (13.10)$$

where e_k is an $n \times 1$ unit vector with "1" as kth component and zeros else,

as the permutation matrix that generates the interesting reordered structure matrix

$$[\tilde{A}] = P'[A]P \qquad (13.1)$$

Example 13.2: Consider again the system depicted in Fig. 13.1. The adjacency matrix N results immediately from [A],

	1	2	3	4	5	6	7	8	9	10	11	12	13	14	15	16
1	1				1											
2				1			1	1				1			1	
3		1		1		1	1			1				1		
4																
5									1		1					
6													1	1	1	
7				1			1						1			
8		1					1						1			
9									1					1		
10					1									1		
11	1															
12		1														
13				1												
14	1				1									1		
15		1					1									
16				1									1			

= N

The connectability matrix [R] becomes

[R] =

	1	2	3	4	5	6	7	8	9	10	11	12	13	14	15	16
1	L				L				L		L			L		
2		L		L			L	L				L	L		L	
3	L	L		L	L	L	L	L	L	L	L	L	L	L	L	
4																
5	L				L				L		L			L		
6	L	L		L	L		L	L	L		L	L	L	L	L	
7				L			L						L			
8		L		L			L	L				L	L		L	
9	L				L				L		L			L		
10	L				L				L		L			L		
11	L				L				L		L			L		
12		L		L			L	L				L	L		L	
13				L												
14	L				L				L		L			L		
15		L		L			L	L				L	L		L	
16				L									L			

The symmetrical structure matrix $[S] = [R] \wedge [R']$ is

[S] =

	1	2	3	4	5	6	7	8	9	10	11	12	13	14	15	16
1	L				L				L		L			L		
2		L						L				L			L	
3																
4																
5	L				L				L		L			L		
6																
7							L									
8		L						L				L			L	
9	L				L				L		L			L		
10																
11	L				L				L		L			L		
12		L						L				L			L	
13																
14	L				L				L		L			L		
15		L						L				L			L	
16																

[S] provides all equivalence classes of vertices. Here they are given by the vertex-sets $\{1, 5, 9, 11, 14\}$, $\{2, 8, 12, 15\}$, $\{7\}$ and six isolated vertices 3, 4, 6, 10, 13, 16.

The structure matrix $[T] = [R] \wedge [\overline{R'}]$ and the integer vector w become

$[T] =$

	1	2	3	4	5	6	7	8	9	10	11	12	13	14	15	16
1																
2				L			L						L			
3	L	L		L	L	L	L	L	L	L	L	L	L	L	L	
4																
5																
6	L	L		L	L		L	L	L		L	L	L	L	L	
7				L									L			
8				L			L						L			
9																
10	L				L				L		L			L		
11																
12				L			L						L			
13				L												
14																
15				L			L						L			
16																

$$w = (3\ 2\ 0\ 9\ 3\ 1\ 6\ 2\ 3\ 1\ 3\ 2\ 8\ 3\ 2\ 0)$$

The reordering steps explained above give the following serial succession of w-components

$$w_3 = w_{16} < w_6 = w_{10} < w_{14} = w_9 = w_1 = w_{11} = w_5 < w_{15} = w_8 = w_{12} = w_2 < w_7 < w_{13} < w_4$$

Now the equivalence classes of strongly connected vertices can be appropriately enumerated:

$$K_1 = \{3\},\ K_2 = \{16\},\ K_3 = \{6\},\ K_4 = \{10\},\ K_5 = \{14, 9, 1, 11, 5\}$$
$$K_6 = \{15, 8, 12, 2\},\ K_7 = \{7\},\ K_8 = \{13\},\ K_9 = \{4\}.$$

The permutation matrix **P** is

$$P = (e_3, e_{16}, e_6, e_{10}, e_{14}, e_9, e_1, e_{11}, e_5, e_{15}, e_8, e_{12}, e_8)$$

From this results

$$[\tilde{A}] = P'[A]P$$

as indicated in Fig. 13.1.c.

It should be noted that the permutation matrix P is not uniquely determined in most cases.
First, transpositions between different equivalent classes supplying the same value for the components of the row vector w are possible. As for the example system, the vertices 16 and 3 as well as the vertices 6 and 10 could be transposed.
Second, the order between vertices of one and the same equivalence class does not play any role so far.

13.3 Some properties of irreducible structure matrices

Now, we are going to describe the details of allowable transitions within an equivalence class of strongly connected vertices.
Let [Q] be an irreducible square structure matrix and G([Q]) be the associated digraph. From the definition of irreducible structure matrices it is clear that an arbitrarily chosen vertex i can be reached from every vertes j of G([Q]), i.e. there are paths from j to i within G([Q]). Now we take into consideration the lengths of all paths from j to i and denote the set of all these lengths by L_{ij}. Let d_j be the greatest common divisor of all the integers contained in L_{jj}. It can be shown by a number-theoretic argumentation (see Kemeny and Snell 1960) that, in case of $d_j > 1$, there holds $d_j = d_k = d$ for every vertex k of G([Q]).
Further, each integer $l_{ij} \in L_{ij}$ may be represented as $l_{ij} = q \cdot d + t_{ij}$ where q is the integer quotient of l_{ij} and d, and t_{ij} is associated remainder. In algebraic terms, $l_{ij} \equiv t_{ij} (\text{mod.} d)$ with $0 \leq t_{ij} < d$.

From this it is easily seen that the vertices of the digraph G([Q]) may be divided into d cyclic subclasses. The transitions within G([Q]) run periodically over all cyclic subclasses. Therefore, we formulate

Definition 13.1

Irreducible matrices with $d > 1$ are called periodic irreducible matrices. In case of $d = 1$ an irreducible matrix is called aperiodic.

Provided that the vertices are appropriately enumerated then periodic irreducible matrices have a typical structure (see Fig. 13.2). An arbitrarily chosen vertex may be denoted by 1. All vertices j with $t_{1j} = 0$ form the first cyclic subclass, all vertices with $t_{1j} = 1$ form the second cyclic subclass, ..., all vertices with $t_{1j} = d-1$ form the last cyclic subclass. The number of vertices within different

cyclic subclasses may be different from each other.
If one enumerates the vertices in such a manner that the first, the second, ..., the last subclass are taken into account successively, then the structure matrix [Q] has the typical shape illustrated in Fig. 13.2 for $d = 5$.

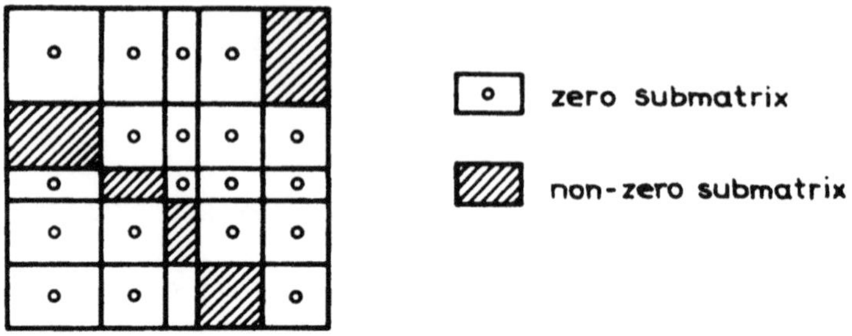

Fig. 13.2

There is a simple graph-theoretic test to determine the <u>index d of periodicity</u>:

<u>Theorem 13.1</u>

Consider the lengths of all cycles within the digraph G([Q]) of an irreducible structure matrix [Q]. The greatest common divisor of those lengths is equal to the index d of periodicity.

<u>Example 13.3</u>: The digraph G([A]) depicted in Fig. 13.1 contains three non-trivial equivalence classes of strongly connected vertices, $K_5 = \{1, 5, 9, 11, 14\}$, $K_6 = \{2, 8, 12, 15\}$, and $K_7 = \{7\}$.
The corresponding strongly connected subgraphs are circumscribed by dotted lines in Fig. 13.1.b. The irreducible matrices associated with K_5 and K_7 are aperiodic matrices of order 5 and 1, respectively. K_6 is associated with a periodic irreducible matrix of order 4 with $d = 2$ cyclic subclasses. These subclasses are $\{2\}$ and $\{8, 12, 15\}$.

Finally, let us prove a Lemma that is of interest in many contexts.

<u>Lemma 13.1</u>

Let [Q] be an irreducible $n \times n$ structure matrix, [R] an $n \times 1$ non-zero structure matrix, and I the $n \times n$ unit matrix. Then for almost all admissible realizations $(Q, R) \in [Q, R]$ there is valid

$$\text{rank } (Q - \lambda I, R) = n \quad \text{for all scalars } \lambda \neq 0. \tag{13.11}$$

Proof: Almost all admissible matrices $Q \in [Q]$ have the properties
(a) the non-vanishing main diagonal elements differ from each other,
(b) the non-vanishing eigenvalues are simple.
Otherwise a matrix $Q \in [Q]$ is said to be atypical. For atypical matrices Q the relation (13.11) need not be valid.
Let (Q,R) be numerically given where Q is assumed not to be atypical.
If $\lambda \neq 0$ is no eigenvalue of Q we obtain rank $(Q - \lambda I) = n$, and (13.11) is obviously fulfilled.
If $\lambda \neq 0$ is an eigenvalue of Q then (b) implies

$$\text{rank } (Q - \lambda I) = n-1.$$

Furthermore, (a) implies that only one main diagonal element of $(Q - \lambda I)$ may vanish at the most.
We remove one column of $(Q - \lambda I)$ and replace it by the column R:
If the kth main diagonal element of $(Q - \lambda I)$ vanishes then we eliminate the kth column, otherwise an arbitrarily chosen column of $(Q - \lambda I)$ may be replaced by R. The modified matrix is denoted by $\bar{Q}$.
Now, (13.11) is true because rank $\bar{Q} = n$ holds for almost all $R \in [R]$.
This statement can be shown as follows.
The n vertices of the digraph $G([Q])$ are strongly connected. This property remains true for the digraph $G([Q - \lambda I])$. Deletion of the kth column of $[Q - \lambda I]$ means elimination of all the edges in $G([Q - \lambda I])$ with the initial vertex k. According to the new column R of $\bar{Q}$ there are newly introduced edges in $G([\bar{Q}])$ from vertex k to at least one vertex j.
Let $j \neq k$. Since all n vertices of $G([Q - \lambda I])$ are strongly connected there exists a "backward" path from vertex j to vertex k which remains unchanged during the kth column replacement process.
In case of $j = k$ there exists a self-cycle attached at vertex k in $G([\bar{Q}])$.
Consequently, the structure graph $G([\bar{Q}])$ contains at least one cycle that coincides with the vertex k and, possibly, some other vertices.
Mark such a cycle.
Consider now all those vertices of $G([\bar{Q}])$ that are not involved in the cycle just marked. Those vertices are involved in other cycles within $G([\bar{Q}])$. In particular, they are initial and final vertices of self-cycles. Thus we have obtained at least one set of vertex-disjoint cycles that consists of n edges and involves all the n vertices of $G([\bar{Q}])$.

At this point we call back to mind the fact (see Appendix, Theorem A2.1) that such a cycle family corresponds to a non-vanishing term of the determinant det $\bar{Q}$. It cannot be excluded that there exist a few non-vanishing terms of det $\bar{Q}$. But there is no numerical cancellation of these terms for almost all $R \in [R]$. Hence the matrix $\bar{Q}$ has full rank in the structural sense (see Definition 12.2), which implies the desired result (13.11). ▲

14 Structural controllability, structural observability and structural completeness

Roughly speaking, a dynamic system is said to be "controllable" if its state vector (11.5) can be caused, by an appropriate manipulation of system inputs, to behave in desirable manner.
For linear time-invariant plants modelled by (11.8), (11.9) precise versions of this concept have been discussed in all texts on modern control theory. All the known controllability criteria involve only the matrices A and B. For that reason, more theoretically oriented authors prefer the phrase "the pair (A,B) is controllable ..." instead of "the plant (11.8), (11.9) is controllable ...", see Wonham 1974 and others.
"Structural controllability" concerns only the structure matrix pair [A,B] introduced by Definition 12.1.

Definition 14.1

A class of systems given by its structure matrix pair [A,B] is said to be structurally controllable (for short, s-controllable) if there exists at least one admissible realization $(A,B) \in [A,B]$ being controllable in the usual numerical sense.

Seemingly, Definition 14.1 of structural controllability appears to be not in accordance with Definition 12.2 of structural properties in general. This first impression is false. Later on, it will be recognized that controllability of one admissible pair (A,B) implies controllability of almost all admissible pairs within the class given by [A,B].

All information contained in [A,B] is reflected in the digraph G([Q]) associated with the $(n+m) \times (n+m)$ structure matrix

$$[Q] = \begin{pmatrix} [A] & [B] \\ 0 & 0 \end{pmatrix} \qquad (14.1)$$

14.1 Input-connectability and structural controllability

An obvious precondition of controllability is that the system inputs are able to influence all state variables. Said in graph-theoretic terms, there must exist paths from input vertices to all state vertices.

Definition 14.2

A class of systems is said to be <u>input-connectable</u> (or input-reachable) if in the digraph G([Q]) defined by (14.1) there is, for each state vertex, a path from at least one of the input vertices to the chosen state vertex.

Input-connectability, however, is not sufficient for s-controllability.

Example 14.1: Consider Fig. 14.1.

$$[Q] = \begin{pmatrix} [A] & [B] \\ 0 & 0 \end{pmatrix} = \left(\begin{array}{cc|c} 0 & 0 & L \\ 0 & 0 & L \\ \hline 0 & 0 & 0 \end{array}\right)$$

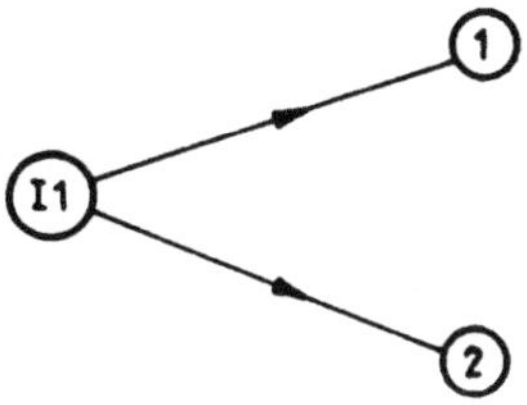

Fig. 14.1

No admissible matrix pair

$$(A,B) = \begin{pmatrix} 0 & 0 & b_1 \\ 0 & 0 & b_2 \end{pmatrix}$$

is controllable in the usual numerical sense. Therefore, [A,B] cannot be s-controllable.

Input-connectability has a remarkable implication which we formulate as

Lemma 14.1

If a class of systems characterized by the structure matrix pair [A,B] is input-connectable, then there holds

$$\text{rank } (A - \lambda I, B) = n \quad \text{for all scalars } \lambda \neq 0 \qquad (14.2)$$

for almost all admissible matrices $(A,B) \in [A,B]$.

<u>Proof:</u> If $\lambda \neq 0$ is no eigenvalue of A, then $(A - \lambda I)$ has full rank, and (14.2) holds obviously. The other case, $\lambda \neq 0$ is an eigenvalue of A, requires a more detailed consideration.
Assume the state variables to be enumerated as described in Section 13. Consider the hyper-rows of [A] which correspond to the equivalence

classes. The entries of at least one hyper-row, which are not located within the corresponding main diagonal block, are all zero. (The last hyper-row has this property at any rate.)
Mark all the hyper-rows of [A] holding this property and mark the corresponding hyper-rows of [B]. It is easy to recognize that the pair [A,B] represents a class of input-connectable systems if and only if none of the marked hyper-rows of [B] has only zero entries.
The following properties are true for almost all admissible matrices $A \in [A]$:

(a) every non-vanishing eigenvalue of A is simple,

(b) every non-vanishing eigenvalue of A is an eigenvalue of one and only one of the quadratic submatrices given by the main diagonal blocks of A.

In order to verify (14.2) we try to find gradually a non-singular $n \times n$ submatrix of $(A - \lambda I, B)$ given by a set of n independent entries.
Each non-zero main diagonal block of [A] may be interpreted as an irreducible structure matrix [Q] in the sense of Lemma 13. 1. Instead of [R] in Lemma 13.1 we shall use the next non-zero column within the same hyper-row of [A,B] which lies to the right of the main diagonal block being considered here.
Provided $\lambda \neq 0$ is an eigenvalue of A then, see (a) and (b), it can be assumed that λ is a simple eigenvalue of exactly one main diagonal submatrix A_{ii}. The dependent column of $A_{ii} - \lambda I$ will be removed and replaced by the next non-zero column of the ith hyper-row of [A,B] which is seen to the right. This column will be denoted by R_i and the modified square submatrix by $\bar{A}_{ii}$. According to Lemma 13.1 both the matrix $(A_{ii} - \lambda I, R_i)$ and the matrix $\bar{A}_{ii}$ have full rank.
Now the possible contribution deliverable from the ith hyper-row to the desired regular $n \times n$ submatrix of $(A - \lambda I, B)$ has been exhausted. All the columns of $(A - \lambda I, B)$, which were used to form $\bar{A}_{ii}$, must not be applied in the sequel.
If R_i does not belong to B, then R_i stems from a column $a_{.j}$ of A.
The vertex j in G([A]) belongs to an equivalence class, say K_j.
The corresponding main diagonal submatrix $(A_{jj} - \lambda I)$ is non-singular.
Instead of the column $a_{.j}$, which cannot be used any longer, we look for the next non-zero column which lies to the right within the jth hyper-row. This column is denoted by R_j. In the same way as in Lemma 13.1 it can be proved that both $(A_{jj} - \lambda I, R_j)$ and the modified square submatrix $\bar{A}_{jj}$ have full rank.
The column where R_j stems from must not be used in the sequel....
The next modified main diagonal submatrix will be $\bar{A}_{kk}$....
This process is continued until the newly added column belongs to B, i.e. the last modified main diagonal submatrix $\bar{A}_{pp}$ arises from $(A_{pp} - \lambda I, R_p)$ where R_p stems from B.

The desired regular $n \times n$ submatrix of $(A - \lambda I, B)$ is yielded by the submatrices $\bar{A}_{ii}$, $\bar{A}_{jj}$, ..., $\bar{A}_{pp}$ together with all the other main diagonal submatrices $(A_{rr} - \lambda I)$ which were not changed during the choice process described above. This completes the proof. ▲

<u>Example 14.2</u>: We shall illustrate the simple basic idea of the foregoing proof by an example (see Fig. 14.2).

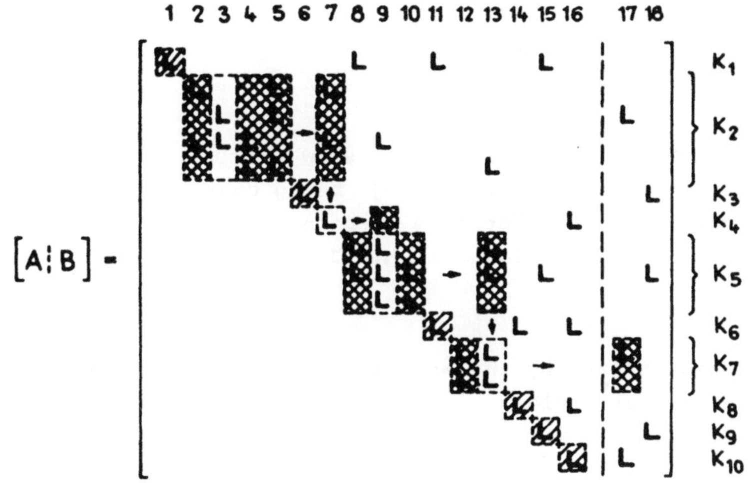

Fig. 14.2

The structure matrix $[A - \lambda I, B]$ has dimension 16×18.
There are 10 main diagonal blocks of order 1, 4, 1, 1, 3, 1, 2, 1, 1, 1.
It is assumed that $\lambda \neq 0$ is an eigenvalue of the second main diagonal submatrix of order 4.
The steps of the choice process described in the proof of Lemma 14.1 are marked by arrows. The non-zero parts of the non-singular 16×16 submatrix obtained have been shaded. The modified submatrices $\bar{A}_{22}$, $\bar{A}_{44}$, $\bar{A}_{55}$ and $\bar{A}_{77}$ are doubly shaded, The unchanged main diagonal submatrices $(A_{33} - \lambda I)$, $(A_{66} - \lambda I)$, $(A_{88} - \lambda I)$, $(A_{99} - \lambda I)$, $(A_{10,10} - \lambda I)$ are simply shaded.

14.2 Criteria of structural controllability

The following criterion of s-controllability is well-known (see Shields and Pearson 1976, Glover and Silverman 1976, Davison 1977, Franksen et al. 1979 and many others).

Theorem 14.1

A class of systems characterized by the $n \times (n+m)$ structure matrix pair $[A,B]$ is s-controllable if and only if
(a) it is input-connectable, and
(b) s-rank $[A,B] = n$. (14.3)

According to Definition 14.1, we have to show that there is at least one admissible matrix pair $(A,B) \in [A,B]$ being controllable in the traditional numerical sense. For this purpose, we remember the known criterion of controllability found by M.L.J. Hautus in 1969.

Lemma 14.2

A system characterized by the matrix pair (A,B) is controllable if and only if there holds

$$\text{rank } (A - \lambda I,\ B) = n \quad \text{for all scalars } \lambda . \tag{14.4}$$

Proof of Theorem 14.1: As a preliminary step, enumerate the state-variables as described in Section 13.
First, we shall prove the necessity of the stated conditions (a) and (b).
If, in contradiction with condition (a), the system is not input-connectable, then the entries of at least one hyper-row of $[A,B]$, which are not located within its main diagonal block, say $[A_{11}]$, are all zero. Now, for each $(A,B) \in [A,B]$, we can draw the following conclusion: Let λ be an eigenvalue of the main diagonal block $A_{11} \in [A_{11}]$. Then λ is also an eigenvalue of $A \in [A]$ and rank $(A - \lambda I, B) < n$, since the rows within the hyper-row under consideration are linearly dependent. According to Lemma 14.2, each admissible system $(A,B) \in [A,B]$ cannot be controllable.

If, in contradiction with condition (b), the structural rank of $[A,B]$ is less than n, then for each matrix pair $(A,B) \in [A,B]$ there holds rank $(A,B) < n$. Hence the controllability criterion (14.4) is violated for $\lambda = 0$. Thus, no admissible system can be controllable.

We continue by proving the sufficiency, i.e., if both the conditions (a) and (b) are met, then the class of systems $[A,B]$ will be shown to be structurally controllable.

Condition (a) implies that for almost all $(A,B) \in [A,B]$ the matrix $(A - \lambda I, B)$ has full rank for every scalar $\lambda \neq 0$. This may immediately be seen by application of Lemma 14.1.
From condition (b) follows that, for almost all $(A,B) \in [A,B]$, the matrix (A,B) has also full rank. Hence, Hautus' controllability criterion (Lemma 14.2) is satisfied for almost all admissible $(A,B) \in [A,B]$. This completes the proof. ▲

Condition (a) of Theorem 14.1 is easy to check. We have to look for paths which connect each state vertex with one of the input vertices. This is a standard task of algorithmic graph theory.

The determination of s-rank required by condition (b) is more difficult. Although several combinatorial algorithms have been described for many years (see, for example, Ford and Fulkerson 1962, Kaufmann1968, Franksen et al. 1979, Johnston et al. 1984), some authors (see Davison 1977) have recommended repeated application of numerical standard algorithms that deliver the rank of numerically given matrices. This concept though it will produce satisfactory results in most cases is hardly compatible with the spirit of the structural approach. One should investigate structural properties with the aid of appropriate tools using merely structural information on the system under consideration. Only when we have proved that the system meets the interesting property in the structural sense then, in a second step of investigation, the corresponding numerical property that presumes the knowledge of numerical data should be checked.

In order to state a purely graph-theoretic criterion of s-controllability we need a special notion not explained in the graph-theoretic appendix.

<u>Definition 14.3</u>

Consider the digraphs $G(Q_j)$ associated with the square matrices Q_j defined by (11.13), (11.14), (11.15), (11.16) for $j = 1,2,3,4$. A given <u>cycle family</u> in $G(Q_j)$ is said to be <u>of width w</u> if this cycle family touches exactly w state vertices.

Theorem 14.2

A class of systems characterized by the $n \times (n+m)$ structure matrix $[A,B]$ is s-controllable if and only if the digraph $G([Q_1])$, see (11.13), meets both the following conditions:

(a) For each state vertex in $G([Q_1])$ there is at least one path from one of the m input vertices to the chosen state vertex.

(b) There is at least one cycle family of width n in $G([Q_1])$.

Proof: As condition (a) is nothing else than the Definition 14.2 of input-connectability, condition (a) of Theorem 14 2 and condition (a) of Theorem 14.1 are equivalent.

Assume now condition (b) of Theorem 14.2 to be met.
Choose a cycle family of width n in $G([Q_1])$. Mark the entries of $[Q_1]$ that correspond to edges of the chosen cycle family. As this cycle family touches every one of the n state vertices, it contains n edges whose final vertices constitute the set of all state vertices. These n edges correspond to n marked entries of the structure matrix $[A,B]$ located both in different rows and in different columns. This implies immediately (comp. Definition 12.3)

$$\text{s-rank } [A,B] = n,$$

i.e., condition (b) of Theorem 14.1 is satisfied.

It remains to show that condition (b) of Theorem 14.1 implies condition (b) of Theorem 14.2.
For this purpose, we mark a set of n indeterminate entries of $[A,B]$ located both in different rows and in different columns. These marked entries correspond to n edges in $G([Q_1])$ whose final vertices are the n state vertices.
Next, consider the subset S of these n edges whose initial vertices are input-vertices. Let p denote the number of elements of S. If $p = 0$ then S is empty, and the desired implication holds obviously. Assume $p \geq 1$. The p edges of S correspond to p indeterminate entries of $[B]$ located at $(k_1,i_1),\ldots,(k_p,i_p)$. Further, there are p columns of $[A]$ that do not contain marked entries. Let their column indices be $j_1,\ldots,j_p$. Now let us supplement the n entries of $[Q_1]$ marked hitherto by p further marked entries, namely the entries $(i_1,j_1),\ldots,(i_p,j_p)$ of $[E]$. The resulting set of n+p marked entries is associated with a set of n+p edges of $G([Q_1])$ that forms a cycle family of width n.
This completes the proof. ▲

Example 14.3: To illustrate the main idea of the second part of the foregoing proof let us consider an example with $n = 5$ state-variables and $m = 3$ inputs.

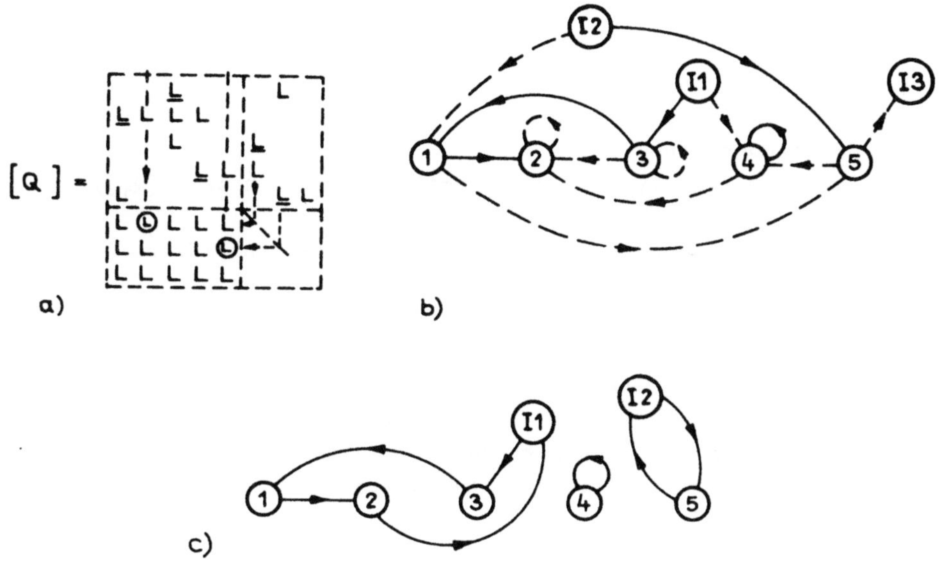

Fig. 14.3

Fig. 14.3.a shows the structure matrix $[Q_1]$. Fig. 14.3.b shows the associated digraph where the [E]-edges have been omitted.
Five entries of [A,B] located both in different rows and in different columns have been underlined. There holds $p = 2$, and $(k_1,i_1) = (3:1)$, $(k_2,i_2) = (5;2)$, $j_1 = 2$, $j_2 = 5$.
The desired cycle family of width $n = 5$ (see Fig. 14.3.c) has been obtained with the aid of two state feedback edges associated with the encircled entries of [E].

Remark: It should be kept in mind that an adjacent edge pair associated with $[b_{ki}]$ and $[e_{ij}]$ can play the same role as one edge associated with $[a_{kj}]$.

There holds s-rank[A,B] < n if and only if there is no cycle family of width n in $G([Q_1])$. If there are two or more cycle families of width n in $G([Q_1])$, then, for some admissible realizations $(A,B) \in [A,B]$ numerical cancellation can happen in such a way that $\text{rank}(A,B) < n$ despite of s-rank[A,B] = n. However, if there exists exactly one cycle family of width n in $G([Q_1])$ such a numerical cancellation is impossible. A few authors (see, in particular, Mayeda and Yamada 1979) have recommended the adverb "strongly" in order to characterize the impossibility of numerical cancellations. Thus, one obtains statements of the following kind.

<u>Corollary 14.1</u>

A class of systems characterized by the structure matrix pair [A,B] is strongly s-controllable if and only if the digraph $G([Q_1])$ meets both the following conditions:

(a) Input-connectability

(b) There is exactly one cycle family of width n in $G([Q_1])$.

The author of this monograph would like to prefer the adverb "simply" instead of "strongly" because all possibilities of helpful structural redundancy are here excluded by definition. This means, the practising engineer has to regard "strong s-controllability" as a relatively weak variant of s-controllability.

At this point, the relations between R.E. Kalman's well-known criterion of (numerical) controllability and the criteria of s-controllability proved above are to be discussed.

<u>Lemma 14.3</u>

A numerically given matrix pair (A,B) is controllable if and only if

$$\text{rank}\ (B,\ AB,\ A^2B, \ldots,\ A^{n-1}B) = n. \tag{14.5}$$

It has been conjectured (see Momen 1982 or Raske 1982) that the corresponding s-rank condition

$$\text{s-rank}\ ([B],\ [AB],\ [A^2B], \ldots,\ [A^{n-1}B]) = n \tag{14.6}$$

is necessary and sufficient for s-controllability.

Because of (12.3) it is clear that (14.6) is a necessary condition for s-controllability. Its sufficiency, however, cannot be proved.

<u>Counterexample 14.4</u>: Consider a class of systems with $n = 3$, $m = 1$, governed by the state equations

$$\dot{x} = \begin{pmatrix} \dot{x}_1 \\ \dot{x}_2 \\ \dot{x}_3 \end{pmatrix} = \begin{pmatrix} 0 & a_{12} & 0 \\ a_{21} & a_{22} & a_{23} \\ 0 & a_{32} & 0 \end{pmatrix} \begin{pmatrix} x_1 \\ x_2 \\ x_3 \end{pmatrix} + \begin{pmatrix} 0 \\ b \\ 0 \end{pmatrix} u = Ax + Bu \tag{14.7}$$

The non-vanishing entries of (A,B) are assumed to be unrelated.
The determinant

$$\det(B,AB,A^2B) = \det \begin{pmatrix} 0 & a_{12}b & a_{12}a_{22}b \\ b & a_{22}b & (a_{21}a_{12}+a_{22}^2+a_{23}a_{32})b \\ 0 & a_{32}b & a_{32}a_{22} \end{pmatrix}$$

$$= -b^3(a_{12}a_{32}a_{22} - a_{32}a_{12}a_{22}) \equiv 0$$

vanishes for all admissible pairs (A,B).
There is no admissible pair $(A,B) \in [A,B]$ that fulfills Kalman's controllability condition (14.5). Henceforth, according to Definition 14.1, the class of systems characterized by [A,B] cannot be s-controllable.
On the other hand, condition (14.6) is met,

$$\text{s-rank}([B],[AB],[A^2B]) = \text{s-rank} \begin{pmatrix} 0 & L & L \\ L & L & L \\ 0 & L & L \end{pmatrix} = 3 = n.$$

Let us try to explain the reasons why the conjecture mentioned above cannot be verified. The class of structurally equivalent systems is defined by the assumption that the non-vanishing entries of [A,B] may be fixed independently in order to get admissible realizations (A,B). From this, one cannot conclude that the entries of $(B,AB,A^2B,\dots,A^{n-1}B)$ might be independently chosen. Hence, Eq. (14.6) does not imply that $\text{rank}(B,AB,A^2B,\dots,A^{n-1}B) = n$ for at least one admissible pair (A,B). Even if the entries of [A,B] are assumed to be unrelated, there may arise structural dependencies between the columns of $(B,AB,\dots,A^{n-1}B)$. Such dependencies are seemingly not easy to recognize graph-theoretically in spite of the fact that the entries of $[A^iB]$ have an immediate graph-theoretic interpretation.

Lemma 14.4

The entry $[A^{k-1}B]_{ji}$ is non-zero if and only if there exists a path of length k from the input-vertex Ii to the state-vertex j within the digraph G([Q]) defined by (14.1).

It should be realized that the k state-vertices touched by the path introduced in Lemma 14.4 are not necessarily distinct. From this results the difficulty concerning the determination of the maximal rank of all admissible matrices $(B,AB,A^2B,\dots,A^{n-1}B)$.

Kalman's criterion (14.5) embraces more than both the necessary condition of input-connectability
(otherwise there are zero rows in $(B, AB, \dots, A^{n-1}B)$)
and the necessary condition of $\mathrm{rank}(A,B) = n$
(consider a single-input system with A = unit matrix and $B = (1, 1, \dots, 1)'$).

Let us now take advantage of the fact that input-connectability is much easier to check than the condition (b) in Theorems 14.1 and 14.2.

For $i = 1, 2, \dots, m$, let B_i be an $n \times i$ matrix consisting of the first i column vectors of B, and let A_i be the $n \times n$ matrix obtained from A replacing by zero lines those rows and columns that belong to state-vertices not connected with at least one input-vertex in $G([Q_1])$. Consider successively the digraphs $G([Q_{1,i}])$ where

$$Q_{1,i} = \begin{pmatrix} A_i & B_i \\ E & O \end{pmatrix} \tag{14.8}$$

are square matrices of order $n+i$.

A state-vertex k is not input-connected if and only if the kth row of (A_i, B_i) vanishes identically for all $i = 1, 2, \dots, m$.

The digraph $G([Q_{1,m}])$ can be used to formulate a modified criterion of s-controllability.

<u>Theorem 14.3</u>

A class of systems characterized by $[A,B]$ is s-controllable if and only if there exists a cycle family of width n in the digraph $G([Q_{1,m}])$, see (14.8).

14.3 Structural observability and structural completeness

It is well-known that observability and controllability are "dual" concepts, i.e., any statement about controllability has its direct counterpart concerning observability:

Input-connectability must be replaced by output-connectability,
the structure matrix pair [A,B] must be replaced by the structure matrix pair $\begin{bmatrix} C \\ A \end{bmatrix}$ or, equivalently, $\begin{bmatrix} A \\ C \end{bmatrix}$,
the digraph $G([Q_1])$ defined by (11.13) must be replaced by $G([Q_2])$ defined by (11.14).

Definition 14.4

A class of systems characterized by its structure matrix pair $\begin{bmatrix} C \\ A \end{bmatrix}$ is said to be structurally observable (for short, s-observable) if there exists at least one admissible realization $\begin{pmatrix} C \\ A \end{pmatrix} \in \begin{bmatrix} C \\ A \end{bmatrix}$ being observable in the usual numerical sense.

Definition 14.5

A class of systems is said to be output-connectable (or output-reachable) if in the digraph $G([Q_2])$ defined by (11.14) there is, for every state-vertex, a path from this state-vertex to at least one of the output-vertices.

In this context, the purely graph-theoretic criterion of s-observability corresponding to Theorem 14.2 reads as follows:

Theorem 14.4

A class of systems characterized by the $(r+n) \times n$ structure matrix $\begin{bmatrix} C \\ A \end{bmatrix}$ is s-observable if and only if the digraph $G([Q_2])$, see (11.14), meets both the following conditions:

(a) For each state-vertex in $G([Q_2])$ there is at least one path from that state-vertex to at least one of the r output-vertices

(b) There is at least one cycle family of width n in $G([Q_2])$.

Definition 14.6

A class of systems being both s-controllable and s-observable is said to be structurally complete (for short, s-complete).

In order to test s-completeness, let us consider the digraph $G([Q_3])$ defined by (11.15).

Theorem 14.5

A class of systems characterized by its structure matrices [A], [B] and [C] is s-complete if and only if the digraph $G([Q_3])$ meets both the following conditions:

(a) For each state-vertex in $G([Q_3])$ there is both a path from the chosen state-vertex to one of the r output-vertices and a path from one of the m input-vertices to the chosen state-vertex.

(b) There is at least one cycle family of width n in $G([Q_3])$.

Proof: Condition (a) means nothing else than both output-connectability and input-connectability.

If condition (b) is fulfilled, let us mark the indeterminate entries of $[Q_3]$ that are associated with all edges of a fixed cycle family of width n. This cycle family contains n edges whose final vertices are the n state-vertices. These n edges are associated with n marked entries of the submatrix [A,B] and ensure that s-rank[A,B] = n.
The chosen cycle family contains also a subset of n edges whose initial vertices are the n state-vertices. These n edges are associated with n marked entries of the submatrix $\begin{bmatrix} C \\ A \end{bmatrix}$ and ensure that s-rank$\begin{bmatrix} C \\ A \end{bmatrix}$ = n.

Thus, the conditions (a) and (b) imply s-controllability as well as s-observability. Consequently, both the conditions together are sufficient for s-completeness.

It remains to show that condition (b) is necessary for s-completeness. If the class of systems characterized by [A], [B], [C] is s-complete, then there exist admissible numerical matrices A, B, C which determine a system that is both controllable and observable in the usual numerical sense. Choose such an admissible matrix triple A, B, C. Then an admissible realization $F \in [E]$ can be chosen in such a way that the closed-loop system matrix $A + BFC$ is non-singular (Theorem 2 in Wang and Davison 1972). The non-singularity implies the existence of a cycle family of width n in $G([Q_3])$. This statement follows immediately from Theorem 31.4 proved below. In this way the proof of Theorem 14.5 has been completed. ▲

Example 14.5: Consider the class of systems with n = 6 state-variables, m = 2 inputs and r = 3 outputs described above by Eq. (12.1). We shall test if these systems are s-complete or not.
In order to apply Theorem 14.5 we have to investigate the digraph $G([Q_3])$. It has been shown in Fig. 12.1.
Obviously, both the conditions of input-connectability and of output-connectability are met. Moreover, there are many cycle families of width n = 6 within $G([Q_3])$, for example, the cycle family sketched in Fig. 14.4.

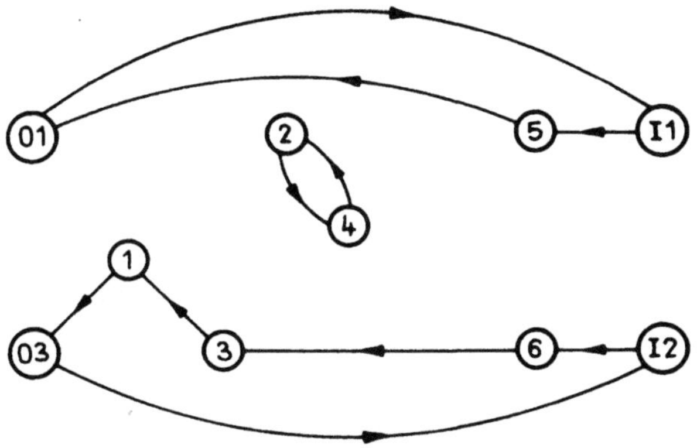

Fig. 14.4

Thus, we have verified that the class of systems under consideration is structurally complete.

Lastly, we shall reformulate the conditions stated in Theorem 14.5. Condition (a) says that $G([Q_3])$ is strongly connected. This may be interpreted as irreducibility condition for the closed-loop system structure matrix [A + BEC]. Each cycle family of width n in $G([Q_3])$ may also be regarded as a cycle family of length n in G([A + BEC]). Thus, one can reformulate Theorem 14.5 as follows:

<u>Corollary 14.2</u>

A class of systems characterized by its structure matrices [A], $[B] \neq 0$, and $[C] \neq 0$ is s-complete if and only if
(a) [A + BEC] is irreducible, and
(b) s-rank[A + BEC] = n.

Similarly, the criteria of s-controllability and s-observability may be reformulated.

<u>Corollary 14.3</u>

A class of systems characterized by its structure matrices [A], $[B] \neq 0$ is s-controllable if and only if [A + BE] is irreducible and has full structural rank.

<u>Corollary 14.4</u>

A class of systems characterized by its structure matrices [A], $[C] \neq 0$ is s-observable if and only if [A + EC] is irreducible and has full structural rank.

15 Do structural properties hold generically?

The notion of genericity has been referred to classes of systems whose parameters need not be fixed at certain actual values. This means the class under consideration may be regarded as an admissible domain in a parameter space. Any actual system characterized by fixed parameters values appears as an element of the admissible domain.

> Definition 15.1
>
> An interesting property holds generically for the class under consideration if the set of actual systems for which the property is really valid forms a dense subset of the admissible domain in the parameter space (comp. Definition 12.2).

If we regard all the non-zero entries of the structure matrices [A], [B], [C], see Definition 12.1, as independently varying parameters then the structural properties discussed above hold generically.
This can be easily seen: Let the number of indeterminate entries of [A], [B], and [C] be d altogether. If we now interpret the set of all admissible systems as a d-dimensional domain of the Euclidean vector space $\mathbb{R}^d$, then the subset of admissible systems for which the property under consideration is violated is formed by the intersection of D and an algebraic subset.
The notion "algebraic subset" is taken from Topology: The intersection of hyperfaces each defined by an algebraic equation is said to be an algebraic subset. In case of controllability, for example, the algebraic subset would be defined by the set of admissible (A,B) with rank(A,B) < n despite of s-rank[A,B] = n.
Consequently, the complementary subset of admissible systems which have the interesting property is dense in D.

Unfortunately, the basic assumption of Definition 12.1 cannot be justified in many practical examples. Mostly, the number of independent parameters which are varying for physical, technological, economic or social reasons is smaller than the number of non-zero entries of [A], [B], and [C]. Then, the really admissible domain must be considered to be a subset S of the domain D defined by all mathematically admissible realizations of the structure matrices [A], [B], and [C]. Generally, S is not dense in D. Thus, if the class of systems under consideration is defined by its changeable physical parameters it cannot be concluded that structural properties (in the sense of this monograph) hold generically. Let us discuss these relations in detail with the aid of a simple example treated extensively in Ackermann 1983.

A loading bridge (see Fig. 15.1) may be modelled as follows. The system input is the force u that accelerates the travelling crab, the system output is the position y of the load.

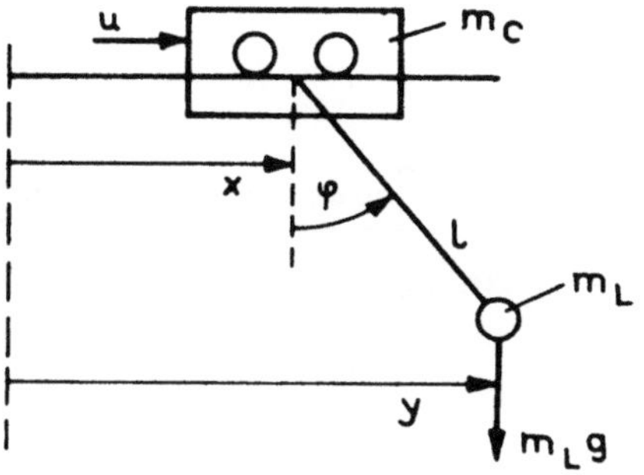

Fig. 15.1

There are four essential physical parameters:

m_C = mass of the travelling crab,
m_L = mass of load,
l = length of cable,
g = acceleration due to gravity.

Using the variables

x = position of the travelling crab,
φ = angle of the cable,

the well-known laws of point-mass mechanics yield the following non-linear equations of motion:

$$\begin{aligned} (m_L + m_C)\ddot{x} + m_L l(\ddot{\varphi}\cos\varphi - \dot{\varphi}^2 \sin\varphi) &= u \\ \ddot{x}\cos\varphi + l\ddot{\varphi} - g\sin\varphi &= 0 \,. \end{aligned} \tag{15.1}$$

After introduction of the state-variables

$$x = \begin{pmatrix} x_1 \\ x_2 \\ x_3 \\ x_4 \end{pmatrix} = \begin{pmatrix} x \\ \dot{x} \\ \varphi \\ \dot{\varphi} \end{pmatrix}$$

one obtains the state-space description

$$x = f(x,u) = \begin{pmatrix} x_2 \\ \dfrac{u + (g\cos x_3 + l\,x_4^2)\, m_L \sin x_3}{m_C + m_L \sin^2 x_3} \\ x_4 \\ -\dfrac{u\cos x_3 + (g + l x_4^2 \cos x_3) m_L \sin x_3 + g m_C \sin x_3}{l(m_C + m_L \sin^2 x_3)} \end{pmatrix} \tag{15.2}$$

$$y = c'x = (1,\ 0,\ 1,\ 0)\ x\ .$$

Linearizing these equations about the equilibrium position $x_3 = \varphi = 0$, $x_4 = \dot{\varphi} = 0$ one obtains a system description of the form (11.8), (11.9):

$$\dot{x} = \left.\frac{\partial f}{\partial x}\right|_{\substack{x_3=0 \\ x_4=0}} x + \left.\frac{\partial f}{\partial u}\right|_{\substack{x_3=0 \\ x_4=0}} u = \begin{pmatrix} 0 & 1 & 0 & 0 \\ 0 & 0 & a_{23} & 0 \\ 0 & 0 & 0 & 1 \\ 0 & 0 & a_{43} & 0 \end{pmatrix} x + \begin{pmatrix} 0 \\ b_2 \\ 0 \\ b_4 \end{pmatrix} u \tag{15.3}$$

$$y = (1,\ 0,\ c_3,\ 0)'x$$

where

$$a_{23} = \frac{m_L}{m_C}\,g, \quad a_{43} = -\frac{m_L + m_C}{m_C}\,\frac{g}{l}\ , \quad b_2 = \frac{1}{m_C}, \quad b_4 = \frac{-1}{m_C l}, \quad c_3 = 1.$$

The linearized system (15.3) may be represented by the corresponding weighted digraph $G(Q_o)$, see (11.12), shown in Fig. 15.2.

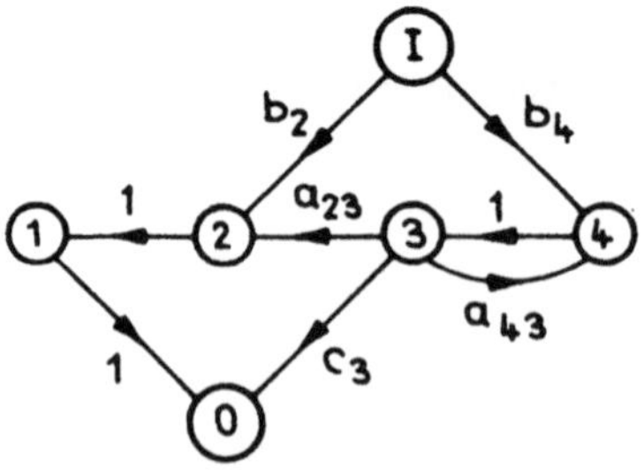

Fig. 15.2

The number of edges in this digraph is equal to the number of independent non-fixed entries within the structure matrices [A], [B], and [C].

As for this example, we have 8 parameters which, according to the basic presumption of the structural approach outlined in Section 12, are assumed to be non-fixed and independently varying. In other words, the domain D is an eight-dimensional part of the $\mathbb{R}^8$.

The example under consideration shows that 3 parameters are precisely fixed at 1. The other five edge weights are considered to be non-fixed. But there exist interdependencies between these five coefficients. Provided that we have to investigate loading bridges which are not portable from, for example, the earth to the moon then the acceleration g is constant and there remain three physical parameters, namely m_C, m_L and l. This means, all admissible actual systems are located on a three-dimensional subset S of the eight-dimensional domain D which is considered in the framework of the structural approach explained in Section 12.
Thus, we have the following situation:
On the one hand, the totality of actual systems which are admissible for physical reasons is located on a subset S of a domain D.
On the other hand, a property such as controllability which has been shown to hold structurally is valid for "almost all" structurally admissible systems which correspond to elements of D.
Unfortunately, from this cannot be concluded that the property is valid for at least one actual system corresponding to an element of S.

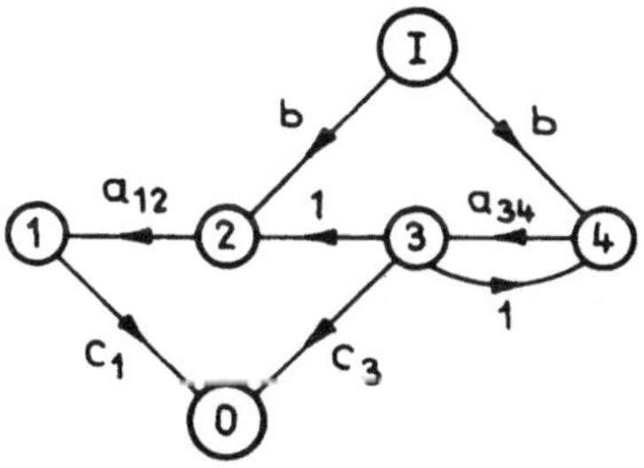

Fig. 15.3

Fig. 15.3 shows the graph representation of a linear system that is structurally equivalent to the loading bridge, compare Fig. 15.2.
We assume that the five parameters a_{12}, a_{34}, b, c_1, c_3 are independently changeable.
Though the system is s-controllable (apply, for example, Theorem 14.1) there exist no numerical values of a_{12}, a_{34}, b, c_1, c_3 for which the corresponding actual system becomes controllable (the rank condition cannot be met).

Let us now draw the conclusions.

1. If a property such as controllability does not hold structurally, then there exists no actual system within the structurally defined class for which that property holds numerically.

2. If a property holds structurally then this property is valid generically with respect to the parameters not fixed at zero by the structure in sense of Definition 12.1.

3. The number of independent physical parameters is often smaller than the number of entries not fixed at zero. Therefore, a property that holds structurally need not be valid generically with respect to the physical system parameters. If the worst comes to the worst, there is no physically admissible system having the property under consideration.

Being fully aware of this situation, some authors (Corfmat and Morse 1976, Anderson and Hong 1982, Linnemann 1982, Hosoe, Hayakawa and Aoki 1984 et al.) aimed to base on a more realistic modelling of physical and technological uncertainties inherent in all natural systems than the modelling based on Definition 12.1. Such a more realistic modelling should take into consideration that those entries of A, B, C known to be constant may be non-zero (in particular, the constant value 1 appears frequently) and changeable entries are not necessarily independent of one another. Because the uncertainties depend on physical parameters, the authors cited above suggested investigating parameter-dependent systems.

To say it more definitely, Corfmat and Morse considered matrices

$$A = A_o + \sum_{i=1}^{k} B_i P_i C_i, \qquad B = B_o + \sum_{i=1}^{k} B_i P_i D_i \tag{15.4}$$

where all matrices have fixed entries except the P_i, while all entries of each of the P_i are indeterminate and unrelated.

Anderson and Hong considered matrices

$$A = A_o + \sum_{i=1}^{k} p_i A_i, \qquad B = B_o + \sum_{i=1}^{k} p_i B_i \tag{15.5}$$

where the $A_o, A_1, \ldots, A_k$; $B_o, B_1, \ldots, B_k$ are thought of as fixed matrices, and the $p_1, \ldots, p_k$ as free real parameters.

Then they derived conditions on the fixed matrices such that the pair (A,B) is controllable for "almost all" parameter values.

In this pamphlet, we shall neither continue those investigations on parameter-dependent systems nor report in detail on the results obtained by others. In the following Chapters , we use the digraph-approach in order to tackle problems of controller synthesis. We aim to take full advantage of digraphs the abilities of which have by no means exhausted until now.

Chapter 2. Digraph approach to controller synthesis based on static state feedback

21 Pole placement by static state feedback

21.1 Problem formulation

We consider closed-loop systems obtained by applying linear static state feedback

$$u(t) = F\,x(t) \tag{21.1}$$

to the plant

$$\dot{x}(t) = A\,x(t) + B\,u(t) \tag{21.2}$$

where $x(t) \in \mathbb{R}^n$, $u(t) \in \mathbb{R}^m$, rank $B = m \leq n$.

From (21.1) and (21.2) we get, by Laplace transform,

$$U(s) = F\,X(s) \tag{21.1'}$$

$$s\,X(s) - x(0) = A\,X(s) + B\,U(s) \tag{21.2'}$$

These equations can equally be written as

$$\begin{pmatrix} s\,I_n - A & -B \\ -F & I_m \end{pmatrix} \begin{pmatrix} X(s) \\ U(s) \end{pmatrix} = \begin{pmatrix} x(0) \\ 0 \end{pmatrix} \tag{21.3}$$

where I_n and I_m are unit matrices of dimension n and m, respectively.
The determinant of the closed-loop system matrix

$$M(s) = \begin{pmatrix} s\,I_n - A & -B \\ -F & I_m \end{pmatrix} \tag{21.4}$$

is called the closed-loop characteristic polynomial

$$\det M(s) = s^n + p_1 s^{n-1} + \dots + p_{n-1} s + p_n \tag{21.5}$$

The roots of the equation

$$s^n + p_1 s^{n-1} + \dots + p_{n-1} s + p_n = 0 \tag{21.6}$$

are called poles of the closed-loop system.

As the matrices A, B, F are assumed to be real, it is clear that the coefficients $p_1, \ldots, p_n$ have also to be real and any feasible complex poles must of course appear in conjugate pairs.

Lemma 21.1.

There holds the identity

$$\det M(s) = \det (s I_n - A - B F) \tag{21.7}$$

Proof: Applying the well-known rules for evaluating the determinants of block matrices,

$$\det \begin{pmatrix} P & R \\ S & T \end{pmatrix} = \det P \det(T - S P^{-1}R) = \det T \det(P - R T^{-1}S), \tag{21.8}$$

one obtains

$$\det M(s) = \det \begin{pmatrix} s I_n - A & -B \\ -F & I_m \end{pmatrix} = \det I_m \det(s I_n - A -(-B)I_m^{-1}(-F))$$

$$= \det(s I_n - A - B F), \qquad \text{q.e.d.}$$

The problem of arbitrary pole placement by state feedback (21.1) is equivalent to the assignment of the coefficients $p_1, p_2, \ldots, p_{n-1}, p_n$ of the closed-loop characteristic polynomial (21.5) to arbitrarily chosen real values.

Definition 21.1

A plant (21.2) is said to have the property of arbitrary pole assignability under static state feedback (21.1) if given any real polynomial

$$p(s) = s^n + p_1 s^{n-1} + \ldots + p_{n-1}s + p_n$$

there exists a real $m \times n$ matrix F such that

$$p(s) = \det M(s) = \det(s I_n - A - B F).$$

21.2 Graph-theoretic characterization and algebraic reinterpretation of the closed-loop characteristic-polynomial coefficients

Consider the (n+m) × (n+m) matrix

$$Q = \begin{pmatrix} A & B \\ F & 0 \end{pmatrix} \tag{21.9}$$

and its associated digraph G(Q) applying the second graph-theoretic characterization of square matrices explained in Appendix A1.3.
Based on the notion of width of cycle families (see Definition 14.3) we state

Theorem 21.1

The coefficients p_i, $1 \leq i \leq n$, of the characteristic polynomial (21.5) are determined by the cycle families of width i within the graph G(Q), see (21.9).
Each cycle family of width i corresponds to one summand of p_i.
The numerical value of the summand results from the weight of the corresponding cycle family. This value must be multiplied by a sign factor $(-1)^d$ if the cycle family under consideration consists of d disjoint cycles.
In particular,

p_1 results from all cycles of width 1, and the common sign factor is -1;

p_2 results from all cycles of width 2, each with sign factor -1, and all disjoint pairs of cycles of width 1, each pair with sign factor +1;

p_3 results from all cycles of width 3, each with sign factor -1, all disjoint pairs with one cycle of width 2 and one cycle of width 1, each pair with sign factor +1, and all disjoint triples of cycles of width 1, each triple with sign factor -1;

etc.

Example 21.1: Consider a plant described by the equations

$$\begin{pmatrix} \dot{x}_1 \\ \dot{x}_2 \\ \dot{x}_3 \end{pmatrix} = \begin{pmatrix} a_{11} & a_{12} & 0 \\ 0 & 0 & 0 \\ a_{31} & a_{32} & 0 \end{pmatrix} \begin{pmatrix} x_1 \\ x_2 \\ x_3 \end{pmatrix} + \begin{pmatrix} b_{11} & 0 \\ 0 & b_{22} \\ 0 & 0 \end{pmatrix} \begin{pmatrix} u_1 \\ u_2 \end{pmatrix} \tag{21.10}$$

Here we have n = 3, m = 2.

The matrix Q introduced in (21.9) becomes

$$Q = \left(\begin{array}{ccc|cc} a_{11} & a_{12} & 0 & b_{11} & 0 \\ 0 & 0 & 0 & 0 & b_{22} \\ a_{31} & a_{32} & 0 & 0 & 0 \\ \hline f_{11} & f_{12} & f_{13} & 0 & 0 \\ f_{21} & f_{22} & f_{23} & 0 & 0 \end{array}\right) \qquad (21.11)$$

The associated digraph $G(Q)$ is shown in Fig. 21.1.

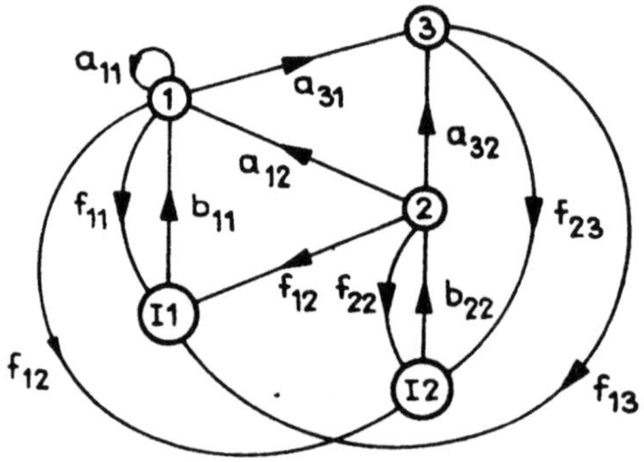

Fig. 21.1

Fig. 21.2 shows the cycle families of widths 1, 2, and 3.

According to Theorem 21.1 one obtains

$$p_1 = -a_{11} - f_{11}b_{11} - f_{22}b_{22}$$

$$\begin{aligned} p_2 = & -f_{23}a_{32}b_{22} - f_{21}a_{12}b_{22} - f_{13}a_{31}b_{11} + a_{11}f_{22}b_{22} \\ & -f_{12}b_{22}f_{21}b_{11} + f_{11}b_{11}f_{22}b_{22} \end{aligned} \qquad (21.12)$$

$$\begin{aligned} p_3 = & \; a_{11}f_{23}a_{32}b_{22} - f_{23}a_{31}a_{12}b_{22} + f_{11}b_{11}f_{23}a_{32}b_{22} \\ & -f_{13}a_{32}b_{22}f_{21}b_{11} - f_{23}a_{31}b_{11}f_{12}b_{22} + f_{22}b_{22}f_{13}a_{31}b_{11} \end{aligned}$$

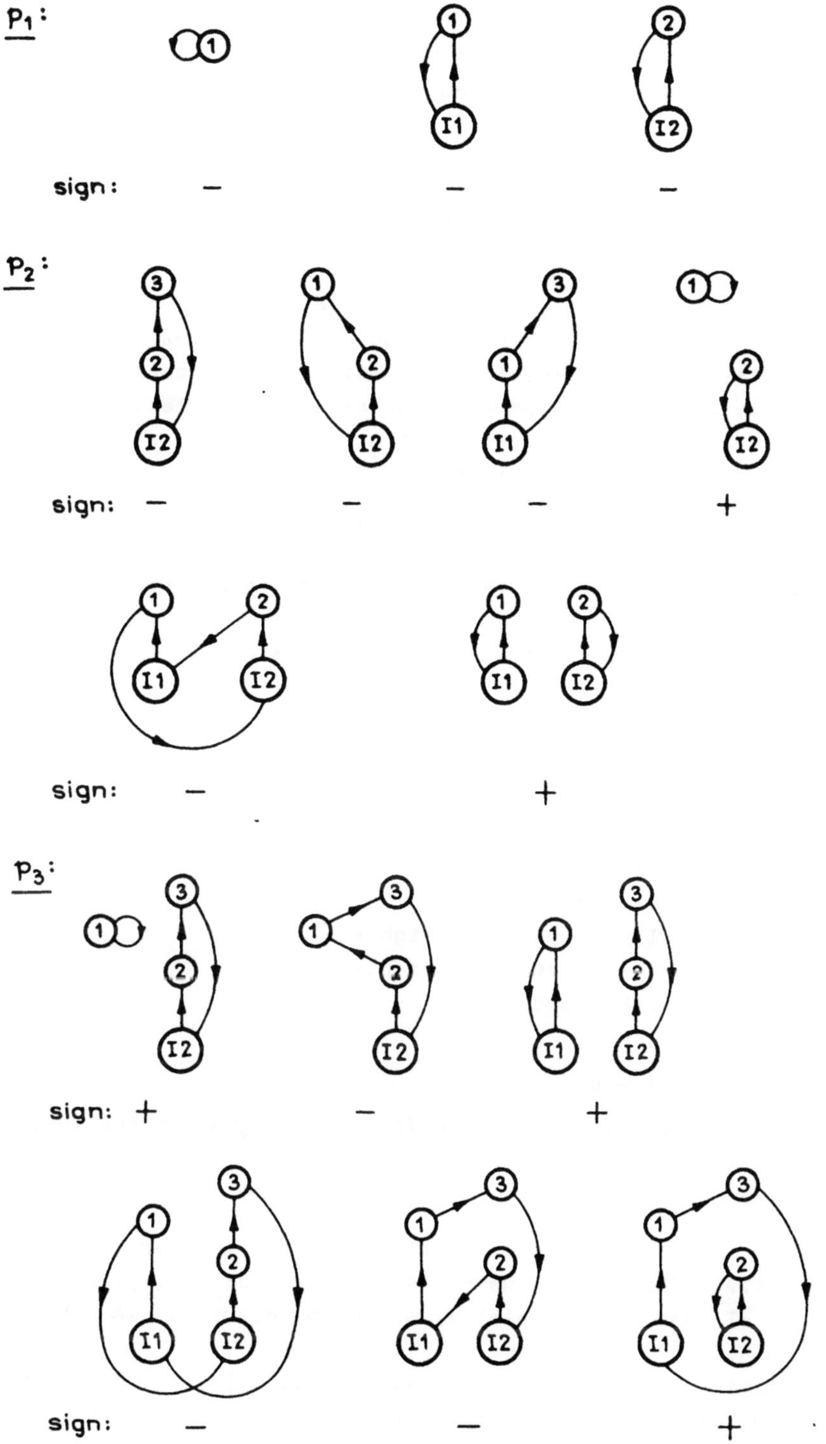
P1:
1
1
I1
2
I2
sign: − − −
P2:
3
2
I2
1
2
I2
3
1
I1
1
2
I2
sign: − − − +
1
2
I1
I2
1
2
I1
I2
sign: − +
P3:
1
3
2
I2
3
1
2
I2
1
I1
3
2
I2
sign: + − +
1
3
2
I1
I2
3
1
2
I1
I2
3
1
2
I1
I2
sign: − − +

Fig. 21.2

Proof of Theorem 21.1:

The basic idea is to apply Theorem A2.1, see Appendix A2, to the square matrix $M(s)$ defined by (21.4) and to collect all the terms of $\det M(s)$ with the same factor s^i, $1 \leq i \leq n$.

Each main diagonal element $s - a_{ii}$, $1 \leq i \leq n$, of $M(s)$ may be reflected in $G(M(s))$ by two self-cycles affixed to the state vertex i, one self-cycle with weight s and the other with weight $-a_{ii}$. The second self-cycle is omitted if $a_{ii} = 0$.

Each summand $p_i s^{n-i}$ on the right-hand side of (21.5) consists of those terms of $\det M(s)$ whose corresponding cycle families in $G(M(s))$ contain exactly $n-i$ self-cycles with weight s. The coefficient p_i is represented by a sum whose summands correspond to cycle families which involve all the m input vertices and i state vertices, and which do not contain any self-cycle with weight s.

Let us consider in detail one arbitrarily chosen cycle family in $G(M(s))$ which contributes to $p_i s^{n-i}$.

If the chosen cycle family contains g ($\neq 0$) feedback edges, then this family consists of

g feedback edges which correspond to entries of $-F$,
g input edges which correspond to entries of $-B$,
$m-g$ edges which correspond to entries of I_m (i.e. self-cycles with weights 1),
$n-i$ self-cycles with weights s, and the remaining
$i-g$ edges which correspond to entries of $-A$.

The total number l of disjoint cycles within the cycle family under consideration results from

$n-i$ = number of self-cycles with weights s,
$m-g$ = number of self-cycles with weights 1, and
d = number of further cycles.

That is

$$l = n + m + d - i - g.$$

According to Theorem A2.1, the sign factor belonging to the chosen cycle family will be

$$(-1)^{n+m-l} = (-1)^{i+g-d}.$$

Now, let us convert the graph $G(M(s))$ into the simpler graph $G(Q)$ defined by (21.9). This means the n+m self-cycles of $G(M(s))$ which correspond to sI_n and I_m are deleted, while the matrices $-A$, $-B$, $-F$ are replaced by A, B, F, respectively.

The cycle family of $G(M(s))$ investigated in detail above is converted into a cycle family of $G(Q)$ consisting of d individual cycles, altogether formed by g feedback edges, g input edges and $i-g$ state edges.

The replacement of $-A$, $-B$, $-F$ by A, B, F may be taken into account by an additional sign factor

$$(-1)^{2g+i-g} = (-1)^{i+g}.$$

For the p_i-summand under investigation we have therefore a total sign factor of

$$(-1)^{i+g-d}(-1)^{i+g} = (-1)^d.$$

Thus, the coefficient p_i, $1 \leq i \leq n$, of the closed-loop characteristic polynomial may indeed be obtained with the aid of the cycle families of width i in $G(Q)$ as stated in Theorem 21.1. ▲

Let

$$p = (p_1, p_2, \ldots, p_n)' \tag{21.13}$$

be the column vector encompassing the coefficients of the closed-loop characteristic polynomial (21.5). Each coefficient can be represented as a sum

$$p_i = p_i^o + h_i(F) \tag{21.14}$$

where $p_1^o, p_2^o, \ldots, p_n^o$ are the coefficients of the characteristic polynomial of the open-loop system (21.2), i.e.

$$\det(s\, I - A) = s^n + p_1^o s^{n-1} + \ldots + p_{n-1}^o s + p_n^o \tag{21.15}$$

and $h_i(F)$ is a multilinear form of feedback gains.

From Theorem 21.1 it can now be easily seen how the coefficients p_i depend on the feedback gains:

Theorem 21.2

The vector (21.13) depends on the feedback gains as follows,

$$\begin{aligned} p = p^o &+ \sum_{i_1=1}^{m} \sum_{j_1=1}^{n} p^{i_1j_1} f_{i_1j_1} \\ &+ \sum_{i_1} \sum_{<\, i_2} \sum_{j_1} \sum_{\neq\, j_2} p^{i_1j_1 \cdot i_2j_2} f_{i_1j_1} f_{i_2j_2} + \ldots \\ &+ \sum \sum_{i_1<i_2<\ldots<i_q} \sum \sum \sum_{j_1 \neq j_2 \neq \ldots \neq j_q} \sum p^{i_1j_1 \cdot i_2j_2 \cdots i_qj_q} f_{i_1j_1} f_{i_2j_2} \cdots f_{i_qj_q} \end{aligned} \tag{21.16}$$

where $q \leq m$. ($\sum\limits_{i_1<i_2<\ldots<i_q}\sum \cdots \sum$ and $\sum\limits_{j_1 \neq j_2 \neq \ldots \neq j_q}\sum \cdots \sum$ mean summation over all q-tupels $(i_1, i_2, \ldots, i_q)$ with $1 \leq i_1 < i_2 < \ldots < i_q \leq m$ and over all q-tupels $(j_1, j_2, \ldots, j_q)$ of pairwise different integers with $1 \leq j_1, j_2, \ldots, j_q \leq n$.)

The ith component of the vector equation (21.16) corresponds to all cycle families of width i in $G(Q)$, see (21.9).

In p^{o} those cycle families are gathered that do not contain feedback edges;

in $p^{i_1 j_1} f_{i_1 j_1}$ those cycle families are gathered that contain exactly one feedback edge, namely the edge leading from state vertex j_1 to input vertex Ii_1;

in $p^{i_1 j_1 \cdot i_2 j_2} f_{i_1 j_1} f_{i_2 j_2}$ those cycle families are gathered that contain exactly two feedback edges, namely the edges leading from j_1 to Ii_1 and from j_2 to Ii_2;

etc.

Example 21.2: For the example system considered above, the relations (21.12) may be rewritten, according to (21.16), in the following way:

$$\begin{pmatrix} p_1 \\ p_2 \\ p_3 \end{pmatrix} = -\begin{pmatrix} a_{11} \\ 0 \\ 0 \end{pmatrix} - \begin{pmatrix} b_{11} \\ 0 \\ 0 \end{pmatrix} f_{11} - \begin{pmatrix} 0 \\ a_{31}b_{11} \\ 0 \end{pmatrix} f_{13} - \begin{pmatrix} 0 \\ a_{12}b_{22} \\ 0 \end{pmatrix} f_{21} + \begin{pmatrix} -b_{22} \\ a_{11}b_{22} \\ 0 \end{pmatrix} f_{22}$$

$$+ \begin{pmatrix} 0 \\ -a_{32}b_{22} \\ (a_{11}a_{32}-a_{31}a_{12})b_{22} \end{pmatrix} f_{23} - \begin{pmatrix} 0 \\ b_{11}b_{22} \\ 0 \end{pmatrix} f_{12}f_{21} + \begin{pmatrix} 0 \\ b_{11}b_{22} \\ 0 \end{pmatrix} f_{11}f_{22} \qquad (21.17)$$

$$+ \begin{pmatrix} 0 \\ 0 \\ b_{11}a_{32}b_{22} \end{pmatrix} f_{11}f_{23} - \begin{pmatrix} 0 \\ 0 \\ a_{32}b_{22}b_{11} \end{pmatrix} f_{13}f_{21} - \begin{pmatrix} 0 \\ 0 \\ a_{31}b_{11}b_{22} \end{pmatrix} f_{23}f_{12} + \begin{pmatrix} 0 \\ 0 \\ b_{22}a_{31} \end{pmatrix} f_{22}f_{13}$$

Based on the one-to-one correspondence between cycle families and determinants, which has been explained in the Appendix A2.1, we are able to give an algebraic reinterpretation of all the coefficients occurring in (21.16). For this purpose, we denote by

$$M \begin{matrix} i_1 i_2 \ldots i_k \\ j_1 j_2 \ldots j_k \end{matrix} \qquad (21.18)$$

a minor of order k of a matrix M, more exactly, the determinant of the $k \times k$ submatrix formed by the common entries of the row vectors $i_1, i_2, \ldots, i_k$ and the column vectors $j_1, j_2, \ldots, j_k$ of M, comp. Eq.(A2.5).

Further, we denote by e_i' the ith unit row vector, i.e.

$$e_i' = \begin{pmatrix} \overset{1}{0} & \overset{2}{0} & \overset{\dots}{\dots} & \overset{i-1}{0} & \overset{i}{1} & \overset{i+1}{0} & \overset{\dots}{\dots} & \overset{n}{0} \end{pmatrix} \qquad (21.19)$$

<u>Theorem 21.3</u>

The coefficients occurring in (21.16) may be computed as follows: For $1 \leq \gamma \leq n$,

$$p_\gamma^o = \sum \sum_{1 \leq k_1 < k_2 < \dots < k_\gamma \leq n} \dots \sum (-A)^{k_1 k_2 \dots k_\gamma}_{k_1 k_2 \dots k_\gamma} \qquad (21.20)$$

$$p_\gamma^{i_1 j_1} = \sum \sum_{1 \leq k_1 < k_2 < \dots < k_\gamma \leq n} \dots \sum \begin{pmatrix} -A & b_{i_1} \\ e_{j_1}' & 0 \end{pmatrix}^{k_1 k_2 \dots k_\gamma, n+1}_{k_1 k_2 \dots k_\gamma, n+1} \qquad (21.21)$$

if $q \geq 2$ then $p_\gamma^{i_1 j_1 \cdot i_2 j_2 \cdots i_q j_q} = 0$ for $1 \leq \gamma \leq q-1$ and

$$p_\gamma^{i_1 j_1 \cdots i_q j_q} = \sum_{1 \leq k_1 < \dots < k_\gamma \leq n} \dots \sum \begin{pmatrix} -A & b_{i_1} & \dots & b_{i_q} \\ e_{j_1}' & 0 & \dots & 0 \\ \vdots & \vdots & \vdots & \vdots \\ e_{j_q}' & 0 & \dots & 0 \end{pmatrix}^{k_1 \dots k_\gamma, n+1 \dots n+q}_{k_1 \dots k_\gamma, n+1 \dots n+q} \qquad (21.21)$$

for $q \leq \gamma \leq n$.

Readers who would like to read a purely algebraic proof of Theorem 21.3 should consult the proof of Theorem 31.1 in Chapter 3. There, elementary algebraic arguments are used to prove the related statements for the more general case of output feedback.

Now, let us demonstrate an interesting relation between Kalman's controllability matrix (comp. Lemma 14.3) and a matrix made-up of the coefficients p_γ^{ij} $(1 \leq \gamma \leq n,\ 1 \leq i \leq m,\ 1 \leq j \leq n)$.

Theorem 21.4

For $i = 1, 2, \dots, m$ there holds the identity

$$(b_i, Ab_i, A^2b_i, \dots, A^{n-1}b_i)\begin{pmatrix} 1 & p_1^o & p_2^o & \cdots & p_{n-1}^o \\ 0 & 1 & p_1^o & \cdots & p_{n-2}^o \\ 0 & 0 & 1 & \cdots & p_{n-3}^o \\ \vdots & \vdots & \vdots & & \vdots \\ 0 & 0 & 0 & \cdots & 1 \end{pmatrix} = -\begin{pmatrix} p_1^{i1} & p_2^{i1} & p_3^{i1} & \cdots & p_n^{i1} \\ p_1^{i2} & p_2^{i2} & p_3^{i2} & \cdots & p_n^{i2} \\ p_1^{i3} & p_2^{i3} & p_3^{i3} & \cdots & p_n^{i3} \\ \vdots & \vdots & \vdots & & \vdots \\ p_1^{in} & p_2^{in} & p_3^{in} & \cdots & p_n^{in} \end{pmatrix} \quad (21.22)$$

where p_γ^o, $1 \leq \gamma \leq n$, are the open-loop characteristic-polynomial coefficients, see (21.15).

Proof: Consider those summands of the closed-loop characteristic polynomial (21.5) that involve the feedback gain f_{ij} and no further feedback gains. They are given by

$$\det\begin{pmatrix} sI_n - A & -b_i & 0 \\ -f_{ij}e_j' & 0 & 0 \\ 0 & 0 & I_{m-1} \end{pmatrix} = f_{ij} \det\begin{pmatrix} sI_n - A & b_i \\ e_j' & 0 \end{pmatrix}$$

$$= f_{ij} \sum_{\gamma=1}^{n} p_\gamma^{ij} s^{n-\gamma} . \quad (21.23)$$

The adjoint matrix Q_{adj} to any $n \times n$ matrix Q is defined by

$$Q\, Q_{adj} = Q_{adj}\, Q = I_n \det Q.$$

Employing a known representation for the matrix $(sI_n - A)_{adj}$ (see, for example, Gantmacher 1966, Ch.4, §3),

$$(sI_n - A)_{adj} = I_n s^{n-1} + (A + p_1^o I_n)s^{n-2} + (A^2 + p_1^o A + p_2^o I_n)s^{n-3}$$

$$+ \dots + (A^{n-2} + p_1^o A^{n-3} + \dots + p_{n-2}^o I_n)s \quad (21.24)$$

$$+ (A^{n-1} + p_1^o A^{n-2} + \dots + p_{n-1} I_n),$$

one obtains with (21.23), (21.8) and (21.24)

$$\sum_{\gamma=1}^{n} p^{ij} s^{n-\gamma} = \det\begin{pmatrix} sI_n - A & b_i \\ e_j' & 0 \end{pmatrix} = -e_j'(sI_n - A)_{adj}\, b_i$$

$$= -e_j' b_i s^{n-1} + e_j'(A + p_1^o I_n) b_i s^{n-2} + \dots + e_j'(A^{n-1} + p_1^o A^{n-2} + \dots + p_{n-1}^o I_n) b_i .$$

Comparing the summands of the same power of s on both sides, one gets

$$-p_1^{ij} = e_j' b_i$$

$$-p_2^{ij} = e_j'(b_i, Ab_i)\begin{pmatrix} p_1^o \\ 1 \end{pmatrix}$$

$$-p_3^{ij} = e_j'(b_i, Ab_i, A^2 b_i)\begin{pmatrix} p_2^o \\ p_1^o \\ 1 \end{pmatrix} \tag{21.25}$$

$$-p_n^{ij} = e_j'(b_i, Ab_i, \ldots, A^{n-2}b_i, A^{n-1}b_i)\begin{pmatrix} p_{n-1}^o \\ p_{n-2}^o \\ \vdots \\ p_1^o \\ 1 \end{pmatrix}$$

For $j = 1, 2, \ldots, n$, all these equations can be compressed into the matrix equation (21.22) that was to be proved.

▲

Finally, it should be noticed that the matrix equation (21.22) can also be understood using purely graph-theoretic arguments. For this purpose we multiply both sides of (21.22) by e_j' from the left and by e_k from the right,

$$-1 \cdot e_j' A^{k-1} b_i + p_1^o \cdot e_j' A^{k-2} b_i + \ldots + p_{k-2}^o e_j' A b_i + p_{k-1}^o \cdot e_j' b_i = -1 \cdot p_k^{ij} \tag{21.26}$$

All the terms occurring in (21.26) admit of a graph-theoretic interpretation. They are subgraphs within the digraph

$$G\begin{pmatrix} A & b_i \\ e_j' & 0 \end{pmatrix} \tag{21.27}$$

The coefficients p_γ^o ($\gamma = 1, 2, \ldots, k-1$) result from the cycle families of width γ without feedback edges (comp. Theorem 21.2), the expressions $e_j' A\, b_i^{\varkappa}$ ($\varkappa = 0, 1, \ldots, k-1$) may be interpreted as the sum of weights of all the paths of length $\varkappa+1$ from input vertex Ii to state vertex j supplemented by a feedback edge (with weight 1) from state vertex j to input vertex Ii (comp. Lemma 14.3), the right-hand side $-1 \cdot p_k^{ij}$ is the (sign-weighted) sum of weights of all the cycle families of width k containing one feedback edge (with weight 1) from j to Ii (comp. Theorem 21.2).

We are going to prove Eq. (21.26) graph-theoretically.

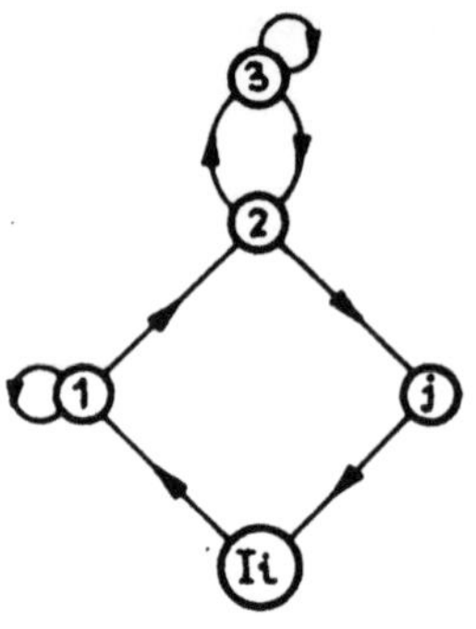

$e_j' A^7 b_i$

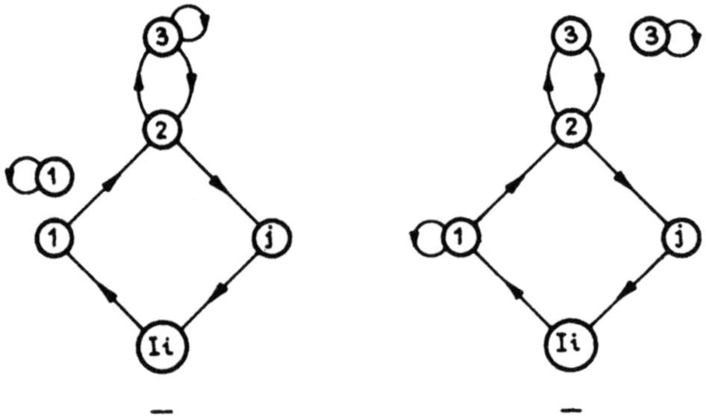

$p_1^0 e_j' A^6 b_i$

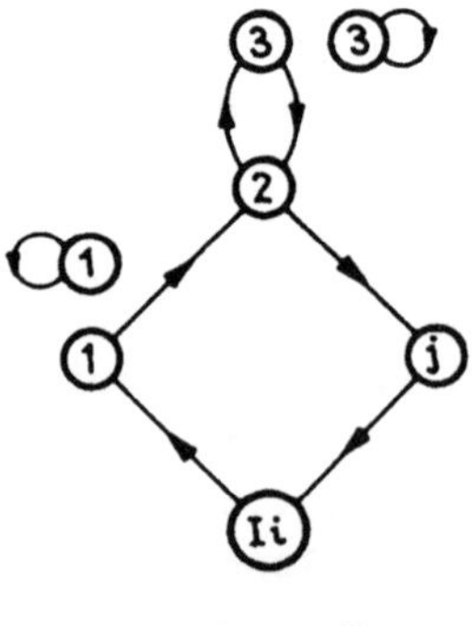

$p_2^0 e_j' A^5 b_i$

Fig. 21.3

Choose a closed path contributing to $e_j^{\prime}A^{k-1}b_i$ and look for all the subgraphs of it that contribute to further terms of (21.26).
If the chosen closed path of length k+1 is a cycle then it contributes to the right-hand side of (21.26) while it does not contain subgraphs contributing to $e_j^{\prime}A^{\varkappa}b_i$ for $\varkappa < k-1$. Taking into account the sign rule stated in Theorem 21.1 the equality between the contributions on both sides of (21.26) is verified.
If the chosen closed path of length k+1 makes a "detour" then this path necessarily contains at least one cycle contributing to some $p_{\varkappa}^{o}$ and a shorter closed subpath contributing to $e_j^{\prime}A^{\varkappa}b_i$ for at least one $\varkappa < k-1$. The chosen closed path cannot contain cycle-families of width k. Hence, it gives no contribution to the right-hand side of (21.26). The signs of the non-vanishing coefficients $p_{\varkappa}^{o}$ ensure the mutual numerical cancellation of the non-zero terms on the left-hand side of (21.26).

<u>Example 21.3</u>: Fig. 21.3 shows an example for k = 8. The closed path of length 8 contributing to $e_j^{\prime}A^{7}b_i$ involves 4 state vertices. It contains two self-cycles that separately contribute to p_1^{o}. Regarded as a pair these self-cycles form a cycle family of width 2 that contributes to p_2^{o}.
There is no contribution to the right-hand side of (21.26).
There are four non-vanishing summands contributing to the left-hand side of (21.26). Each summand has the same absolute value, two of them with negative and two with positive sign. Therefore, they cancel each other numerically.

Until now we have shown the equality of both sides of (21.26) considering separately each closed path that contributes to $e_j^{\prime}A^{k-1}b_i$.
Even more the equality holds for the totality of all the closed paths of length k+1 which contribute to $e_j^{\prime}A^{k-1}b_i$.

▲

21.3 Pole placement for single-input systems

If the system (21.1), (21.2) has $m = 1$ input, then F becomes a row vector f' and B becomes a column vector b.

Assume the system to be structurally controllable, comp. Theorem 14.1 and Theorem 14.2. This implies

$$\text{s-rank}\begin{pmatrix}[A] & [b] \\ [f'] & 0\end{pmatrix} = n+1 \qquad (21.28)$$

or, in graph-theoretic terms, there is at least one cycle family of length $n+1$ forming a spanning subgraph of the digraph

$$G\begin{pmatrix}[A] & [b] \\ [f'] & 0\end{pmatrix} \qquad (21.29)$$

Choose such a cycle family of length $n+1$ and delete its feedback edge. Thus we obtain a spanning subgraph consisting of a simple path and, possibly, disjoint cycles. The property of input-connectability ensures that all of these cycles can be "reached" from the simple path just obtained.

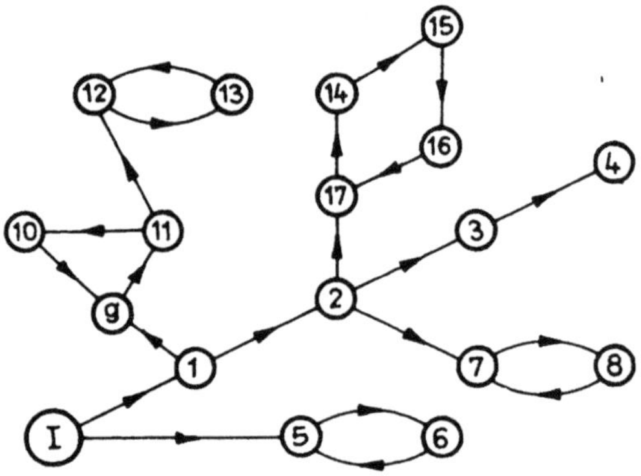

Fig. 21.4

Fig. 21.4 illustrates these features typical of s-controllable single-input systems. Here we have $n = 17$ state vertices. The edge sequence $I \rightarrow 1 \rightarrow 2 \rightarrow 3 \rightarrow 4$ forms a simple path from which all state vertices can be reached.

At this point we recall the historically first characterization of s-controllability published by C.T. Lin in 1974. Using a more flowery language, saying "stem" and "bud" insted of "simple path" and "cycle", respectively, he denoted a spanning subgraph of (21.29) having the structure illustrated in Fig.21.4 as a "cactus". (More exactly: A 'cactus' associated with [A,b] is a spanning input-connected graph consisting of a simple path with the initial vertex I, p (≥ 0) vertex-disjoint cycles and p distinguished edges each of which connects exactly one cycle with the path or with another cycle.) As a consequence, Lin formulated his main result as follows:

Lemma 21.2

The pair [A,b] is s-controllable if and only if its associated digraph is spanned by a cactus.

We shall here prove another graph-theoretic characterization of s-controllability for single-input systems.

Lemma 21.3

The pair [A,b] is s-controllable if and only if there exist within the digraph (21.29) n cycle families such that each of these cycle families contains another feedback edge and has a different width.

Proof: (a) Sufficiency:
Assume there exist n different cycle families as specified in Lemma 21.3. Consider such a set of cycle families. Each of the n feedback edges involved starts from another state vertex j and leads to the input vertex, where $j = 1, 2, \ldots, n$. Each feedback edge belongs to a cycle consisting of a simple path from the input vertex to the state vertex j and the feedback edge under consideration. This implies the input-connectedness of every state vertex $j = 1, 2, \ldots, n$.
By assumption, there is a cycle family of width n in the digraph (21.29). Thus both the conditions for s-controllability stated in Theorem 14.2 are fulfilled.

(b) Necessity:
If the pair [A,b] is s-controllable then such a set of n cycle families as specified in Lemma 21.3 can be constructed as follows:
Choose in the given digraph (21.29) a cycle family of width n, containing a feedback edge, and, additionally, a minimal set of input and state edges such that the input-connectedness of all state vertices remains valid. In the sequel, we consider the subgraph $\bar{G}$ made-up by the chosen edges and all feedback edges.
The digraph $\bar{G}$ is nothing else than a cactus in the sense of Lin, comp. Lemma 21.2, completed by n feedback edges. It should be realized that the reduced digraph $\bar{G}$ is still associated with an s-controllable pair $[\bar{A},\bar{b}] \subset [A,b]$.
The set of final vertices of all input edges in $\bar{G}$ is denoted by S_1 and its cardinality by s_1. If $s_1 > 1$ then $s_1 - 1$ of the S_1-vertices belong to cycles in $\bar{G}$. Denote one of these $s_1 - 1$ state vertices by 1.
If $s_1 = 1$ then S_1 consists of one state vertex which is denoted by 1.

In both cases, the input edge (I,1) together with the feedback edge (1,I) constitute a cycle family of width 1.
Now we try to continue the path from I to 1. The set of final vertices ($\neq 1$) of all state edges starting in 1 is denoted by S_2, its cardinality by s_2. If $s_2 > 1$ then s_2-1 of the S_2-vertices belong to cycles in $\overline{G}$ not yet touched previously. Denote one of these s_2-1 state vertices by 2. If $s_2 = 1$ then S_2 consists of one state vertex which is denoted by 2. In both cases, the edge sequence $\{(I,1), (1,2), (2,I)\}$ constitutes a cycle family of width 2.
Again we try to continue the path $I \rightarrow 1 \rightarrow 2$. The set of final vertices ($\notin \{1, 2\}$) of all state edges starting in 2 is denoted by S_3, its cardinality by s_3. If $s_3 > 1$ then s_3-1 of the S_3-vertices belong to cycles in $\overline{G}$ not yet touched previously. Denote one of these s_3-1 state vertices by 3. If $s_3 = 1$ then S_3 consists of one state vertex which is denoted by 3. The edge sequence $\{(I,1), (1,2), (2,3), (3,I)\}$ constitutes a cycle family of width 3. ...
After a finite number of steps, say j_1 steps, it will be impossible to continue the path $I \rightarrow 1 \rightarrow 2 \rightarrow \ldots \rightarrow j_1$.
Now we look for a state set S_j with the maximal index j, $1 \leq j < j_1$, in which at least one state vertex has not been denoted yet. Let the maximal index be j_2. The state vertices contained in

$$\{S_{j_1}, S_{j_1-1}, \ldots, S_{j_2+1}\}$$

are spanned by a cycle family that constitutes a permanent part of all further cycle families to be constructed in the following.

If there are $\overline{s}_{j_2} > 1$ vertices in S_{j_2} not denoted as yet then $\overline{s}_{j_2}-1$ vertices of S_{j_2} belong to cycles not touched previously. Denote one of these $\overline{s}_{j_2}-1$ vertices by j_1+1. If $\overline{s}_{j_2} = 1$ then denote the corresponding vertex of S_{j_2} by j_1+1.
The edge sequence $\{(I,1), (1,2), \ldots, (j_2-1, j_2), (j_2, j_1+1), (j_1+1, I)\}$ forms a cycle. This cycle together with the cycle family spanned by the state sets $\{S_{j_1}, \ldots, S_{j_2+1}\}$ used previously constitute a cycle family of width j_1+1.

Next we try to continue the process of path prolongation starting from the path $I \rightarrow 1 \rightarrow \ldots \rightarrow j_2-1 \rightarrow j_2 \rightarrow j_1+1$ and looking for final vertices ($\notin \{1, 2, \ldots, j_2, \ldots, j_1, j_1+1\}$) of state edges starting in (j_1+1). ...
The construction process outlined yields n different cycle families whose width is enlargened successively by one. The cycle family of width i, $1 \leq i \leq n$, obtained in this manner is uniquely associated with a state vertex denoted by i during the sketched process.

For the sake of illustration, the cactus of Fig. 21.4 has been drawn again in Fig. 21.5.

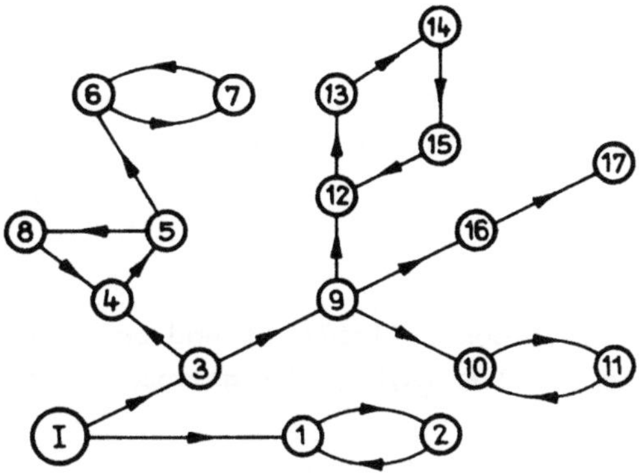

Fig. 21.5

The state vertices, however, have been reordered according to the construction process of this proof.
Thus, the necessity of the proposition stated in Lemma 21.3 has been shown.

▲

For single-input systems, the identity (21.22) may be written as follows:

$$-(b,Ab,A^2b,\dots,A^{n-1}b)\begin{pmatrix} 1 & p_1^o & p_2^o \dots p_{n-1}^o \\ 0 & 1 & p_1^o \dots p_{n-2}^o \\ 0 & 0 & 1 \ \dots p_{n-3}^o \\ \vdots & \vdots & \vdots \\ 0 & 0 & 0 \ \dots \ 1 \end{pmatrix} = \begin{pmatrix} p_1^{11} & p_2^{11} & p_3^{11} \dots p_n^{11} \\ p_1^{12} & p_2^{12} & p_3^{12} \dots p_n^{12} \\ p_1^{13} & p_2^{13} & p_3^{13} \dots p_n^{13} \\ \vdots & \vdots & \vdots \\ p_1^{1n} & p_2^{1n} & p_3^{1n} \dots p_n^{1n} \end{pmatrix} = P^1 \quad (21.30)$$

The entries of the $n \times n$ matrix P^1 can be interpreted graph-theoretically (comp. Theorem 21.2):

> p_w^{1j} ($w = 1,\dots,n$; $j = 1,\dots,n$) is the (sign-weigthed) sum of weights of all those cycle families of width w within the digraph (21.29), that contain one feedback edge (with weight 1), from state vertex j to the input.

On the other hand, there holds the algebraic relation (comp. Theorem 21.3)

$$p_w^{1j} = \sum_{1 \le k_1 < k_2 < \ldots < k_w \le n} \sum \ldots \sum \begin{pmatrix} -A & b \\ e_j' & 0 \end{pmatrix} \begin{matrix} k_1k_2\ldots k_w,n+1 \\ \\ k_1k_2\ldots k_w,n+1 \end{matrix} \qquad (21.31)$$

The matrix P^1 introduced in (21.30) allows nice characterizations of both structural controllability and numerical controllability of single-input systems.

<u>Lemma 21.4</u>

The pair $[A,b]$ is s-controllable if and only if, for at least one admissible pair $(A,b) \in [A,b]$, the matrix P^1 meets the following two conditions:

(a) P^1 contains no zero row vector.

(b) The nth column vector of P^1 does not vanish.

<u>Proof</u>: Condition (a) is equivalent to input-connectability. Because of (21.31) it can be easily seen that condition (b) is equivalent to

$$\text{s-rank}[A,b] = n. \qquad (21.32)$$

Thus, Lemma 21.4 is an immediate consequence of Theorem 14.1.

▲

<u>Lemma 21.5</u>

The pair (A,b) is controllable if and only if the matrix P^1 has full rank, i.e.

$$\text{rank } P^1 = n \qquad (21.33)$$

or, equivalently,

$$\det P^1 \neq 0. \qquad (21.34)$$

<u>Proof</u>: The identity (21.30) implies

$$\text{rank}(b, Ab, A^2b, \ldots, A^{n-1}b) = \text{rank } P^1.$$

Now the statement of Lemma 21.5 is obvious taking into account Kalman's controllability criterion (14.5).

▲

An important consequence of the foregoing considerations is

<u>Theorem 21.5</u>

A single-input system has the property of arbitrary pole assignability under static state feedback if and only if the system is controllable. (This statement has been well-known for many years.)

Let

$$p^o = (p_1^o, p_2^o, \dots, p_n^o)' \quad \text{and} \quad p = (p_1, p_2, \dots, p_n)'$$

denote the column vectors encompassing the coefficients of the open-loop characteristic polynomial and of the closed-loop characteristic polynomial, respectively, then the feedback gains providing the desired pole assignment are uniquely determined and may be computed as

$$f' = (p - p^o)' (P^1)^{-1} \tag{21.35}$$

Proof: For single-input systems, Eq.(21.16) reads as

$$p = p^o + \sum_{j=1}^{n} p^{1j} f_j = p^o + (P^1)' f \tag{21.36}$$

Eq.(21.36) can be uniquely solved for f if and only if rank $P^1 = n$. This is equivalent to controllability (see Lemma 21.5).
Under the assumption (21.33) the feedback gains forming the column vector f are given by (21.35). This completes the proof.

▲

Example 21.4: Consider the plant

$$\begin{pmatrix} \dot{x}_1 \\ \dot{x}_2 \\ \dot{x}_3 \end{pmatrix} = \begin{pmatrix} a_{11} & a_{12} & 0 \\ 0 & 0 & 0 \\ a_{31} & a_{32} & 0 \end{pmatrix} \begin{pmatrix} x_1 \\ x_2 \\ x_3 \end{pmatrix} + \begin{pmatrix} 0 \\ b_2 \\ 0 \end{pmatrix} u \tag{21.37}$$

and the associated digraph

$$G \left(\begin{array}{ccc|c} a_{11} & a_{12} & 0 & 0 \\ 0 & 0 & 0 & b_2 \\ a_{31} & a_{32} & 0 & 0 \\ \hline f_1 & f_2 & f_3 & 0 \end{array} \right) \tag{21.38}$$

drawn in Fig. 21.6.

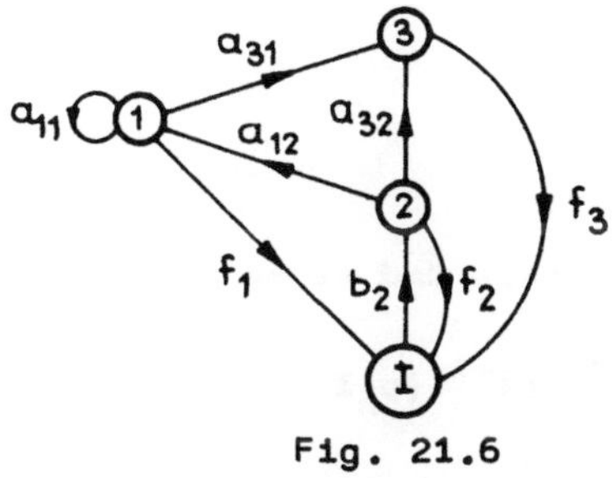

Fig. 21.6

The vector p^o and the matrix P^1 can immediately be obtained from Fig. 21.6, using the graphical interpretation of p^o and P^1:

$$p^o = -\begin{pmatrix} a_{11} \\ 0 \\ 0 \end{pmatrix}, \quad (P^1)' = -\begin{pmatrix} 0 & b_2 & 0 \\ b_2a_{12} & b_2(a_{32}-a_{11}) & b_2a_{32} \\ 0 & b_2(a_{12}a_{31}-a_{32}a_{11}) & b_2(a_{12}a_{31}-a_{32}a_{11}) \end{pmatrix} \tag{21.39}$$

The system under investigation has the property of arbitrary pole assignability under static state feedback if and only if

$$\det P^1 = b_2^3a_{12}(a_{12}a_{31}- a_{32}a_{11}) \neq 0 \tag{21.40}$$

Let the numerical parameter values be given by

$$a_{11} = b_2 = 1, \quad a_{12} = a_{31} = a_{32} = -1 \tag{21.41}$$

In this case the necessary and sufficient condition (21.40) is met. Assume the desired closed-loop system poles to be -2, $-1+j$, $-1-j$. Then the corresponding characteristic polynomial is

$$(s + 2)(s + 1 - j)(s +1 + j) = s^3 + 4s^2 + 6s + 4,$$

i.e.

$$p = \begin{pmatrix} 4 \\ 6 \\ 4 \end{pmatrix} \tag{21.42}$$

The desired pole assignment will be achieved if and only if the feedback gains are chosen as solutions of the following system of equations:

$$\begin{pmatrix} 0 & 1 & 0 \\ -1 & -2 & -1 \\ 0 & 2 & 2 \end{pmatrix} \begin{pmatrix} f_1 \\ f_2 \\ f_3 \end{pmatrix} = -\begin{pmatrix} 4 \\ 6 \\ 4 \end{pmatrix} + \begin{pmatrix} -1 \\ 0 \\ 0 \end{pmatrix} = -\begin{pmatrix} 5 \\ 6 \\ 4 \end{pmatrix}$$

One obtains

$$\begin{pmatrix} f_1 \\ f_2 \\ f_3 \end{pmatrix} = \begin{pmatrix} 13 \\ -5 \\ 3 \end{pmatrix} \tag{21.43}$$

<u>Remark</u>: Eq. (21.35) may be regarded as an alternative to Ackermann's formula

$$f' = (0,0,\dots,1)(b,Ab,\dots,A^{n-1}b)^{-1} \, (p_nI + p_{n-1}A + \dots + p_1A^{n-1} + A^n)$$

$$= (p_n,p_{n-1},\dots,p_1,1) \begin{pmatrix} e' \\ e'A \\ \vdots \\ e'A^n \end{pmatrix} \quad \text{where} \quad e' = (0,0,\dots,1)(b,Ab,\dots,A^{n-1}b)^{-1}$$

21.4 Pole placement for multi-input systems

Assume the system to be structurally controllable, comp. Theorem 14.1 and Theorem 14.2.

First, let us try to generalize Lemma 21.2 for multi-input systems.

Consider the digraph

$$G\begin{pmatrix} A & B \\ F & 0 \end{pmatrix} \tag{21.44}$$

and choose a cycle family of width n. This cycle family forms a subgraph G' of (21.44). Complete G' by a minimal set of further edges of (21.44) such that each state vertex becomes input-connected in the resulting subgraph G". Possibly, there are certain input vertices involved in G", but not involved in G'. Denote the subset of input vertices of G" not involved in G' by I". For each input vertex Ii of I" we may locally alter the subgraph G" as follows (comp. Fig. 21.7):

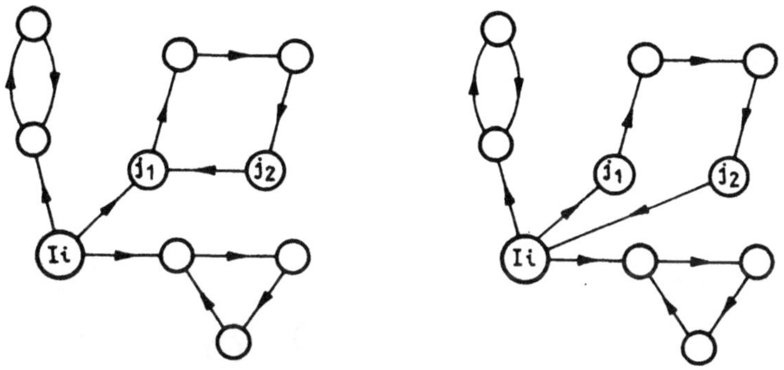

Fig. 21.7

Choose one of the cycles directly connected with Ii. Denote its state vertex adjacent to Ii by j_1. It is the final vertex of a state edge starting in j_2. Delete the state edge (j_2, j_1) and replace it by the feedback edge (j_2, Ii). In this way we obtain a subgraph G"' of (21.44). Let G"' involve exactly $\overline{m}$ feedback edges, $1 \leq \overline{m} \leq m$.
Removal of all these $\overline{m}$ feedback edges supplies us a further subgraph $\overline{G}$. It consists of a set of $\overline{m}$ disjoint "cacti" (see Lemma 21.2), each rooted in another input vertex and, together, involving all n state vertices of the digraph (21.44).

Thus we have substantiated the following generalization of Lemma 21.2; comp. Lin 1977:

Lemma 21.6

The $n \times (n+m)$ structure matrix pair [A,B] is s-controllable if and only if its associated digraph (21.44) contains a set of $\bar{m}$ ($\leq m$) disjoint cacti, each of them rooted in another input vertex and, together, touching all n state vertices.

Example 21.5: Fig. 21.8.a exhibits the digraph $G\begin{pmatrix} A & B \\ 0 & 0 \end{pmatrix}$ for an example system with $n = 21$ and $m = 4$.
A possible set of $\bar{m} = 3$ disjoint cacti with the properties indicated in Lemma 21.6 is shown in Fig. 21.8.b.

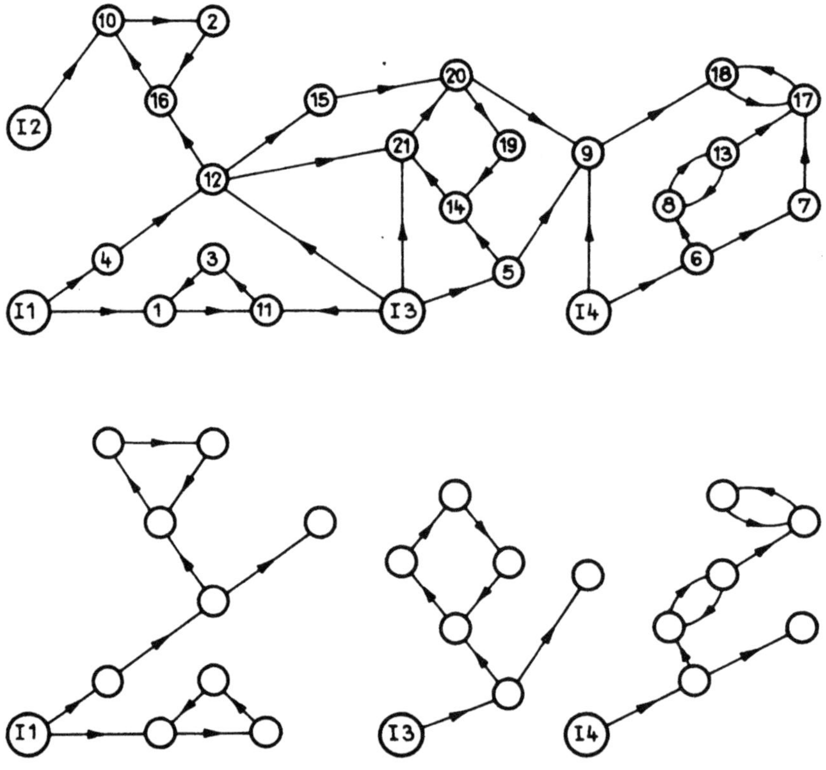

Fig. 21.8

Assume the systems under investigation to be structurally controllable. Then there holds

Theorem 21.6

Structural controllability of a class of systems [A,B] implies the arbitrary pole assignability by static state feedback for almost all actual systems $(A,B) \in [A,B]$.

Proof (constructive): The task of pole assignment may be accomplished as follows:

Step 1: Consider the digraph (21.44), and choose a subgraph consisting of $\bar{m}$ disjoint cacti as indicated in Lemma 21.6. Enumerate these cacti arbitrarily from 1 to $\bar{m}$. (There are $\bar{m}$! possible different enumerations.)

Step 2: Choose an $m \times n$ feedback structure matrix $[\bar{F}]$ whose $\bar{m}-1$ non-vanishing elements correspond to feedback edges leading from the final vertex of the path of cactus j to the root of cactus j+1 ($j = 1,\dots,\bar{m}-1$).
Choose a unit structure vector $[g]$ such that the input vertex associated with the column structure vector $[Bg]$ is the root of cactus 1.
The matrix $[\bar{F}]$ and the vector $[g]$ make sure the s-controllability of the single-input pair $[A + B\bar{F}, Bg]$. Moreover, for almost all admissible $(A,B) \in [A,B]$, $\bar{F} \in [\bar{F}]$, $g \in [g]$, the pair $(A + B\bar{F}, Bg)$ is controllable in the traditional numerical sense.

Step 3: Choose admissible $(A,B,\bar{F},g) \in [A,B,\bar{F},g]$ such that the pair $(A + B\bar{F}, Bg)$ becomes controllable.
Set up the system of equations

$$p - \bar{p}^{o} = (\bar{P}^{1})' f \tag{21.45}$$

where p encompasses the characteristic polynomial coefficients defined by the desired pole locations of the closed-loop system,
$\bar{p}^{o}$ is determined by the cycle families in $G(A + B\bar{F})$,
$\bar{P}^{1}$ is determined by the cycle families in

$$G\begin{pmatrix} A + B\bar{F} & Bg \\ 11\dots1 & 0 \end{pmatrix}$$

or, in algebraic terms (comp. (21.31)),

$$\bar{p}^{1j} = \sum_{1\le k_1 <} \sum_{k_2 <} \dots \sum_{\dots < k_w \le n} \begin{vmatrix} -(A+B\bar{F}) & Bg \\ e_j' & 0 \end{vmatrix}^{k_1\dots k_w, n+1}_{k_1\dots k_w, n+1} \tag{21.46}$$

From Lemma 21.5 it is known that

$$\operatorname{rank} \bar{P}^{1} = n \tag{21.47}$$

if and only if $(A + B\bar{F}, Bg)$ is controllable.
Hence, Eq. (21.45) can be solved for f for almost all $(A,B) \in [A,B]$.

Step 4: The overall feedback matrix

$$F = \bar{F} + gf' \tag{21.48}$$

provides the desired pole placement.

▲

It should be noted that, by contrast with the single-input case, the feedback matrix F given by (21.48) is not uniquely determined neither by the structure matrix [A,B] nor by an admissible realization of the plant (A,B). The numerical values of the non-vanishing entries of $\bar{F}$ may be chosen independently both of the actual pair and of the desired pole locations. The numerical values of the n-vector f depend on A, B, $\bar{F}$, g, and on the prescribed pole places.

Example 21.6: Consider again the example system introduced above as Example 21.5. After completing the subgraph consisting of $\bar{m} = 3$ disjoint cacti (see Fig. 21.8) by two appropriately chosen feedback edges there results a single large cactus (Fig. 21.9.a).
The edge sequences $15 \rightarrow I3 \rightarrow 5$ and $9 \rightarrow I4 \rightarrow 6$ play the same role as state edges $15 \rightarrow 5$ and $9 \rightarrow 6$, respectively, see Fig.21.9.b.
The $\bar{F}$-entries different from $\bar{f}_{3,15}$ and $\bar{f}_{4,9}$ may be put equal to zero.
The whole system characterized by the system matrix

$$A + B\bar{F} \tag{21.49}$$

may be controlled from the single input I1.
Choosing $g = (1\ 0\ 0\ 0)'$ the pair $[A + B\bar{F}, Bg]$ is s-controllable.
In the digraph

$$G\begin{pmatrix} A + B\bar{F} & Bg \\ f' & 0 \end{pmatrix} \tag{21.50}$$

there exist $n = 21$ cycle families such that each of them contains another feedback edge and has a different width. The state vertices have been reordered accordingly, comp. the proof of Lemma 21.3 and see Fig. 21.9.c.

The proof of Theorem 21.6 gives evidence for the following statement:

Corollary 21.1

In case of controllable systems under static state feedback, the determination of the feedback gains ensuring any desired pole placement demands essentially the resolution of the system (21.45) of linear algebraic equations.

So far we have discussed the digraph approach to a known result proved algebraically by W.M. Wonham as early as in 1967.

> A plant (21.2) has the property of arbitrary pole assignability under static state feedback (21.1) if and only if the pair (A,B) is controllable.

Besides, it has been explained how an appropriate feedback matrix F can

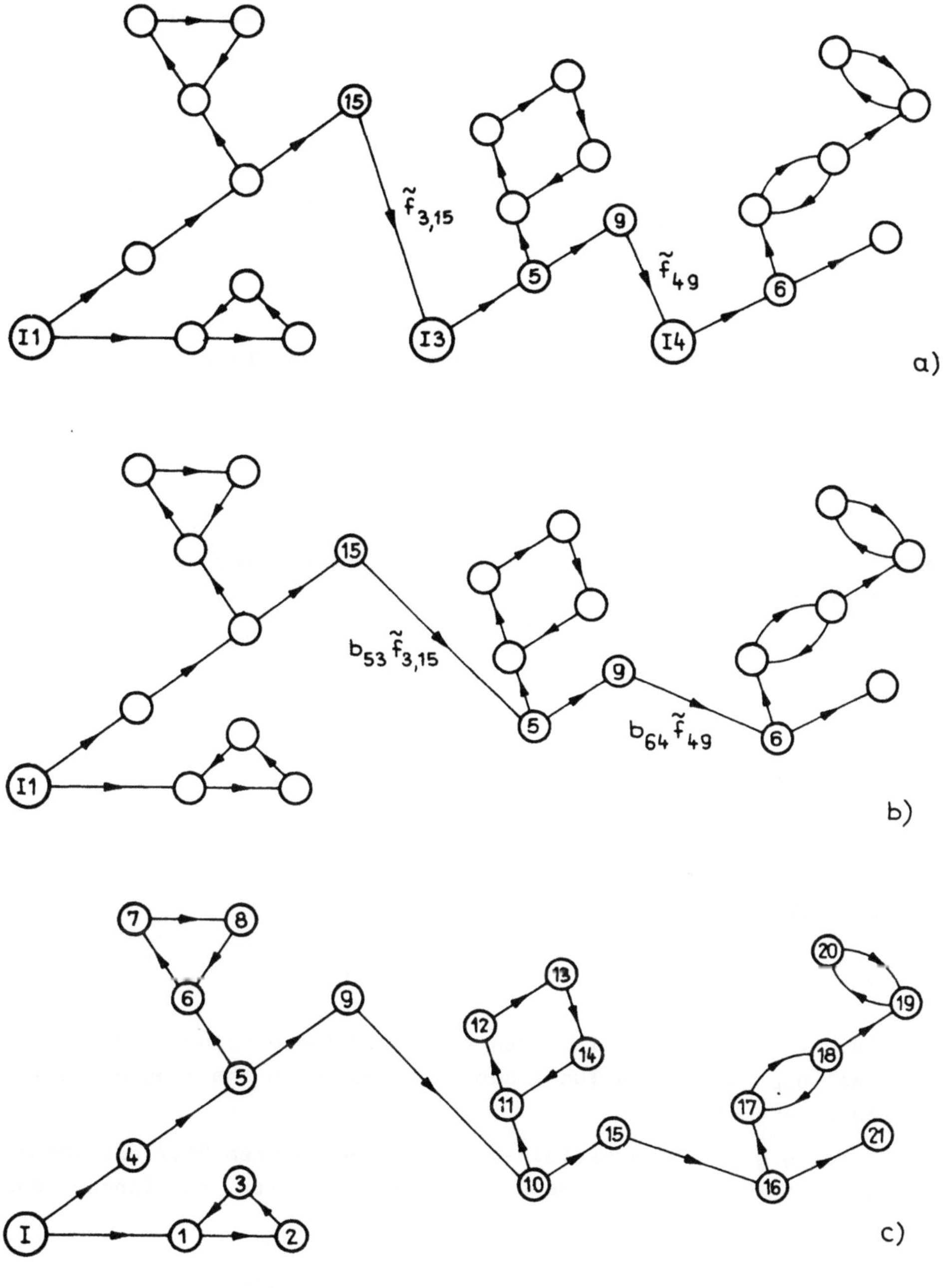

Fig. 21.9

be found. However, for multi-input systems specification of the closed-loop poles does not uniquely define F. In the following let us identify the degrees of freedom offered by static state feedback beyond specification of the closed-loop poles.

21.5 Determination of all the feedback matrices which provide the desired pole placement

As above, let $p = (p_1, p_2, \dots, p_{n-1}, p_n)'$ be a column vector encompassing the coefficients of the closed-loop characteristic polynomial. If the matrices A and B are numerically fixed, then p may be regarded as a function of F,

$$p = g(F) \tag{21.51}$$

Each real $m \times n$ matrix F may be regarded as an element of a real Euclidean vector space $\mathbb{R}^{m \cdot n}$. Eq. (21.51) defines a smooth mapping

$$g: \mathbb{R}^{m \cdot n} \longrightarrow \mathbb{R}^n \tag{21.52}$$

In case of controllable pairs (A,B) the mapping g is surjective, since each $p \in \mathbb{R}^n$ is a possible image. (Surjective mappings are sometimes called to be "onto".)

We use $T_F\,\mathbb{R}^{m \cdot n}$ and $T_{g(F)}\,\mathbb{R}^n$ to denote, respectively, the tangent space to $\mathbb{R}^{m \cdot n}$ at F and the tangent space to $\mathbb{R}^n$ at g(F).
The elements of $T_F\,\mathbb{R}^{m \cdot n}$ and $T_{g(F)}\,\mathbb{R}^n$ are denoted by dF and dp, respectively. The differential of g at the point $F \in \mathbb{R}^{m \cdot n}$ is a linear mapping from $T_F\,\mathbb{R}^{m \cdot n}$ to $T_{g(F)}\,\mathbb{R}^n$,

$$g_{*F}: T_F\,\mathbb{R}^{m \cdot n} \longrightarrow T_{g(F)}\,\mathbb{R}^n \tag{21.53}$$

or, expressed in more traditional terms,

$$dp = g_{*F}(dF) \tag{21.54}$$

Provided the pair (A,B) is given then both the mapping g and its differential g_{*F} appear as functions that depend on the feedback gains in a transparent way.

The mapping g has been explained above, see Theorem 21.2 and Theorem 21.3. As an immediate consequence of those results, the linear mapping (21.54) may be represented as follows:

$$dp = g_{*F}(dF) = \sum_{i_1=1}^{m} \sum_{j_1=1}^{n} \Big(p^{i_1 j_1} + \sum_{\substack{i_2 \\ > i_1}} \sum_{\substack{j_2 \\ \neq j_1}} p^{i_1 j_1 \cdot i_2 j_2} f_{i_2 j_2} \tag{21.55}$$
$$+ \dots + \underbrace{\sum \dots \sum}_{\substack{i_2 < \dots < i_q \\ > i_1}} \underbrace{\sum \dots \sum}_{\substack{j_2 \neq \dots \neq j_q \\ \neq j_1}} p^{i_1 j_1 \cdot i_2 j_2 \cdots i_q j_q} f_{i_2 j_2} \cdots f_{i_q j_q} \Big) df_{i_1 j_1}$$

Definition 21.2

A point $F \in \mathbb{R}^{m \cdot n}$ is said to be a regular point of the mapping (21.51) if

$$\text{rank } g_{*F} = n \tag{21.56}$$

Otherwise, F is said to be a non-regular point.

Theorem 21.7

For controllable pairs (A,B), the set of regular F-points is open and dense in $R^{m \cdot n}$.
The complementary set of non-regular F-points forms an algebraic subset $M_1 \subset R^{m \cdot n}$ that can be explicitly computed,

$$M_1 = \left\{F \in \mathbb{R}^{m \cdot n}: \text{rank } g_{*F} < n\right\}$$

$$= \left\{F \in \mathbb{R}^{m \cdot n}: \det (g_{*F})(g_{*F})' = 0\right\}$$

$$= \bigcap_{(j)} \left\{F \in \mathbb{R}^{m \cdot n}: (g_{*F})\begin{matrix} 1\ 2\ \dots\ n \\ j_1 j_2\ \dots j_n \end{matrix} = 0\right\} \tag{21.57}$$

where $(g_{*F})\begin{matrix} 1\ 2\ \dots\ n \\ j_1 j_2\ \dots j_n \end{matrix}$ is the determinant of the $n \times n$ submatrix formed by the column vectors $j_1, j_2, \dots, j_n$ of g_{*F}, and $\bigcap_{(j)}$ symbolizes the intersection of subsets $\{F \in \mathbb{R}^{m \cdot n}: \dots\}$ each defined by a sequence $1 \leq j_1 < j_2 < \dots < j_n \leq m \cdot n$.

Proof: The existence of regular F-points is obvious because the smooth mapping g is onto. Relation (21.57) is an immediate consequence of Definition 21.2. Eq. (21.55) implies that the minors occurring in (21.57) are polynomials with the feedback gains as arguments. Hence, M_1 is an algebraic subset of $\mathbb{R}^{m \cdot n}$ which can be computed as stated in Theorem 21.7.

▲

At this point it should be stressed that the controllability of (A,B) does not imply regularity of g(F) for each $F \in \mathbb{R}^{m \cdot n}$.

Example 21.7: Consider the example system introduced by Eq. (21.10). The mapping g has been given by Eq. (21.17). Putting all non-vanishing entries of A and B equal to 1 one obtains the controllable pair

$$(A,B) = \left(\begin{array}{ccc|cc} 1 & 1 & 0 & 1 & 0 \\ 0 & 0 & 0 & 0 & 1 \\ 1 & 1 & 0 & 0 & 0 \end{array}\right) \tag{21.58}$$

the mapping

$$p = \begin{pmatrix} p_1 \\ p_2 \\ p_3 \end{pmatrix} = g(F) = \begin{pmatrix} -1 \\ 0 \\ 0 \end{pmatrix} - \begin{pmatrix} 1 \\ 0 \\ 0 \end{pmatrix} f_{11} - \begin{pmatrix} 0 \\ 1 \\ 0 \end{pmatrix} (f_{13} + f_{21} + f_{23} + f_{12}f_{21} - f_{11}f_{22})$$

$$+ \begin{pmatrix} -1 \\ 1 \\ 0 \end{pmatrix} f_{22} + \begin{pmatrix} 0 \\ 0 \\ 1 \end{pmatrix} (f_{11}f_{23} - f_{13}f_{21} - f_{23}f_{12} + f_{22}f_{13}) \tag{21.59}$$

and its differential (see (21.55))

$$g_{*F} = \begin{pmatrix} -1 & 0 & 0 & 0 & -1 & 0 \\ f_{22} & -f_{21} & 1 & -1-f_{12} & 1+f_{11} & -1 \\ f_{23} & -f_{23} & f_{22}-f_{21} & -f_{13} & f_{13} & f_{11}-f_{12} \end{pmatrix} \tag{21.60}$$

The subset M_1 defined by (21.57) may be obtained by elementary transformations well-known from linear algebra,

$$M_1 = \left\{ F \in \mathbb{R}^6 : \text{rank } g_{*F} < 3 \right\}$$

$$= \left\{ F \in \mathbb{R}^6 : \text{rank} \begin{pmatrix} -1 & 0 & 0 & 0 & 0 & 0 \\ f_{22} & -f_{21} & 1 & -1-f_{12} & 1+f_{11}-f_{22} & -1 \\ f_{23} & -f_{23} & f_{22}-f_{21} & -f_{13} & f_{13}-f_{23} & f_{11}-f_{12} \end{pmatrix} < 3 \right\}$$

$$= \left\{ F \in \mathbb{R}^6 : \text{rank} \begin{pmatrix} -f_{21} & 1 & -1-f_{12} & 1+f_{11}-f_{22} & -1 \\ -f_{23} & f_{22}-f_{21} & -f_{13} & f_{13}-f_{23} & f_{11}-f_{12} \end{pmatrix} < 2 \right\}$$

$$= \left\{ F \in \mathbb{R}^6 : f_{13} = 0,\ f_{23} = 0,\ f_{21} = f_{22},\ f_{11} = f_{12} \right\} \tag{21.61}$$

This means the subset M_1 is a two-dimensional plane in the Euclidean 6-dimensional space.

<u>Definition 21.3</u>

An element $p \in \mathbb{R}^n$ is said to be <u>regular value</u> of the mapping (21.51) if each $F \in g^{-1}(p)$ – with $g^{-1}(p)$ denoting the inverse image of p – is a regular point in $\mathbb{R}^{m \cdot n}$.

An element $p \in \mathbb{R}^n$ is said to be <u>critical value</u> of the mapping (21.51) if there exists a non-regular point $F \in g^{-1}(p)$.

Theorem 21.8

The set N_1 of critical values of the mapping (21.51) is the image of the set M_1 (see (21.57)) of non-regular F-points,

$$N_1 = g(M_1) \tag{21.62}$$

The complementary subset

$$\overline{N}_1 = \mathbb{R}^n \setminus N_1 \tag{21.63}$$

constitutes the set of regular values of the map (21.51).

Proof: The first statement is an immediate consequence of the Definition 21.3 of critical values.
The second statement holds because the mapping (21.51) is surjective.

Example 21.8: For the system of Example 21.7 the set N_1 follows from (21.61) and (21.62) as

$$N_1 = g(M_1) = \begin{pmatrix} -1 \\ 0 \\ 0 \end{pmatrix} - \begin{pmatrix} 1 \\ 0 \\ 0 \end{pmatrix} (f_{11}+f_{21}) \tag{21.64}$$

The subset N_1 is a one-dimensional straight line in the Euclidean 3-dimensional space.

Theorem 21.9

Let $p \in \mathbb{R}^n$ be a regular value of the mapping (21.51).
Then the inverse image

$$M = g^{-1}(p) \tag{21.65}$$

is a smooth manifold of dimension

$$\dim M = n(m - 1) \tag{21.66}$$

Some $n(m-1)$ local coordinates of $\mathbb{R}^{m \cdot n}$ may be used as local coordinates of M.

Proof: Consider a given regular $p \in \mathbb{R}^n$. Choose a point $F \in \mathbb{R}^{m \cdot n}$ such that $p = g(F)$. This is always possible because the mapping g is onto. From Definition 21.2 and Definition 21.3 it is known that $\operatorname{rank} g_{*F} = n$. Now apply the Implicit Function Theorem: There exists a neighbourhood $U \ni F$ that is locally homeomorph to the Euclidean space $\mathbb{R}^{(m-1)n}$. Some $(m-1)n$ entries of the feedback matrix F may be chosen as local coordinates in U, say $(f_{i_1}, \ldots, f_{i_{(m-1)n}})$. The remaining n local

coordinates appear as smooth functions of $(f_{i_1}, f_{i_2}, \ldots, f_{i_{(m-1)n}})$.
This implies M to be a smooth manifold of dimension (m-1)n.

▲

__Corollary 21.2__

Assume the desired poles $s_1^c, \ldots, s_n^c$ of the closed-loop system to be simple and distinct from the eigenvalues of A.
Then the n(m-1)-dimensional manifold of state feedback matrices F each of which provides the pole assignment is given by

$$F = -(w_1, w_2, \ldots, w_n)[(A-s_1^c I)^{-1} Bw_1, (A-s_2^c I)^{-1} Bw_2, \ldots, (A-s_n^c I)^{-1} Bw_n]^{-1} \quad (21.67)$$

where each of the n vectors $w_1, w_2, \ldots, w_n$ represents (m-1) free scalar parameters.

__Proof__: We denote by v_i the eigenvector associated with s_i^c, i.e.

$$(A + BF)v_i = s_i^c v_i \quad (21.68)$$

Putting

$$Fv_i = w_i \quad (21.69)$$

one obtains $(A-s_i^c I)v_i = -Bw_i$.
By assumption, both the matrices $(A-s_i^c I)$ for $i = 1, \ldots, n$ and the modal matrix $(v_1, v_2, \ldots, v_n)$ are non-singular. Thus we can write

$$-(v_1 v_2, \ldots, v_n) = [(A-s^c I)^{-1} Bw_1, (A-s^c I)^{-1} Bw_2, \ldots, (A-s^c I)^{-1} Bw_n] \quad (21.70)$$

Multiplying both sides by F from the left and taking into account (21.69) we get

$$-(w_1, w_2, \ldots, w_n) = F[(A-s_1^c I)^{-1} Bw_1, \ldots, (A-s_n^c I)^{-1} Bw_n] \quad (21.71)$$

From this follows (21.67).
Obviously, the m-vectors w_i may be interpreted as parameter vectors. Any set of parameter vectors $(\alpha_1 w_1, \alpha_2 w_2, \ldots, \alpha_n w_n)$ with non-vanishing scalars $\alpha_1, \alpha_2, \ldots, \alpha_n$ gives the same results as the set $(w_1, w_2, \ldots, w_n)$.
Therefore, we have n(m-1) free parameters.

▲

In an explicit manner Eq. (21.67) was first derived and recognized as "a procedure for computing F" by B.C. Moore in 1976. G. Roppenecker has used this relation in 1981 and later on. There the reader can find many interesting applications, see also O. Föllinger 1986.

22 Disturbance rejection

22.1 Problem formulation and preliminary results

Consider an externally disturbed system mathematically described by the equations

$$\dot{x}(t) = A\, x(t) + B\, u(t) + D\, v(t) \tag{22.1}$$

$$y(t) = C\, x(t) \tag{22.2}$$

where $x(t) \in \mathbb{R}^n$, $u(t) \in \mathbb{R}^m$, and $y(t) \in \mathbb{R}^r$ symbolize the state vector, the input vector, and the output vector as above. The newly introduced vector $v(t) \in \mathbb{R}^q$ represents the disturbances. The $n \times q$ matrix D is assumed to be real and time-invariant.

The problem of complete disturbance rejection is to find a state feedback matrix F such that

$$u(t) = F\, x(t) \tag{22.3}$$

so that the disturbances v(t) have no influence on the outputs y(t).

Fig. 22.1 shows the generalized digraph associated with the square matrix

$$Q_5 = \begin{pmatrix} 0 & C & 0 & 0 \\ 0 & A & B & D \\ 0 & E & 0 & 0 \\ 0 & 0 & 0 & 0 \end{pmatrix} \quad \text{of order } r+n+m+q \tag{22.4}$$

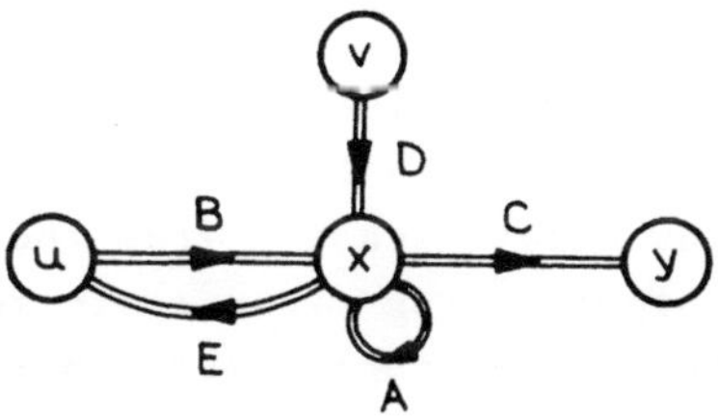

Fig. 22.1

All entries of the $m \times n$ matrix E may be freely chosen.

In Fig. 22.2 a characteristic part of the digraph $G(Q_5)$ has been sketched.

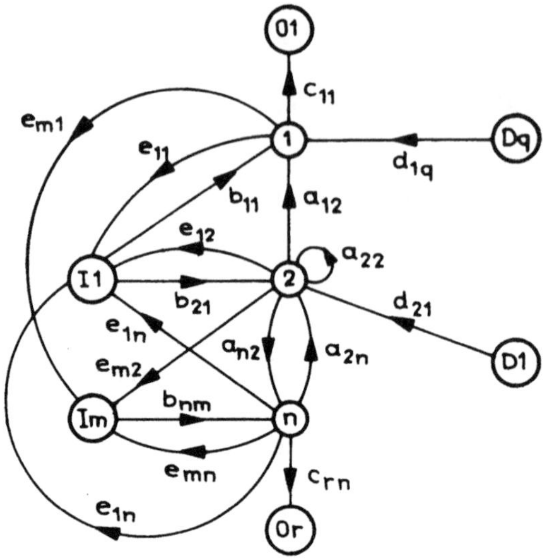

Fig. 22.2

Now we are able to formulate the problem of complete disturbance rejection in graph-theoretic terms.

Map the disturbed plant (22.1), (22.2) into the digraph $G(Q_6)$ associated with

$$Q_6 = \begin{pmatrix} 0 & C & 0 & 0 \\ 0 & A & B & D \\ 0 & 0 & 0 & 0 \\ 0 & 0 & 0 & 0 \end{pmatrix} \qquad (22.5)$$

Complete disturbance rejection means that we are able to compensate simultaneously each of the paths leading from a disturbance vertex to an output vertex by a path leading from the same disturbance vertex to the same output vertex via a state feedback edge. More definitely, each of the simple paths in the digraph $G([Q_6])$ leading from a disturbance vertex to an output vertex must contain a state edge (j,i) that can be compensated by edge pairs $\{(j,Ik), (Ik,i)\}$ with the aid of an admissible feedback matrix $F \in [E]$.

Compensation takes place if and only if

$$a_{ij} = - \sum_k b_{ik} f_{kj} \qquad (22.6)$$

For the example system shown in Fig. 22.2 complete disturbance rejection is impossible because the path $\{(v_q,1), (1,y_1)\}$ cannot be affected

by any state feedback (22.3).
This observation can be generalized at once.

<u>Lemma 22.1</u>

Complete disturbance rejection is impossible if there is in $G(Q_6)$ a path of length 2 from a disturbance vertex to an output vertex, or, said in algebraic terms,
complete disturbance rejection is impossible if there are admissible realizations $C \in [C]$ and $D \in [D]$ such that

$$C\,D = 0 \qquad (22.7)$$

Looking for admissible feedback matrix candidates $F \in [E]$ by which disturbance rejection might be achieved two important observations must be taken into account.
First, feedback edges with non-vanishing weights can produce new paths from disturbance vertices to output vertices.
Second, only those state edges can be compensated whose final vertices are incident with an input edge.

22.2 A necessary condition and a sufficient condition for disturbance rejection

Definition 22.1

The maximal subset of the output-connected state vertices whose output-connectedness cannot be compensated by state feedback is denoted by V_{max}.

The state subset V_{max} may be determined as follows:

First, reverse the edge orientations in the digraph $G([Q_o])$ associated with

$$Q_o = \begin{pmatrix} 0 & C & 0 \\ 0 & A & B \\ 0 & 0 & 0 \end{pmatrix} \quad \text{introduced in (11.12).}$$

Second, pursue each of the paths starting in any output vertices until a state vertex adjacent to an input vertex has been reached.

The set of all state vertices touched during this process constitutes the subset V_{max}.

A formal algorithm for the determination of V_{max} has been written down below (Fig. 22.11).

By V_D we denote the subset of state vertices in $G([Q_6])$, see (22.5), adjacent to disturbance vertices, i.e.

$$V_D = \left\{ i:\ i \in \{1,2,..,n\},\ (Dj,i) \text{ exists for some } j \in \{1,..,q\} \right\} \qquad (22.8)$$

Lemma 22.1 may now be extended to a more comprehensive necessary structural condition for complete disturbance rejection.

Theorem 22.1

For complete disturbance rejection it is necessary that the state subsets V_{max} and V_D are disjoint,

$$V_{max} \cap V_D = \emptyset \ . \qquad (22.9)$$

Proof: Each state vertex of V_{max} must not be adjacent to disturbance vertices. Else disturbance rejection by state feedback would be out of the question. ▲

Assume condition (22.9) to be met. In other words, the disturbances structurally characterized by $[D]$, see (22.1), act directly only on states of

$$\bar{V}_{max} = \{1, 2, ..., n\} \setminus V_{max} \qquad (22.10)$$

In order to decouple from the system outputs the effect of disturbances we have to choose the feedback gain matrix F in such a way that the weights of all paths from disturbance vertices to output vertices vanish in the closed-loop system. For this purpose, we try to compensate all edges of an appropriately chosen cut set of edges in $G([Q_6])$.

Definition 22.2

A disturbance-output cut set, for short, D-O cut set, in the digraph $G([Q_6])$, see (22.5), is defined as a minimal set of state edges whose removal would disconnect all paths leading from one of the disturbance vertices D1,..., Dq to one of the output vertices O1,..., Or.

The attribute "minimal" indicates that no proper subset of those state edges is capable of cutting each of the disturbance-output paths simultaneously.

The set of initial vertices (final vertices) of all edges of a D-O cut set is denoted by T_0 (T_1).

Every D-O cut set induces a natural partitioning of the state vertices as well as of the input vertices. For illustration, see Fig. 22.3.

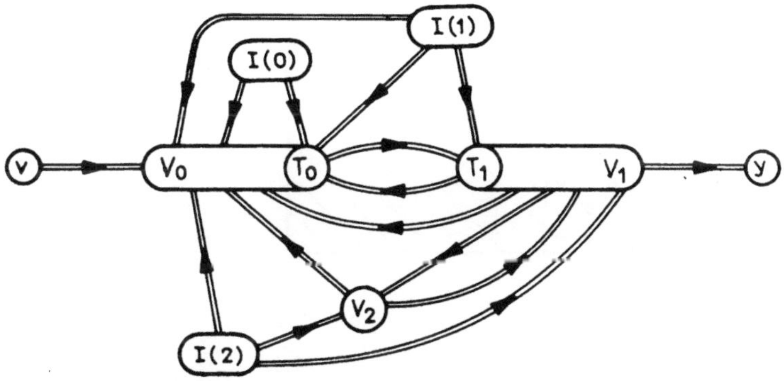

Fig. 22.3

A detailed description of all the subsets of vertices occurring in Fig. 22.3 gives

Definition 22.3

In the digraph $G([Q_6])$, see (22.5), consider a D-O cut set with the initial vertex set T_0 and final vertex set T_1.

The subset of state vertices involved in paths from any disturbance vertices to T_0 is denoted by V_0

The subset of state vertices involved in paths from T_1 to any output vertices is denoted by V_1.
The subset of the remaining state vertices neither contained in V_o nor in V_1 is denoted by V_2.
The subset of inputs from which all paths to any output vertices pass through T_o is denoted by I(0).
The subset of inputs, not contained in I(0), from which all paths to output vertices pass through T_1 is denoted by I(1).
The remaining inputs neither contained in I(0) nor in I(1) form the input subset I(2).

These partitions of the state set and of the input set are associated with a corresponding partitioning of the structure matrices [A] and [B], see Fig. 22.4.

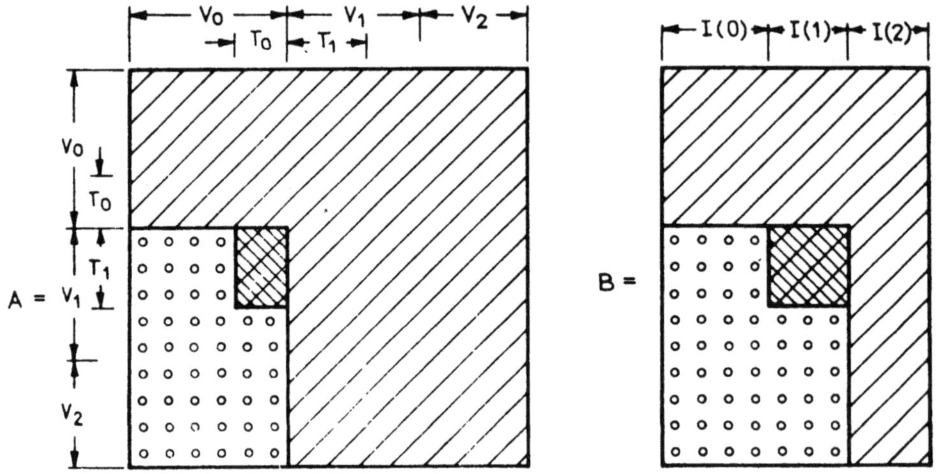

Fig. 22.4

Now we can formulate a sufficient condition for disturbance rejection by state feedback.

Theorem 22.2

A given system (22.1), (22.2) admits of complete disturbance rejection by state feedback (22.3) if there exists a D-O cut set in the digraph G([Q_6]) such that the rank of the associated submatrix $B_{T_1,I(1)}$ formed by the common entries of the T_1-rows and the I(1)-columns of B is equal to the cardinality of T_1, i.e.

$$\text{rank } B_{T_1,I(1)} = \text{card}(T_1). \qquad (22.11)$$

In Fig. 22.4, the submatrix $B_{T_1,I(1)}$ has been marked by double hatching.

<u>Proof of Theorem 22.2</u>: The trivial case in which there are no paths between disturbance vertices and output vertices has been tacitly excluded in the following consideration.
Select a D-O cut set and determine the associated partitioning of the state-variables and of the inputs as indicated by Definition 22.3. Accordingly, the structure matrix [E] can be subdivided into sections as sketched in Fig. 22.5.

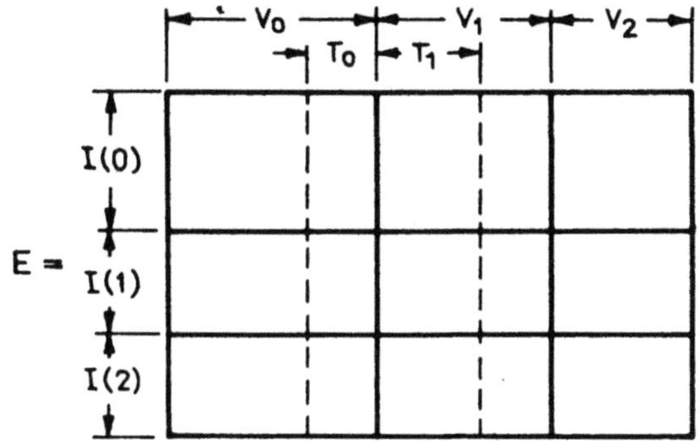

Fig. 22.5

We have to choose a state feedback matrix $F \in [E]$ in such a way that the closed-loop system matrix $A + BF$ has the shape shown in Fig. 22.6.

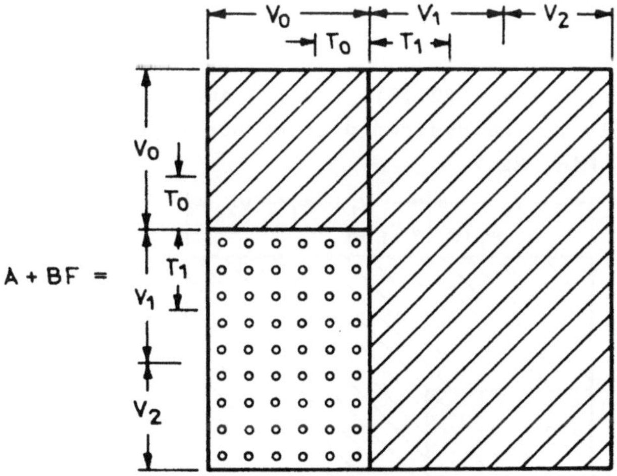

Fig. 22.6

This means

$$B_{T_1, I(1) \cup I(2)} F_{I(1) \cup I(2), V_o \setminus T_o} = 0 \qquad (22.12)$$

$$B_{(V_1 \setminus T_1) \cup V_2, I(2)} F_{I(2), V_o} = 0 \qquad (22.13)$$

$$A_{T_1, T_o} + B_{T_1, I(1) \cup I(2)} F_{I(1) \cup I(2), T_o} = 0 \qquad (22.14)$$

Putting

$$F_{I(1) \cup I(2), V_o \setminus T_o} = 0 \qquad (22.15)$$

$$F_{I(2), V_o} = 0 \qquad (22.16)$$

the equations (22.12) and (22.13) are trivially satisfied whereas Eq. (22.14) turns into

$$A_{T_1, T_o} + B_{T_1, I(1)} F_{I(1), T_o} = 0 \qquad (22.17)$$

or, explicitly written,

$$a_{ji} + \sum_{\substack{k \\ \text{with} \\ Ik \in I(1)}} b_{jk} f_{ki} = 0 \quad \text{for} \quad j \in T_1, \quad i \in T_o \qquad (22.17')$$

The matrix equation (22.17) is solvable if and only if the rank condition (22.11) is met. This completes the proof.

▲

The derived equations (22.15), (22.16), (22.17) imply all the feedback matrices F which supply us the desired disturbance rejection to have a structure as shown in Fig. 22.7.

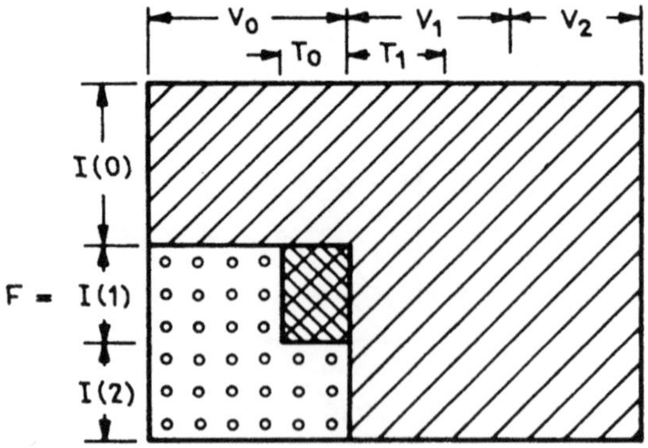

Fig. 22.7

In Fig. 22.7, the submatrix $F_{I(1),T_o}$ has been marked by double hatching.

Those parts of F which have been marked by simple hatching can be freely chosen. These degrees of freedom may be used for other purposes, especially for pole assignment.

The proof of Theorem 22.2 gives us evidence for further statements.

<u>Theorem 22.3</u>

A given system (22.1), (22.2) admits of complete disturbance rejection by state feedback (22.3) if there exists an D-O cut set such that the associated matrix equation

$$A_{V_1 \cup V_2, V_o} + B_{V_1 \cup V_2, I(1) \cup I(2)} F_{I(1) \cup I(2), V_o} = 0 \tag{22.18}$$

or, equivalently, the system of matrix equations (22.12), (22.13), (22.14) is solvable.

The remaining feedback gains not contained in $F_{I(1) \cup I(2), V_o}$ may be freely chosen.

<u>Theorem 22.4</u>

The solvability of (22.18) is sufficient for the rejection of the full variety of disturbances which act directly only on states contained in the subset V_o

It should be noted that V_o is uniquely determined by the chosen D-O cut set but, in most cases, not by the disturbance structure matrix [D] occurring in Eq. (22.1).

Theorem 22.4 says that one and the same state feedback matrix F is capable of rejecting disturbances of different structures. Unfortunately, the variety of rejectable disturbances depends on the choice of the D-O cut set that is strongly influenced by the given disturbance matrix D.

22.3 Compensation of the full variety of rejectable disturbances

Let us turn to the necessary condition (22.9) and ask for feedback matrices F which are able to compensate all disturbances meeting condition (22.9).

For this purpose, we need a little different partition of the state variables and the inputs, comp. Fig. 22.8.

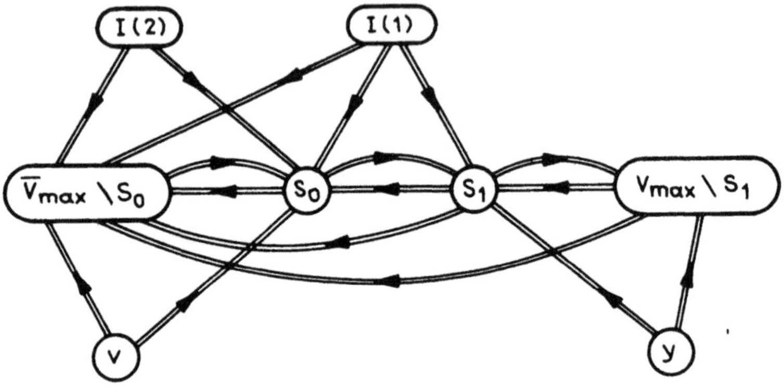

Fig. 22.8

Let be

S_1 the subset of state vertices contained in V_{max} (see Definition 22.1) and adjacent to inputs, i.e.

$$S_1 = \{ i: i \in \{1,2,\dots,n\},\ i \in V_{max},\ (Ik,i) \text{ exists for some } k \in \{1,\dots,m\}\} \tag{22.19}$$

S_o the subset of state vertices not contained in V_{max} and being initial vertices of edges ending in S_1, i.e.

$$S_o = \{ j: j \in \{1,2,\dots,n\},\ j \notin V_{max},\ (j,i) \text{ exists for some } i \in S_1\} \tag{22.20}$$

I(1) the subset of input vertices adjacent to state vertices of S_1, i.e.

$$I(1) = \{Ik: k \in \{1,2,\dots,m\},\ (Ik,i) \text{ exists for some } i \in S_1\} \tag{22.21}$$

I(2) the subset of input vertices not adjacent to vertices of S_1, i.e.

$$I(2) = \{Ik: k \in \{1,2,\dots,m\},\ (Ik,i) \text{ does not exist for each } i \in S_1\}. \tag{22.22}$$

The generalized representation of the digraph $G([Q_6])$ taking into account all of the vertex subsets just introduced and the possible connectivity relations between these subsets has been shown in

Fig. 22.8.
The matrices A, B and F partitioned correspondingly have been sketched in Fig. 22.9.

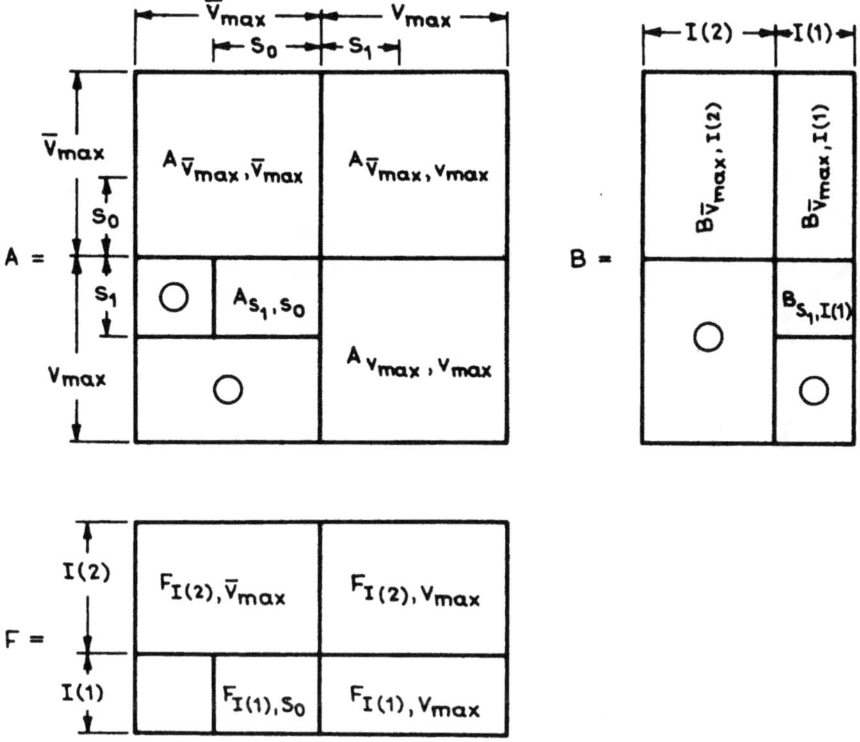

Fig. 22.9

Example 22.1: Fig. 22.10 shows the digraph $G([Q_6])$ for a class of systems given by their structure.
There are $n = 17$ state-variables, $m = 4$ inputs, $r = 3$ outputs and $q = 2$ disturbances.
Here the vertex subsets are

$$V_{max} = \{4, 5, 6, 7, 8, 9, 10, 17\}$$
$$\overline{V}_{max} = \{1, 2, 3, 11, 12, 13, 14, 15, 16\}$$
$$S_1 = \{4, 8, 9\}, \quad S_o = \{11, 12\},$$
$$I(1) = \{I2, I3\}, \quad I(2) = \{I1, I4\},$$
$$V_D = \{13, 15, 16\}.$$

Now we try to choose F in such a way that

$$A_{V_{max},\overline{V}_{max}} + B_{V_{max},I(1)} F_{I(1),\overline{V}_{max}} = 0 \tag{22.23}$$

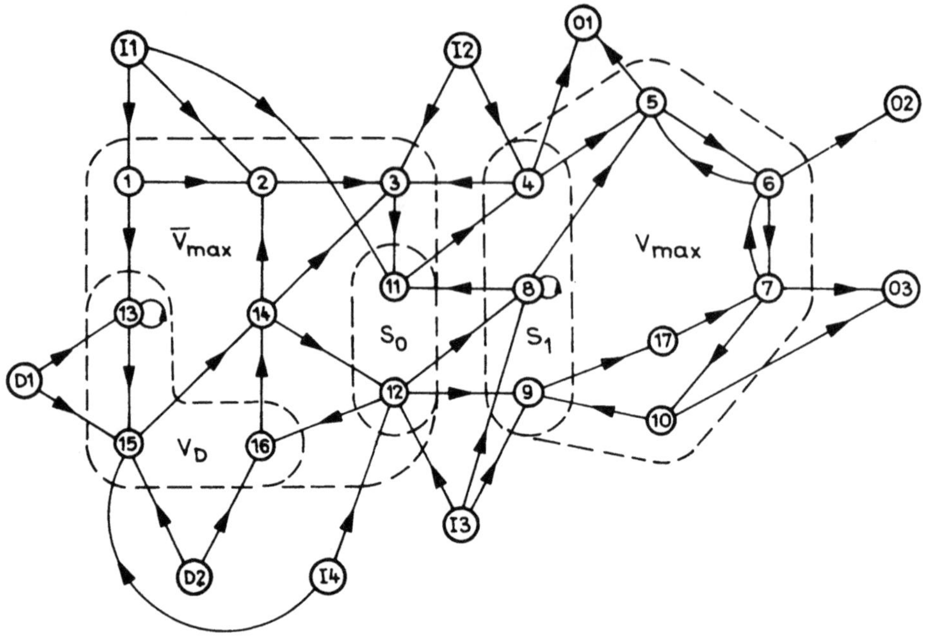

Fig. 22.10

Taking into account the special structure of A and B, see Fig. 22.9, the matrix equation (22.23) splits into

$$B_{S_1,I(1)}\, F_{I(1),\,\overline{V}_{max}\setminus S_o} = 0 \tag{22.24}$$

$$A_{S_1,S_o} + B_{S_1,I(1)}\, F_{I(1),S_o} = 0 \tag{22.25}$$

Thus, we have derived a criterion for disturbance rejection.

<u>Theorem 22.5</u>

For complete rejection of any disturbances meeting the condition (22.9), it is necessary and sufficient that the matrix equations (22.24) and (22.25) can be solved.
This is possible if and only if

$$\operatorname{rank} B_{S_1,I(1)} = \operatorname{card}(S_1) \tag{22.26}$$

The most convenient solution of (22.24) is

$$F_{I(1),\overline{V}_{max}\setminus S_o} = 0 \tag{22.27}$$

Eq. (22.25) is solvable if and only if (22.26) holds.

The choice of the feedback matrices

$F_{I(2),\overline{V}_{max}}$, $F_{I(2),V_{max}}$, $F_{I(1),V_{max}}$ (comp. Fig. 22.9)

does not influence the rejection of the disturbances under investigation.

Example 22.2: Consider the example system sketched in Fig. 22.10. Since

$$V_{max} = \{4, 5, 6, 7, 8, 9, 10, 17\} \text{ and } V_D = \{13, 15, 16\}$$

are disjoint, the necessary condition (22.9) is met.

Here the matrix equation (22.25) reads as follows:

$$\begin{pmatrix} a_{4,11} & 0 \\ 0 & a_{8,12} \\ 0 & a_{9,12} \end{pmatrix} + \begin{pmatrix} b_{4,2} & 0 \\ 0 & b_{8,3} \\ 0 & b_{9,3} \end{pmatrix} \begin{pmatrix} f_{2,11} & f_{2,12} \\ f_{3,11} & f_{3,12} \end{pmatrix} = \begin{pmatrix} 0 & 0 \\ 0 & 0 \\ 0 & 0 \end{pmatrix}$$

Obviously,

$$\operatorname{rank} B_{S_1,I(1)} = \operatorname{rank} \begin{pmatrix} b_{42} & 0 \\ 0 & b_{83} \\ 0 & b_{93} \end{pmatrix} = 2 < 3 = \operatorname{card}(S_1) \qquad (22.28)$$

Complete disturbance rejection with the aid of state feedback for all the disturbances meeting condition (22.9) is impossible because condition (22.26) is not fulfilled.

This does not necessarily imply that rejection of the actual disturbances exhibited in Fig. 22.10 is impossible.

Let us try to apply Theorem 22.2.

Choose the edges (3,11), (14,12) as disturbance-output cut set. This implies the following subdivision of state vertices and inputs (comp. Fig. 22.3 and Fig. 22.8):

$$T_o = \{3, 14\}$$

$$V_o = \{2, 3, 13, 14, 15, 16\}$$

$$T_1 = \{11, 12\}$$

$$V_1 = \{4, 5, 6, 7, 8, 9, 10, 11, 12, 17\}$$

$$V_2 = \{1\}$$

$$I(0) = \emptyset$$

$$I(1) = \{I1, I4\}$$

$$I(2) = \{I2, I3\}$$

The sufficient condition (22.11) is met because of

$$\operatorname{rank} B_{T_1,I(1)} = \operatorname{rank} \begin{pmatrix} b_{11,1} & 0 \\ 0 & b_{12,4} \end{pmatrix} = 2 = \operatorname{card}(T_1).$$

Therefore, feedback matrices F supplying disturbance rejection do exist. Suited feedback matrices may be obtained with the aid of (22.15), (22.16) and (22.17) as follows:

$$f_{ij} = 0 \quad \text{for} \quad i = 1,\ 2,\ 3,\ 4 \quad \text{and} \quad j = 2,\ 13,\ 15,\ 16$$

$$f_{ij} = 0 \quad \text{for} \quad i = 2,\ 3 \quad \text{and} \quad j = 2,\ 3,\ 13,\ 14,\ 15,\ 16$$

$$\begin{pmatrix} a_{11,3} & 0 \\ 0 & a_{12,14} \end{pmatrix} + \begin{pmatrix} b_{11,1} & 0 \\ 0 & b_{12,4} \end{pmatrix} \begin{pmatrix} f_{1,3} & f_{1,14} \\ f_{4,3} & f_{4,14} \end{pmatrix} = \begin{pmatrix} 0 & 0 \\ 0 & 0 \end{pmatrix}$$

Hence,

$$f_{1,3} = -\frac{a_{11,3}}{b_{11,3}}, \quad f_{1,14} = 0, \quad f_{4,3} = 0, \quad f_{4,14} = -\frac{a_{12,14}}{b_{12,4}}$$

while the remaining feedback gains may be freely chosen.

Whence

$$[F] = \begin{array}{c} \\ 1 \\ 2 \\ 3 \\ 4 \end{array} \begin{array}{c} \begin{array}{ccccccccccccccccc} 1 & 2 & 3 & 4 & 5 & 6 & 7 & 8 & 9 & 10 & 11 & 12 & 13 & 14 & 15 & 16 & 17 \end{array} \\ \begin{pmatrix} L & 0 & \underline{L} & L & L & L & L & L & L & L & L & L & 0 & \underline{0} & 0 & 0 & L \\ L & 0 & 0 & L & L & L & L & L & L & L & L & L & 0 & 0 & 0 & 0 & L \\ L & 0 & 0 & L & L & L & L & L & L & L & L & L & 0 & 0 & 0 & 0 & L \\ L & 0 & \underline{0} & L & L & L & L & L & L & L & L & L & 0 & \underline{L} & 0 & 0 & L \end{pmatrix} \end{array}$$

The entries underlined constitute the submatrix $F_{I(1),T_o}$ numerically fixed by condition (22.17).
The other L-entries are indeterminate. Thus, there are 44 feedback gains that need not be fixed in order to ensure complete disturbance rejection. These degrees of freedom should be used for the simultaneous solution of other controller synthesis requirements.

If condition (22.26) is not fulfilled then complete rejection of all disturbances meeting condition (22.9) can still be achieved after augmenting the system by appropriately chosen additional inputs.

Suppose the investigator to be able to introduce

$$\text{card}(S_1) - \text{rank}\ B_{S_1,I(1)}$$

additional inputs acting directly on states of S_1.
Denote the set of additional inputs by $I(a)$.
Let the additional inputs act on S_1 in such a way that

$$\text{rank}\ B_{S_1,I(1)\cup I(a)} = \text{card}(S_1) \tag{22.29}$$

Then the augmented system of linear equations

$$A_{S_1,S_o} + B_{S_1,I(1)\cup I(a)}\ F_{I(1)\cup I(a),S_o} = 0 \tag{22.30}$$

or, explicitly written,

$$a_{ji} + \sum_{\substack{k \\ \text{with} \\ Ik \in I(1) \cup I(a)}} b_{jk} f_{ki} = 0 \quad \text{for} \quad j \in S_1, \quad i \in S_o \qquad (22.30')$$

becomes solvable.
In the augmented system, therefore, complete rejection of any disturbances meeting (22.9) is possible by means of state feedback.

Example 22.3: We continue to investigate the example system discussed before. Because of (22.28) we have to introduce

$$\text{card}(I(a)) = \text{card}(S_1) - \text{s-rank}[B_{S_1,I(1)}] = 3 - 2 = 1$$

additional input. Thus, $I(a) = \{I5\}$.
Let I5 act on the ninth state-variable, see Fig. 22.11.

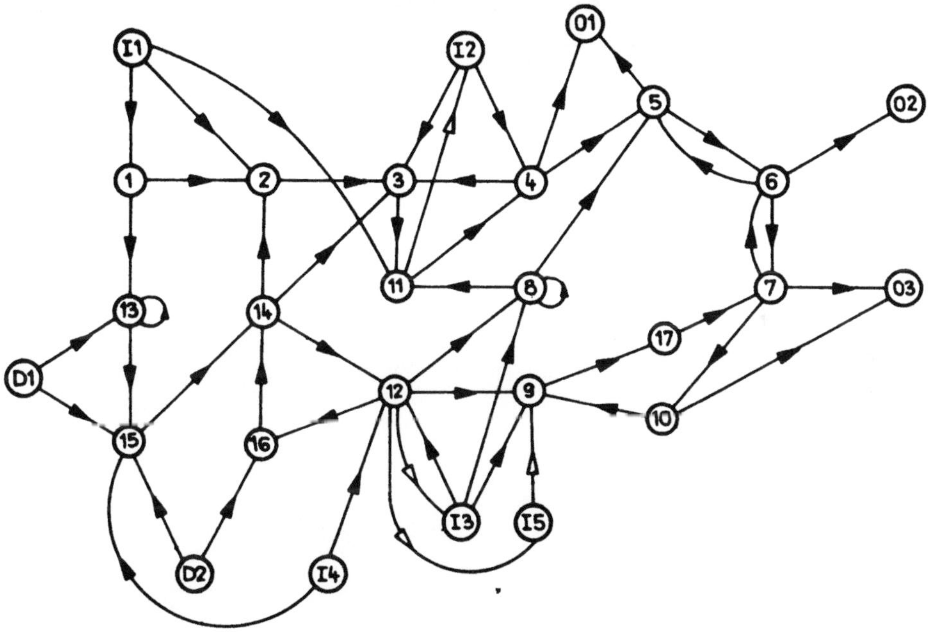

Fig. 22.11

The augmented structure matrix $B_{S_1,I(1)\cup I(a)}$ becomes

$$B_{S_1,I(1)\cup I(a)} = \begin{pmatrix} b_{4,2} & 0 & 0 \\ 0 & b_{8,3} & 0 \\ 0 & b_{9,3} & b_{9,5} \end{pmatrix} \qquad (22.31)$$

Obviously,

$$\text{rank } B_{S_1,I(1)\cup I(a)} = 3.$$

This ensures the solvability of (22.30). For the example under investigation Eq.(22.30) can explicitly be written down:

$$\begin{pmatrix} a_{4,11} & 0 \\ 0 & a_{8,12} \\ 0 & a_{9,12} \end{pmatrix} + \begin{pmatrix} b_{4,2} & 0 & 0 \\ 0 & b_{8,3} & 0 \\ 0 & b_{9,3} & b_{9,5} \end{pmatrix} \begin{pmatrix} f_{2,11} & f_{2,12} \\ f_{3,11} & f_{3,12} \\ f_{5,11} & f_{5,12} \end{pmatrix} = \begin{pmatrix} 0 & 0 \\ 0 & 0 \\ 0 & 0 \end{pmatrix}$$

One gets

$$f_{2,11} = - \frac{a_{4,11}}{b_{4,2}} , \quad f_{3,11} = f_{5,11} = f_{2,12} = 0, \quad f_{3,12} = - \frac{a_{8,12}}{b_{8,3}} ,$$

$$f_{5,12} = \frac{1}{b_{9,5}}(-a_{9,12} + \frac{b_{9,3}}{b_{8,3}} a_{8,12})$$

The edges associated with $b_{9,5}$, $f_{2,11}$, $f_{3,12}$ and $f_{5,12}$ have been marked by open arrows in Fig. 22.11.

22.4 An algorithm for disturbance rejection

A feedback design procedure that ensures complete rejection of any disturbances meeting condition (22.9) can now be briefly stated.

Step 1: Consider the governing system equations (22.1), (22.2) and form the structure matrices [A], [B], [C], [D].

Step 2: Inspect the structure matrices [B], [C], [D] and write down the structure row vectors [b'], $[c^1]$, [d']
where

$$[b']_i = \begin{cases} L & \text{if the ith row of [B] contains at least one L} \\ 0 & \text{else} \end{cases} \tag{22.32}$$

$$[c^1]_i = \begin{cases} L & \text{if the ith column of [C] contains at least one L} \\ 0 & \text{else} \end{cases} \tag{22.33}$$

$$[d']_i = \begin{cases} L & \text{if the ith row of [D] contains at least one L} \\ 0 & \text{else} \end{cases} \tag{22.34}$$

for $i = 1, 2, \ldots, n$.

Form the structure row vector

$$[r^1] = [c^1] \wedge [d'] \tag{22.35}$$

by elementwise conjunction, i.e.

$$[r^1]_i = \begin{cases} L & \text{if both } [c^1]_i = L \text{ and } [d']_i = L \\ 0 & \text{else} \end{cases} \quad \text{for } i = 1, \ldots, n.$$

We have $[r^1] \neq (0, 0, \ldots, 0)$ if and only if the relation (22.7) holds, comp. Lemma 22.1. In this case, complete disturbance rejection is impossible. Otherwise, go on to the next step.

Step 3: Determine the subsets V_{max} and S_1 of state variables. These subsets may be found with the aid of the following algorithm (see Fig. 22.12):

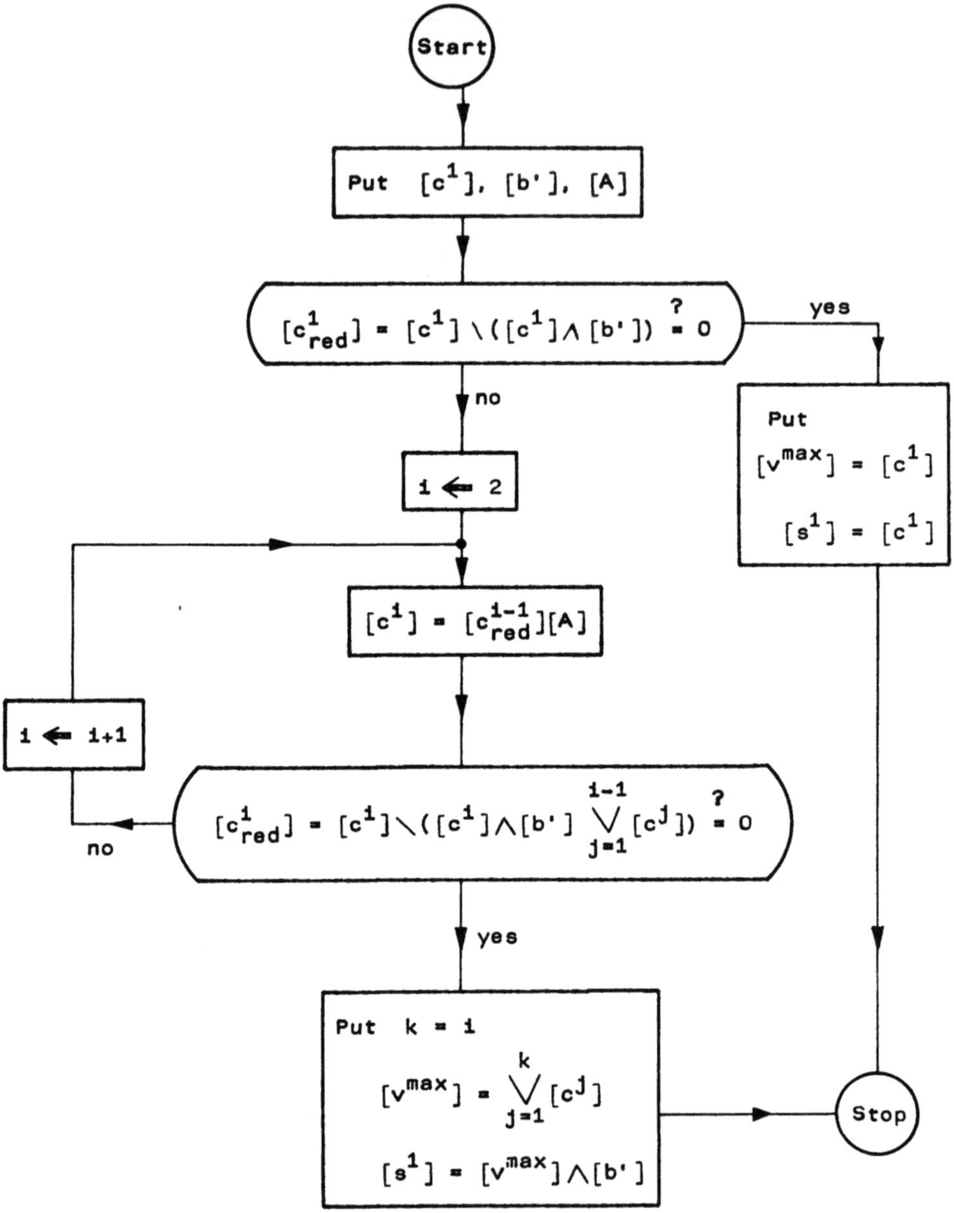

Fig. 22.12

The non-vanishing elements of the structure row vectors $[v^{max}]$ and $[s^1]$ obtained by the outlined algorithm give the desired sets V_{max} and S_1, respectively,

$$V_{max} = \{i: i \in \{1, 2, \ldots, n\}, [v^{max}]_i = L\} \tag{22.36}$$

$$S_1 = \{i: i \in \{1, 2, \dots, n\},\ [s^1]_i = L\}$$ (22.37)

Step 4: Form the structure row vector

$$[r^2] = [v^{max}] \wedge [d']$$ (22.38)

If $[r^2] \neq 0$ then complete disturbance rejection is impossible, comp. Theorem 22.2. Otherwise go on to the next step.

Step 5: Using V_{max} and S_1 determine the state subsets $\overline{V}_{max}$ and S_o and the partition I(1), I(2) of the input set,

$$\overline{V}_{max} = \{i: i \in \{1, 2, \dots, n\},\ i \notin V_{max}\}$$ (22.39)

$$S_o = \{i: i \in \{1, 2, \dots, n\},\ i \notin V_{max},\ [a_{ji}] = L \text{ for } j \in S_1\}$$ (22.40)

$$I(1) = \{Ik: k \in \{1, \dots, m\},\ [b_{ik}] = L \quad \text{for} \quad i \in S_1\}$$ (22.41)

$$I(2) = \{Ik: k \in \{1, \dots, m\},\ Ik \notin I(1)\}$$ (22.42)

Step 6: Consider the partitioning of the matrices A, B, F induced by the state and input subsets determined in the foregoing steps. Possibly after reordering of the state and input sets, the matrices A, B and F are partitioned into submatrices as shown in Fig. 22.9.
Complete rejection of all disturbances with $[r^2] = 0$, see (22.28), makes demands on the feedback gains.

Step 7: Put

$$F_{I(1),\overline{V}_{max} \setminus S_o} = 0$$ (22.43)

Step 8: Solve the matrix equation

$$A_{S_1,S_o} + B_{S_1,I(1)}\ F_{I(1),S_o} = 0$$ (22.44)

comp. (22.25).
Here we have to distinguish between three different situations. First,

$$\text{card}(S_1) = \text{card}(I(1))$$ (22.45)

and the square matrix $B_{S_1,I(1)}$ is non-singular.

Then there exists a unique solution

$$F_{I(1),S_o} = - B^{-1}_{S_1,I(1)} A_{S_1,S_o} \qquad (22.46)$$

Second,

$$\mathrm{card}(S_1) < \mathrm{card}(I(1)) \quad \text{and} \qquad (22.47)$$

$$\mathrm{rank}\ B_{S_1,I(1)} = \mathrm{card}(S_1) \qquad (22.48)$$

Then there exists a $(\mathrm{card}(I(1)) - \mathrm{card}(S_1))$-dimensional set of solutions.

Provided that there are no other aspects which the investigator would like to take into account then he can select a square non-singular submatrix $B_{S_1,I(1)_{sub}}$ and put

$$F_{I(1)_{sub},S_o} = - (B_{S_1,I(1)_{sub}})^{-1} A_{S_1,S_o} \qquad (22.49)$$

$$F_{I(1)\setminus I(1)_{sub},S_o} = 0 \qquad (22.50)$$

Third, if

$$\mathrm{rank}\ B_{S_1,I(1)} < \mathrm{rank}(B_{S_1,I(1)}, A_{S_1,S_o}) \qquad (22.51)$$

then the matrix equation (22.44) cannot be solved. Hence, complete rejection of all the disturbances with $[r^2] = [v^{max}] \wedge [d']$ is impossible. In this case we should check if the actual disturbances given by $[D]$ can be rejected applying Theorem 22.3.

Even if the procedure just written down cannot be executed successfully, the digraph approach gives us hints how to overcome the impossibility of complete disturbance rejection.

Provided that we are caused to finish at Step 2 or at Step 4 then only structural system changes can enable us to get complete disturbance rejection by state feedback.

If we are stopped at Step 8 because inequality (22.51) holds then we should first ask whether this inequality is also true in the structural sense, i.e., does

$$\text{s-rank}[B_{S_1,I(1)}] < \text{s-rank}[B_{S_1,I(1)}, A_{S_1,S_o}]$$

hold? If not, the matrix equation (22.44) becomes solvable after a slight perturbation of some non-identically vanishing entries of $B_{S_1,I(1)}$. Otherwise, we can try to overcome the unsolvability by introduction of appropriately chosen new inputs as described above, compare (22.29).

Furthermore, it should be not forgotten that partial disturbance rejection can also be helpful in many cases. If the necessary condition (22.9) for complete disturbance rejection is not met one should at least compensate those disturbance-output paths which pass through the state subset $\overline{V}_{max}$. For this purpose, the design procedure outlined above may be applied without making essential alterations.

Example 22.4: Let us illustrate all steps of the design procedure using the example system the structure digraph of which has been shown in Fig. 22.10.

Step 1: The structure matrices [A], [B], [D], [C] can be read from Fig. 22.10 as follows:

[A] =

	1	2	3	4	5	6	7	8	9	10	11	12	13	14	15	16	17
1																	
2	L													L			
3		L		L										L			
4											L						
5				L		L		L									
6					L												
7						L											L
8				L				L				L					
9										L		L					
10							L										
11			L					L									
12														L			
13													L				
14															L	L	
15													L				
16												L					
17									L								

[B] =

	I1	I2	I3	I4
1	L			
2	L			
3		L		
4		L		
5				
6				
7				
8			L	
9			L	
10				
11	L			
12			L	L
13				
14				
15				L
16				
17				

[D] =

	D1	D2
1		
2		
3		
4		
5		
6		
7		
8		
9		
10		
11		
12		
13	L	
14		
15	L	L
16		L
17		

[C] =

	1	2	3	4	5	6	7	8	9	10	11	12	13	14	15	16	17
01				L	L												
02						L											
03							L			L							

Step 2: From [B], [C], [D] one obtains the structure row vectors [b'], [c^1], [d'] as

[b'] = (L L L L O O O L L O L L O O L O O)

[c^1] = (O O O L L L L O O L O O O O O O O)

[d'] = (O O O O O O O O O O O O L O L L O)

Step 3: According to the instructions of the flow-chart one obtains

$$[c^1_{red}] = (0\ 0\ 0\ 0\ L\ L\ L\ 0\ 0\ L\ 0\ 0\ 0\ 0\ 0\ 0\ 0)$$

$$[c^2] = (0\ 0\ 0\ L\ L\ L\ L\ L\ 0\ 0\ 0\ 0\ 0\ 0\ 0\ 0\ L)$$

$$[c^2_{red}] = (0\ 0\ 0\ 0\ 0\ 0\ 0\ 0\ 0\ 0\ 0\ 0\ 0\ 0\ 0\ 0\ L)$$

$$[c^3] = (0\ 0\ 0\ 0\ 0\ 0\ 0\ 0\ 0\ L\ 0\ 0\ 0\ 0\ 0\ 0\ 0)$$

$$[c^3_{red}] = (0\ 0\ 0\ 0\ 0\ 0\ 0\ 0\ 0\ 0\ 0\ 0\ 0\ 0\ 0\ 0\ 0)$$

$$[v^{max}] = \bigvee_{i=1}^{3} [c^i] = (0\ 0\ 0\ L\ L\ L\ L\ L\ L\ L\ 0\ 0\ 0\ 0\ 0\ 0\ L)$$

$$[s^1] = [v^{max}] \wedge [b'] = (0\ 0\ 0\ L\ 0\ 0\ 0\ L\ L\ 0\ 0\ 0\ 0\ 0\ 0\ 0\ 0)$$

Hence, $V_{max} = \{4, 5, 6, 7, 8, 9, 10, 17\}$, $S_1 = \{4, 8, 9\}$.

Step 4: $[r^2] = [v^{max}] \wedge [d'] = (0\ 0\ 0\ 0\ 0\ 0\ 0\ 0\ 0\ 0\ 0\ 0\ 0\ 0\ 0\ 0\ 0)$

Step 5: Eqs. (22.39-42) give $\overline{V}_{max} = \{1, 2, 3, 11, 12, 13, 14, 15, 16\}$
$S_o = \{11, 12\}$, $I(1) = \{I2, I3\}$, $I(2) = \{I1, I4\}$.

Step 6 and Step 7: The reordered matrices are

[A] =

	1	2	3	13	14	15	16	11	12	4	8	9	5	6	7	10	17
1																	
2	L				L												
3		L			L					L							
13	L			L													
14						L	L										
15				L													
16									L								
11			L								L						
12					L												
4								L									
8									L	L					L		
9										L					L		
5											L	L			L		
6														L			
7															L		L
10																L	
17													L				

[B] =

	I1	I4	I2	I3
1	L			
2	L			
3			L	
13				
14				
15		L		
16				
11	L			
12		L		L
4			L	
8				L
9				L
5				
6				
7				
10				
17				

[F] = (rows I1, I4, I2, I3)

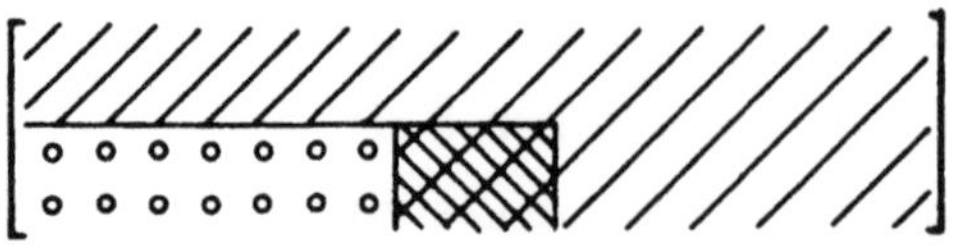

Step 8: The matrix equation (22.44) reads as

$$\begin{pmatrix} a_{4,11} & 0 \\ 0 & a_{8,12} \\ 0 & a_{9,12} \end{pmatrix} + \begin{pmatrix} b_{4,2} & 0 \\ 0 & b_{8,3} \\ 0 & b_{9,3} \end{pmatrix} \begin{pmatrix} f_{2,11} & f_{2,12} \\ f_{3,11} & f_{3,12} \end{pmatrix} = \begin{pmatrix} 0 & 0 \\ 0 & 0 \\ 0 & 0 \end{pmatrix}$$

Generally, this equation cannot be solved because inequality (22.51) holds generically.

Finally, readers are reminded that the disturbance rejection problem has had much attention within the geometric approach to control systems, see, in particular, Wonham 1974 and many more recent publications. The so-called (A,B)-invariant spaces play a key role there. The supremal (A,B)-invariant subspace $\mathcal{V}^*$ contained in the kernel Ker C corresponds to the state subset $\overline{V}_{max}$, while the image Im D corresponds to V_D. The solvability condition

$$\mathcal{V}^* \subset \text{Im } D \tag{22.52}$$

fundamental within the geometric approach corresponds to the necessary condition

$$V_{max} \cap V_D = \emptyset \tag{22.9}$$

within the digraph approach.

The relations between many results on disturbance rejection known from the geometric approach on the one hand and from the digraph approach on the other hand have been considered in detail by Andrei 1985.

23 Digraph approach to noninteraction controls by means of state feedback

For multivariable systems, the design criterion of noninteraction has had much attention for decades. Roughly speaking, the problem is to determine conditions on the controller such that each system input will modify only its corresponding system output and nothing else.

A general method for designing noninteracting controls for multivariable systems described mathematically by transfer function matrices was given by Boksenbom and Hood as early as in 1950. This work plays an important role in early texts on control theory, see for instance Tsien 1954 or Schwarz 1967.

The state space approach to noninteraction or decoupling - a synonymous and now even more usual notation - was paved by Morgan 1964, Rekasius 1965, Falb and Wolovich 1967, Gilbert 1969 et al., and further developed by Wonham and Morse 1970, Silverman and Payne 1971, Cremer 1971, Denham 1973, Kono and Sugiura 1974, Kamiyama and Furuta 1976, Nakamizo and Kobayashi 1980, Dion 1983, Suda and Umahashi 1984 and many others. Based on the digraph representation for large-scale sparse multivariable systems, noninteracting control was investigated by Andrei 1985.

Although the papers cited differ significantly one from another, the basic idea is always the same:
Produce new "action paths" by means of state feedback such that the effects of existing "coupling paths" are compensated.

The digraph approach enables us to develop this underlying principle in a manner particularly transparent.

23.1 An introductory example

Let us reconsider first a classic introductory example for noninteracting controls treated by the pioneers in this field in the early fifties, see Tsien 1954: the problem of controlling a turbopropeller engine seen, however, from the view of digraph approach.

The physical components of a turbopropeller engine are sketched in Fig. 23.1.

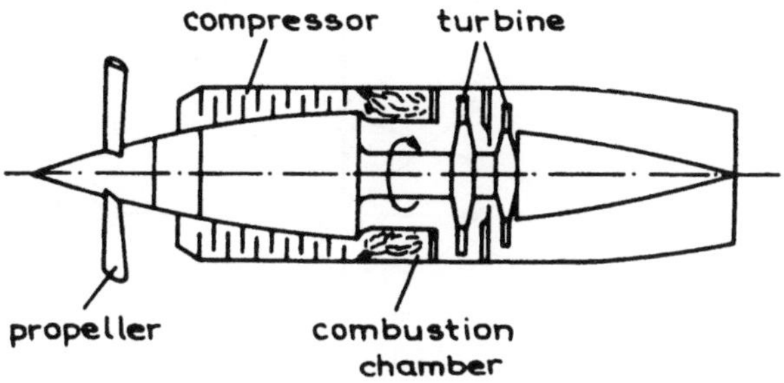

Fig. 23.1

Let the system outputs be

y_1 the deviation of the rotating speed from its normal value at the desired steady-state operating point

y_2 the deviation of the turbine-inlet temperature from its normal value.

It is obvious that one important design criterion for the system control is the independent controllability of these outputs by two appropriately chosen control inputs. A change in rotating speed should not necessarily be combined with a change of turbine-inlet temperature and vice versa.

Near a chosen steady-state operating point there hold the following linearized relations between the system variables (for details, see Tsien 1954):

$$\begin{pmatrix} \dot{x}_1 \\ \dot{x}_2 \\ \dot{x}_3 \end{pmatrix} = \begin{pmatrix} a_{11} & 0 & a_{13} \\ 0 & a_{22} & a_{23} \\ 0 & 0 & 0 \end{pmatrix} \begin{pmatrix} x_1 \\ x_2 \\ x_3 \end{pmatrix} + \begin{pmatrix} b_{11} & 0 \\ b_{21} & b_{22} \\ 0 & b_{32} \end{pmatrix} \begin{pmatrix} u_1 \\ u_2 \end{pmatrix} \tag{23.1}$$

$$\begin{pmatrix} y_1 \\ y_2 \end{pmatrix} = \begin{pmatrix} 1 & 0 & 0 \\ 0 & 1 & 0 \end{pmatrix} \begin{pmatrix} x_1 \\ x_2 \\ x_3 \end{pmatrix} \tag{23.2}$$

where the third state-variable is

x_3 the deviation of the fuel rate from the normal value,

and the inputs are

u_1 the deviation of the propeller blade angle from the normal point,

u_2 the time-derivative of the fuel rate.

Fig. 23.2 shows a digraph reflecting the system equations (23.1) and (23.2).

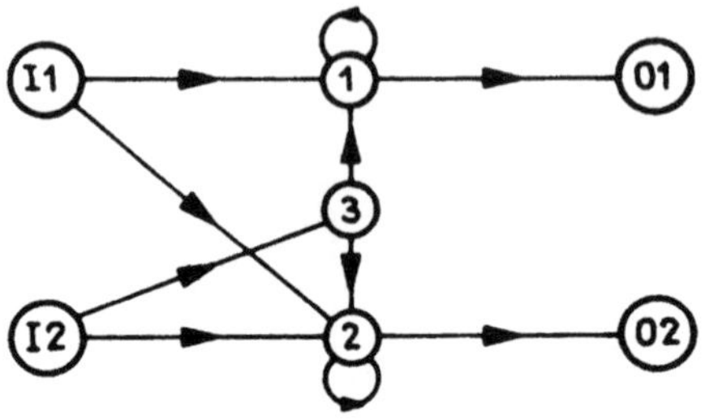

Fig. 23.2

Evidently, the input u_1 exerts an immediate influence on both the outputs simultaneously. The coupling input edge (I1,2) may be compensated by transition from the output vector $u = (u_1, u_2)'$ to a new input vector $\bar{u} = (\bar{u}_1, \bar{u}_2)'$ defined by

$$u = G\,\bar{u} = \begin{pmatrix} 1 & 0 \\ -\dfrac{b_{21}}{b_{22}} & 1 \end{pmatrix} \begin{pmatrix} \bar{u}_1 \\ \bar{u}_2 \end{pmatrix} \tag{23.3}$$

Then we have

$$B\,u = B\,G\,\bar{u} = \begin{pmatrix} b_{11} & 0 \\ b_{21} & b_{22} \\ 0 & b_{32} \end{pmatrix} \begin{pmatrix} 1 & 0 \\ -\dfrac{b_{21}}{b_{22}} & 1 \end{pmatrix} \bar{u} = \begin{pmatrix} b_{11} & 0 \\ 0 & b_{22} \\ -\dfrac{b_{32}b_{21}}{b_{22}} & b_{32} \end{pmatrix} \bar{u}$$

$$= \bar{B}\,\bar{u} \tag{23.4}$$

This simple algebraic manipulation can easily be interpreted graphically. The Eqs. (23.1) to (23.3) are associated with a compound digraph shown in Fig. 23.3.

The paths $\bar{I}1 \rightarrow I1 \rightarrow 2$ and $\bar{I}1 \rightarrow I2 \rightarrow 2$ compensate each other. Their path weigths have the same magnitude and opposite signs. This means there is no action path that really leads from input vertex $\bar{I}1$ to state vertex 2.

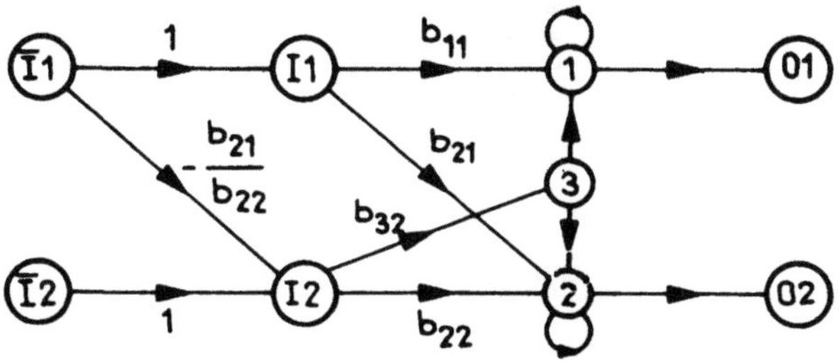

Fig. 23.3

The equality

$$B\,G\,\bar{u} = \bar{B}\,\bar{u} \tag{23.5}$$

comp. (23.4), can be interpreted as elimination of the vertices I1 and I2. Thus Fig. 23.3 is transformed into Fig. 23.4.

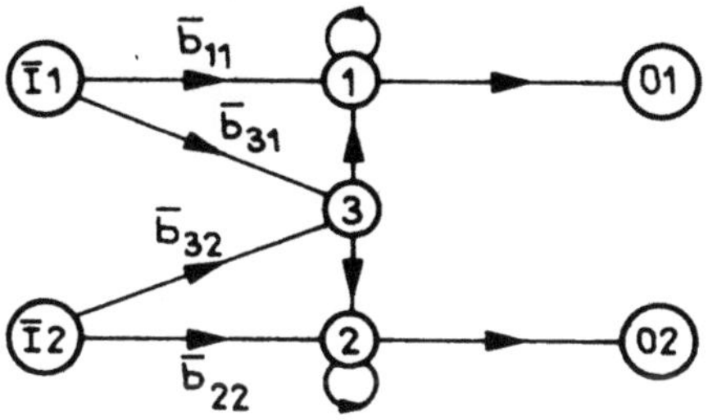

Fig. 23.4

The new input edge weights result from (23.4) as

$$\bar{b}_{11} = b_{11},\; \bar{b}_{31} = -\frac{b_{32}b_{21}}{b_{22}},\; \bar{b}_{32} = b_{32},\; \bar{b}_{22} = b_{22}$$

Although for the example under consideration the undesired coupling input edge has been removed, noninteracting control cannot be carried out yet. The coupling state edge (3,1) causes the input command $\bar{u}_2$ to affect not only the output y_2 but also the output y_1.
The coupling edge (3,1) with weight a_{13} can be compensated by means of a path $3 \rightarrow \bar{I}1 \rightarrow 1$ if its feedback edge $(3,\bar{I}1)$ is weighted by

$$f_{13} = -\frac{a_{13}}{\bar{b}_{11}}\,.$$

This may be achieved by transition from the input command $\bar{u}_1$ to a new input command w_1 defined by

$$\bar{u}_1 = f_{13}x_3 + w_1 \quad \text{with} \quad f_{13} = -\frac{a_{13}}{\bar{b}_{11}} = -\frac{a_{13}}{b_{11}} \tag{23.6}$$

see Fig. 23.5.a.

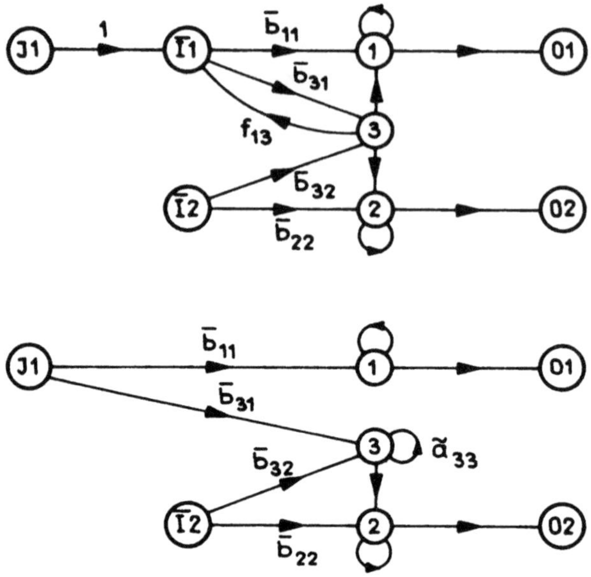

Fig. 23.5

After elimination of vertex $\bar{I}1$ we obtain the digraph of Fig. 23.5.b. The self-cycle attached at vertex 3 has a weight

$$\tilde{a}_{33} = \bar{b}_{31}\, f_{13} = -\frac{b_{32}b_{21}a_{13}}{b_{22}b_{11}} \tag{23.7}$$

It should be realized that the digraphs shown in Fig. 23.5.a and Fig. 23.5.b are indeed input-output equivalent.

Similarly, the coupling state edge (3,2) can be compensated. We use a new input command w_2 defined by

$$\bar{u}_2 = f_{23}\, x_3 + w_2 \quad \text{with} \quad f_{23} = -\frac{a_{23}}{\bar{b}_{22}} \tag{23.8}$$

see Fig. 23.6.a.

Elimination of vertex $\bar{I}2$ leads to Fig. 23.6.b.

The self-cycle attached at vertex 3 has a weight

$$\bar{a}_{33} = \tilde{a}_{33} + f_{23}b_{32} = -\frac{b_{32}}{b_{22}}\left(\frac{b_{21}}{b_{11}}\, a_{13} + a_{23}\right) \tag{23.9}$$

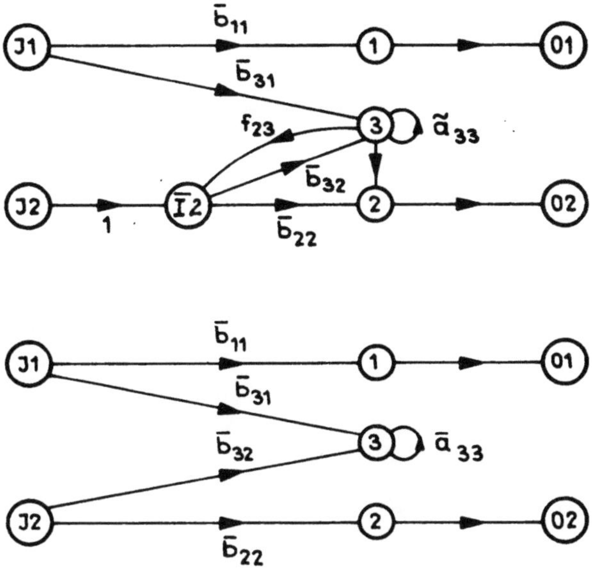

Fig. 23.6

Now he have reached our aim of noninteracting controls of both outputs. The first and the second output can be controlled independently using the inputs

$$w_1 = \bar{u}_1 - f_{13}x_3 = u_1 + \frac{a_{13}}{b_{11}}\, x_3 \qquad (23.10)$$

and

$$w_2 = \bar{u}_2 - f_{23}x_3 = u_2 + \frac{b_{21}}{b_{22}}\, u_1 + \frac{a_{23}}{b_{22}}\, x_3 \qquad (23.11)$$

respectively.
Employing this control law the resulting closed-loop system is governed by the equations

$$\begin{pmatrix} \dot{x}_1 \\ \dot{x}_2 \\ \dot{x}_3 \end{pmatrix} = \begin{pmatrix} a_{11} & 0 & a_{13} \\ 0 & a_{22} & a_{23} \\ 0 & 0 & 0 \end{pmatrix} \begin{pmatrix} x_1 \\ x_2 \\ x_3 \end{pmatrix} + \begin{pmatrix} b_{11} & 0 \\ b_{21} & b_{22} \\ 0 & b_{32} \end{pmatrix} \begin{pmatrix} 1 & 0 \\ -\frac{b_{21}}{b_{22}} & 1 \end{pmatrix} \left[\begin{pmatrix} 0 & 0 & f_{13} \\ 0 & 0 & f_{23} \end{pmatrix} \begin{pmatrix} x_1 \\ x_2 \\ x_3 \end{pmatrix} + \begin{pmatrix} w_1 \\ w_2 \end{pmatrix} \right]$$

$$= \begin{pmatrix} a_{11} & 0 & a_{13} + b_{11}f_{13} \\ 0 & a_{22} & a_{23} + b_{22}f_{23} \\ 0 & 0 & b_{32}\left(\frac{b_{21}}{b_{22}} f_{13} + f_{23}\right) \end{pmatrix} \begin{pmatrix} x_1 \\ x_2 \\ x_3 \end{pmatrix} + \begin{pmatrix} b_{11} & 0 \\ 0 & b_{22} \\ -\frac{b_{32}b_{21}}{b_{22}} & b_{32} \end{pmatrix} \begin{pmatrix} w_1 \\ w_2 \end{pmatrix}$$

$$= \begin{pmatrix} a_{11} & 0 & 0 \\ 0 & a_{22} & 0 \\ 0 & 0 & \frac{b_{32}}{b_{22}}\left(\frac{b_{21}}{b_{11}} a_{13} - a_{23}\right) \end{pmatrix} \begin{pmatrix} x_1 \\ x_2 \\ x_3 \end{pmatrix} + \begin{pmatrix} b_{11} & 0 \\ 0 & b_{22} \\ -\frac{b_{32}b_{21}}{b_{22}} & b_{32} \end{pmatrix} \begin{pmatrix} w_1 \\ w_2 \end{pmatrix} \qquad (23.12)$$

23.2 Problem formulation

As above we start with an internal system description

$$\dot{x} = A\,x + B\,u \tag{23.13}$$

$$y = C\,x \tag{23.14}$$

where $u \in \mathbb{R}^m$, $x \in \mathbb{R}^n$, $y \in \mathbb{R}^r$.
Employing a control law of the form

$$u = G\,v \tag{23.15}$$

$$v = F\,x + w \tag{23.16}$$

there results a closed-loop system

$$\dot{x} = (A + B\,G\,F)\,x + B\,G\,w = \overline{A}\,x + \overline{B}\,w \tag{23.17}$$

$$y = C\,x \qquad (23.14)$$

The associated generalized digraph is shown in Fig. 23.7.

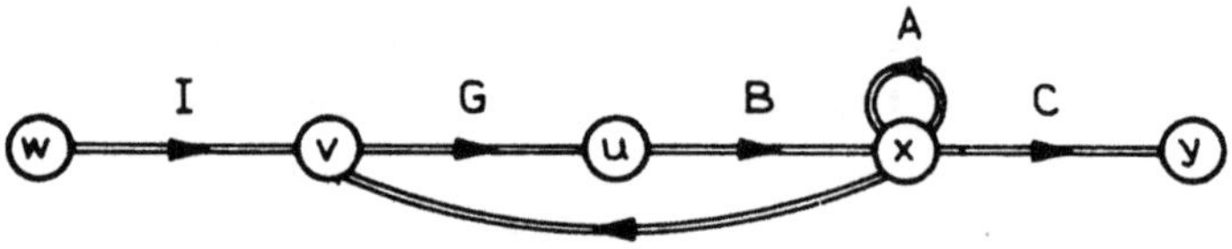

Fig. 23.7

Suppose that the output vector y for the plant (23.13), (23.14) consists of s subvectors

$$y^\sigma = C^\sigma\,x(t) \qquad \sigma \in \{1,2,\ldots,s\} \tag{23.18}$$

where y^σ is an r_σ-vector,

$$\sum_{\sigma=1}^{s} r_\sigma = r \tag{23.19}$$

and C^σ is an $r_\sigma \times n$ matrix.
Therefore, possibly after reordering the outputs, we have

$$C = \begin{pmatrix} C^1 \\ C^2 \\ \vdots \\ C^s \end{pmatrix} \tag{23.20}$$

The objective of decoupling by static state feedback is to determine a matrix pair (G, F) so that, for each $\sigma \in \{1,2,\ldots,s\}$, there is an input subvector w^σ which can control the output subvector y^σ without influencing the remaining output subvectors $y^1,\ldots,y^{\sigma-1},y^{\sigma+1},\ldots,y^s$.

23.3 A necessary condition for decoupling by static state feedback

Analogously to Definition 22.1 and equation (22.19) we state

Definition 23.1

For each output subvector y^{σ}, where $\sigma \in \{1,2,\ldots,s\}$, the maximal subset of σ-output-connected state vertices whose σ-output-connectedness cannot be compensated by state feedback is denoted by V^{σ}_{max}.

The subset of state vertices contained in V^{σ}_{max} and adjacent to input vertices is denoted by S^{σ}_{1}, i.e.

$$S^{\sigma}_{1} = \{i:\ i \in V^{\sigma}_{max},\ (Ik,i) \text{ exists for some } k \in \{1,2,\ldots,m\}\} \qquad (23.21)$$

The state subsets V^{σ}_{max} and S^{σ}_{1} may be determined using the algorithm outlined in Fig. 22.11. Instead of $[c^{1}]$ defined by (22.33) we have to use $[c^{\sigma,1}]$ defined by

$$[c^{\sigma,1}]_i = \begin{cases} L, & \text{if the ith column of } [C^{\sigma}] \text{ contains an } L \\ 0 & \text{else} \end{cases} \qquad (23.22)$$

Then the algorithm supplies V^{σ}_{max} and S^{σ}_{1} instead of V_{max} and S_1, respectively.

The same reasoning from which above resulted Theorem 22.2 leads here immediately to

Theorem 23.1

For decoupling by static state feedback it is necessary that the state vertex subsets V^{σ}_{max} are disjoint, i.e.

$$\bigcap_{\sigma=1}^{s} V^{\sigma}_{max} = \emptyset. \qquad (23.23)$$

The state subset

$$S_1 = \bigcup_{\sigma=1}^{s} S^{\sigma}_{1} \qquad (23.24)$$

induces a natural partitioning of all input vertices into the subset $I(1)$ of input vertices adjacent to S_1 and the subset $I(2)$ of input vertices not adjacent to S_1, see (22.21) and (22.22).

The input subvectors corresponding to $I(1)$ and $I(2)$ are denoted by u^1 and u^2, respectively.

The subset of state vertices complementary to V^{σ}_{max} is denoted by $\overline{V}^{\sigma}_{max}$,

$$\overline{V}^{\sigma}_{max} = \{1,2,\ldots,n\} \setminus V^{\sigma}_{max} \qquad (23.25)$$

In accordance with (22.10) we define

$$\overline{V}_{max} = \{1, 2, \ldots, n\} \setminus \bigcup_{\sigma=1}^{s} V^{\sigma}_{max}. \tag{23.26}$$

Provided that the condition (23.23) is met then the system digraph has a typical structure sketched in Fig. 23.8.

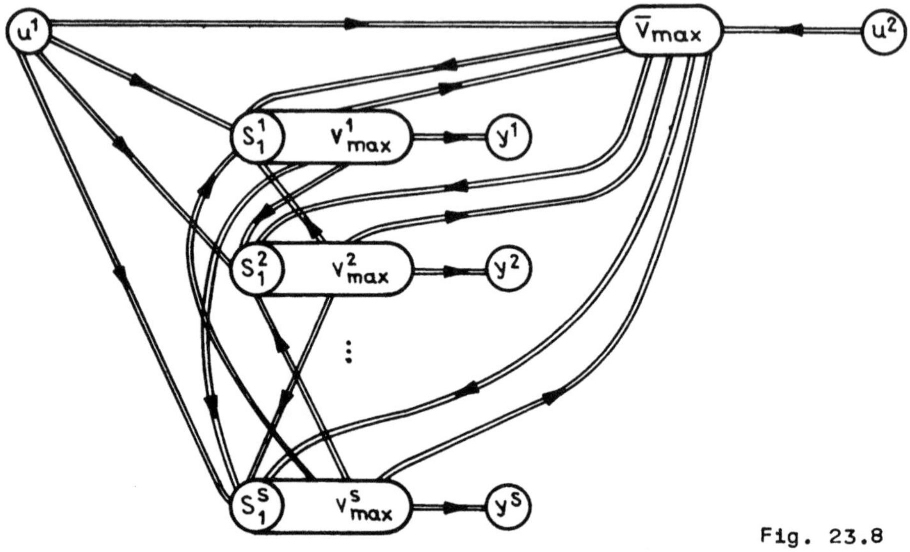

Fig. 23.8

Of course, this structure must be equivalently reflected by the structure of the matrices A, B, and C. This has been illustrated in Fig. 23.9 for the case s = 3.

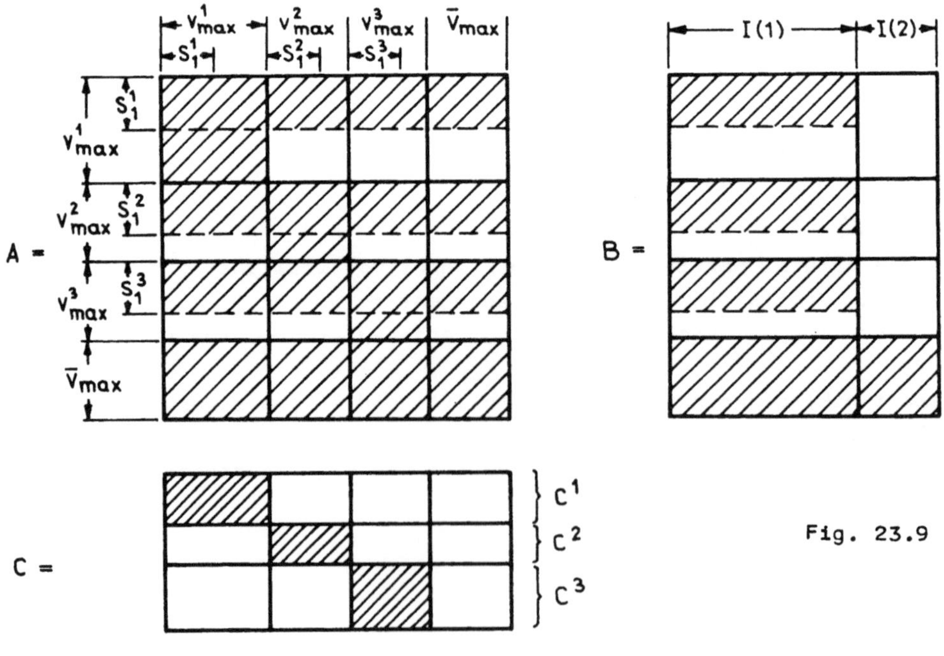

Fig. 23.9

23.4 A sufficient condition for decoupling by static state feedback

Now, we look for matrix pairs (G, F) that provide the desired decoupling. From Fig. 23.8 it is immediately seen that the input submatrix

$$B_{S_1,I(1)}$$

formed by the common entries of the S_1-rows and the I(1)-columns of B will play a key role in this context.

First, we ask for a transformation matrix $G_{I(1),\bar{I}}$ such that

$$B_{S_1,I(1)}\, G_{I(1),\bar{I}} = \bar{B}_{S_1,\bar{I}} \tag{23.27}$$

becomes a quasi-diagonal matrix.

That is, the new input vector v has s subvectors v^σ, $\sigma \in \{1,2,\dots,s\}$, such that the non-diagonal blocks of $\bar{B}_{S_1,\bar{I}}$ vanish,

$$\bar{B}_{S_1^\tau,\bar{I}^\sigma} = 0 \quad \text{for } \tau \neq \sigma \in \{1,2,\dots,s\} \tag{23.28}$$

Here the input vertex subset $\bar{I}^\sigma$ is associated with v^σ.

As far as the original vector u^2 is concerned, there is no need to alter this input subvector. Therefore, we put

$$G_{I(2),\bar{I}} = 0, \quad G_{I(1),I(2)} = 0, \quad G_{I(2),I(2)} = \text{unit matrix} \tag{23.29}$$

For the case of s = 3, the equations (23.27) to (23.29) have been illustrated in Fig. 23.10.

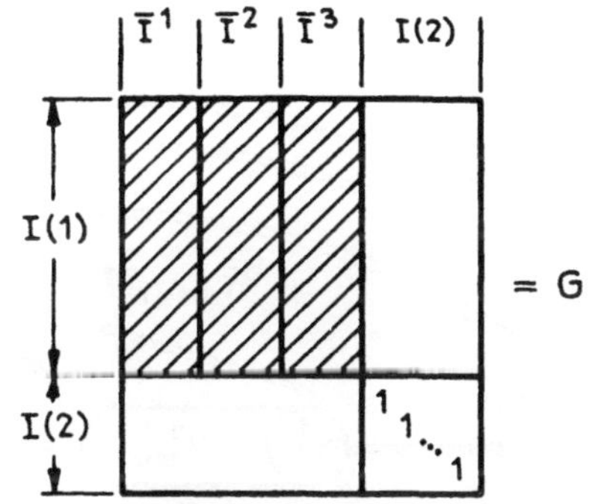

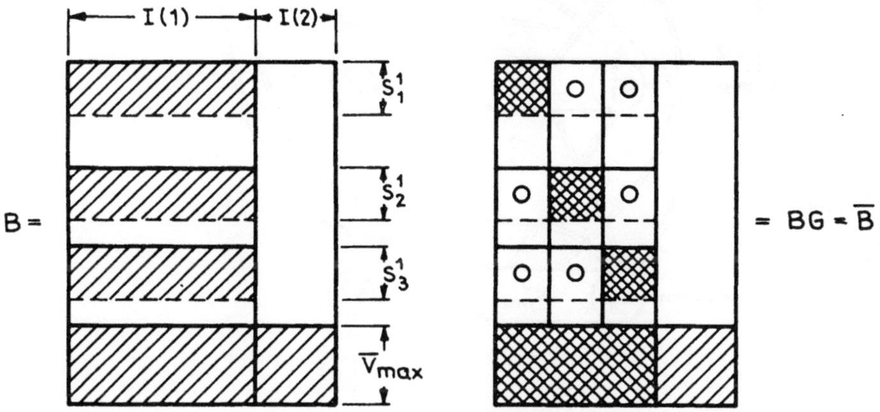

Fig. 23.10

In graph-theoretic terms, the foregoing process of reasoning has been sketched in Fig. 23.11. Both these digraphs are input-output equivalent.

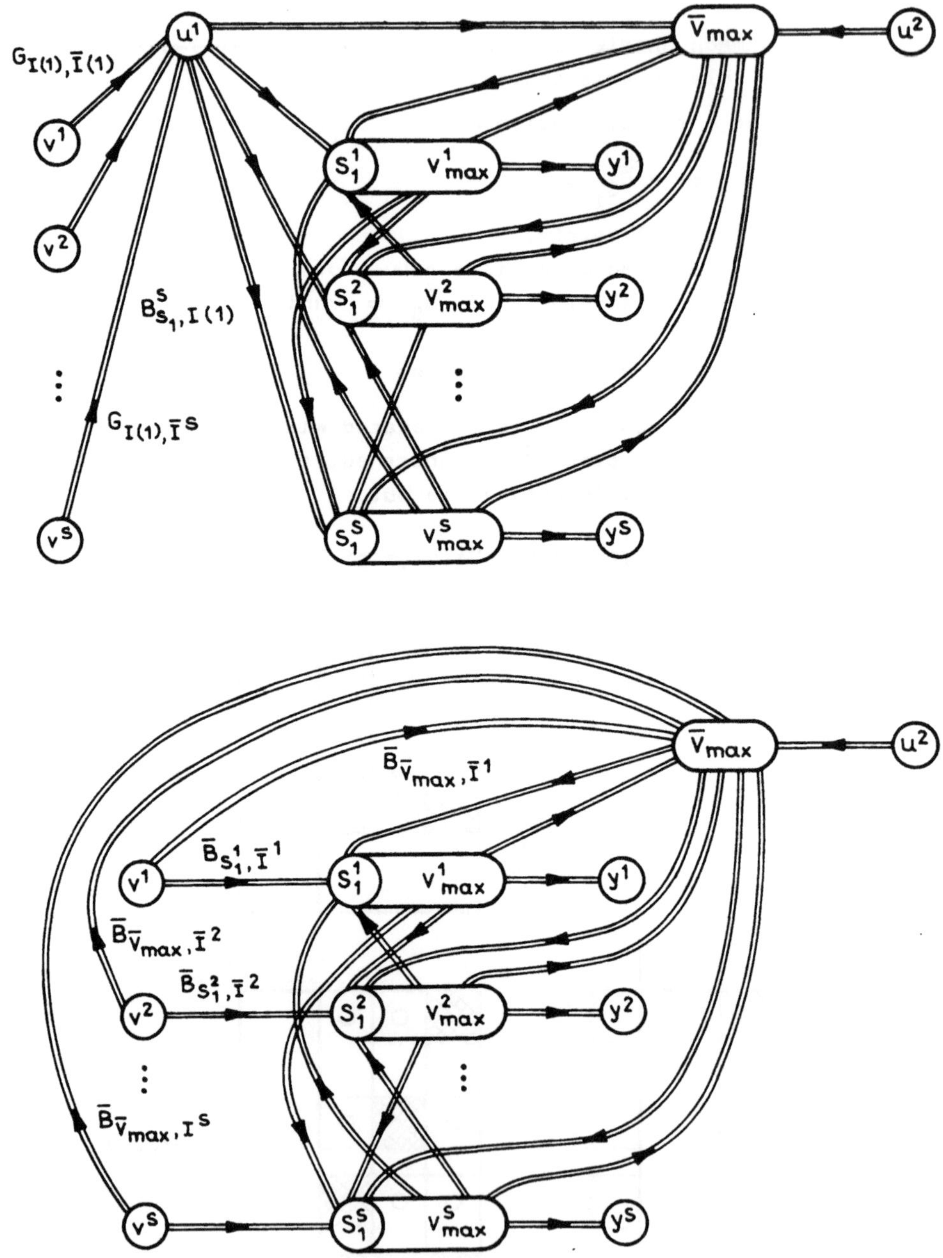

Fig. 23.11

Second, we need state feedback edges which are able to compensate all the state edges whose final vertices belong to one of the subsets S_1^σ for $\sigma \in \{1,2,\dots,s\}$. More exactly, the interesting feedback matrix F should satisfy the demands

$$a_{ji} + \sum_{\substack{k \\ \text{with} \\ \bar{I}k \in \bar{I}^\sigma}} \bar{b}_{jk} f_{ki} = 0 \quad \text{for} \quad j \in S_1^\sigma, \quad i \in \bar{V}^\sigma_{max} \tag{23.30}$$

or, in equivalent vector notation,

$$A_{S_1^\sigma, i} + \bar{B}_{S_1^\sigma, I^\sigma} F_{\bar{I}^\sigma, i} = 0 \quad \text{for each} \quad i \in \bar{V}^\sigma_{max} \tag{23.30'}$$

This relation represents $\mathrm{card}(\bar{V}^\sigma_{max})$ systems of linear algebraic equations with the same coefficients matrix $\bar{B}_{S_1^\sigma, \bar{I}^\sigma}$. They are solvable if and only if

$$\mathrm{rank}\, \bar{B}_{S_1^\sigma, \bar{I}^\sigma} = \mathrm{card}(S_1^\sigma) \tag{23.31}$$

This implies

$$\mathrm{card}(\bar{I}^\sigma) \geq \mathrm{card}(S_1^\sigma) \tag{23.32}$$

We choose $\mathrm{card}(\bar{I}^\sigma) = \mathrm{card}(S_1^\sigma)$ in the sequel.
Assume the equations (23.30) to be solvable.

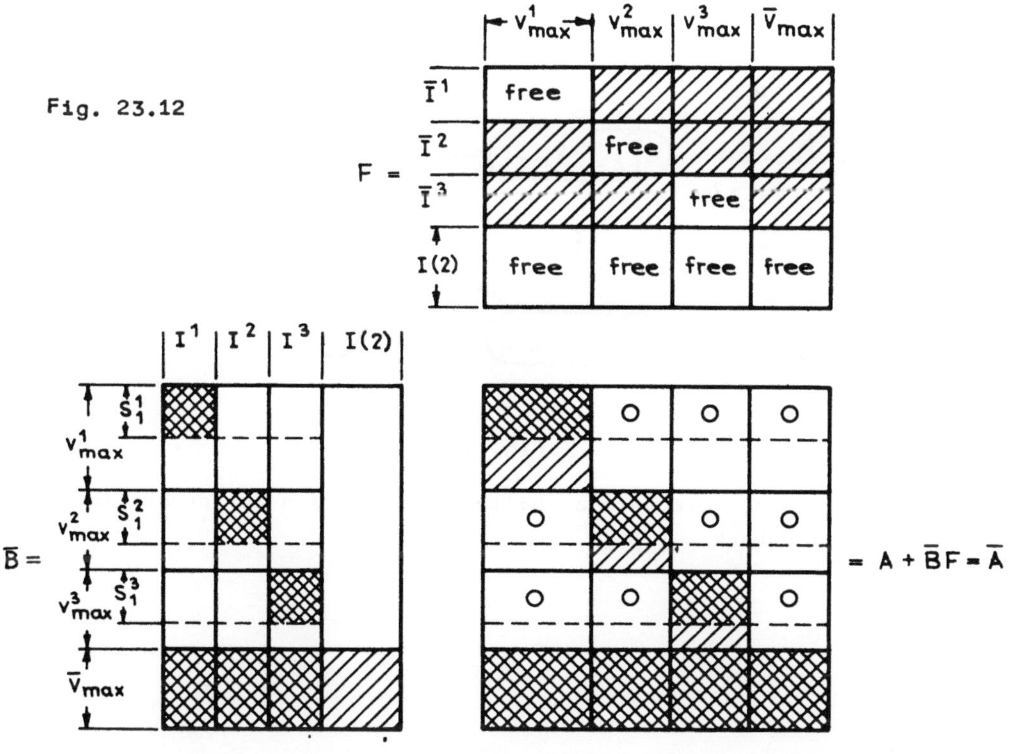

Fig. 23.12 shows the structure of the new input matrix $\overline{B}$ (comp. Fig. 23.10), the feedback matrix F and the resulting decoupled system matrix

$$A + \overline{B} F = \overline{A}.$$

The hatched submatrices of F result from solving the equations (23.30). The entries of all the submatrices marked by "free" may be chosen freely without influencing the structure of $\overline{A}$ exhibited in Fig. 23.12. Those parts of $\overline{B}$ and $\overline{A}$ which are simply hatched agree with the corresponding parts of B and A, respectively, while the doubly hatched parts have been altered during the decoupling process just described.

The obtained decoupled system is governed by the equations

$$\dot{x} = \overline{A}\, x + \overline{B}\, w = \overline{A}\, x + \overline{B} \begin{pmatrix} w^1 \\ \vdots \\ w^s \\ u^2 \end{pmatrix} \qquad (23.33)$$

$$y = C\, x$$

They are associated with a digraph sketched in Fig. 23.13.

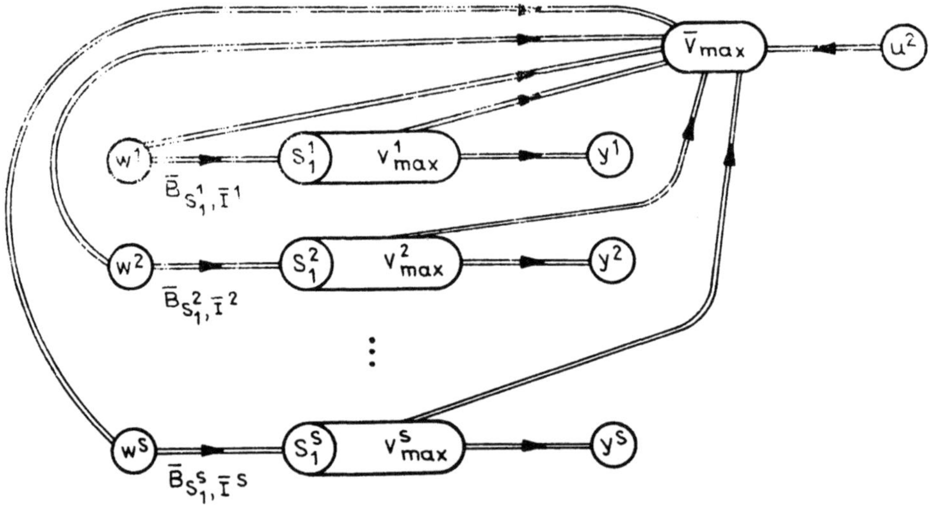

Fig. 23.13

Now, we are able to formulate a sufficient condition for decoupling by static state feedback.

Theorem 23.2

Provided that the necessary condition (23.23) is met then the condition

$$\text{rank } B_{S_1,I(1)} = \text{card}(S_1) \qquad (23.34)$$

is sufficient for decoupling by static state feedback.

Proof (constructive):

1. Consider the multivariable system (23.13), (23.14) and the given partition (23.18) of the output vector into s subvectors $y^1, y^2, \ldots, y^s$.
2. Determine the state subsets V^{σ}_{max} and S^{σ}_1 for $\sigma = 1,2,\ldots,s$ and the input subset $I(1)$.
3. Check the validity of the necessary condition (23.23).
4. Consider the submatrix $B_{S_1,I(1)}$ and check the condition (23.34).

   Both the conditions (23.23) and (23.34) are presumed to be fulfilled in what follows.
5. Form a $\text{card}(S_1) \times \text{card}(S_1)$ quasi-diagonal matrix $\overline{B}_{S_1,\overline{I}}$ consisting of s non-singular square diagonal blocks $\overline{B}_{S_1^{\sigma},\overline{I}^{\sigma}}$ of dimension $\text{card}(S_1^{\sigma}) \times \text{card}(S_1^{\sigma})$.

   The actual choice of the non-vanishing entries may be fitted for the special needs of the actual example under investigation. For principal theoretical reasons, however, only non-singularity of each diagonal block is demanded. Thus, for instance, the $\text{card}(S_1) \times \text{card}(S_1)$ unit matrix may be chosen as $\overline{B}_{S_1,\overline{I}}$.
6. Compute a $\text{card}(I(1)) \times \text{card}(S_1)$ matrix $G_{I(1),\overline{I}}$ solving (23.37) for $G_{I(1),\overline{I}}$. In case of $\text{card}(I(1)) > \text{card}(S_1)$ there are many solutions. Each of them may be used in the sequel.

   The other submatrices of G are given by (23.29).
7. For each $\sigma = 1,2,\ldots,s$ compute $F_{\overline{I}^{\sigma},\overline{V}^{\sigma}_{max}}$ solving (23.30).

   The remaining entries of F may be used for other purposes, in particular, for pole assignment.

   The desired decoupling has been achieved now.
8. The decoupled system has $\overline{B} = B \cdot G$ and $\overline{A} = A + \overline{B} \cdot F$ as input matrix and as system matrix, respectively. For $\sigma = 1,2,\ldots,s$, the new input subvector

$$w^{\sigma} = - F_{\overline{I}^{\sigma},\{1,2,\ldots,n\}} x + (\overline{B}_{S_1^{\sigma},\overline{I}^{\sigma}})^{-1} B_{S_1^{\sigma},I(1)} u^1 \qquad (23.35)$$

   controls the given output subvector y^{σ} without influencing the remaining system outputs, compare Fig. 23.13.

▲

If the sufficient condition (23.34) is not met then, provided the necessary condition (23.23) is valid, decoupling by static state feedback can still always be achieved after augmenting the system by appropriately chosen additional inputs which act directly on states of the subset S_1. In accordance with the analogous argumentation in Section 22 we denote the set of additional inputs by I(a).

To ensure decoupling, we must make certain that

$$\text{rank } B_{S_1, I(1) \cup I(a)} = \text{card}(S_1) \tag{23.36}$$

Then all conclusions can be drawn as above.

At this point it should be underlined that condition (23.34) need not be true to secure decoupling for a given actual system. On the other hand, it seems to be impossible to derive an improved condition of general validity.

Example 23.2: The digraph in Fig. 23.14 represents a system with $m = 2$ inputs, $r = 2$ outputs and $n = 7$ states.

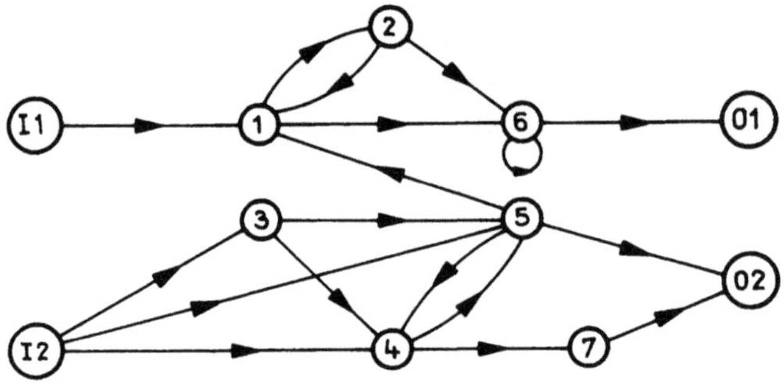

Fig. 23.14

For this example system, the interesting vertex subsets can immediately read from Fig. 23.14 as

$$V^1_{max} = \{1, 2, 6\}$$
$$S^1_1 = \{1\}$$
$$V^2_{max} = \{4, 5, 7\}$$
$$S^2_1 = \{4, 5\}$$
$$\bar{V}_{max} = \{3\}$$

Because of $\quad V^1_{max} \cap V^2_{max} = \emptyset$

the necessary condition (23.23) holds true.

The matrix $B_{S_1,I(1)}$ has the following structure:

$$[B_{S_1,I(1)}] = \begin{matrix} & 1 & 2 \\ 1 & L & 0 \\ 4 & 0 & L \\ 5 & 0 & L \end{matrix}$$

Obviously,

$$\operatorname{rank} B_{S_1,I(1)} \leq 2 < 3 = \operatorname{card}(S_1).$$

Thus, condition (23.34) is not fulfilled.

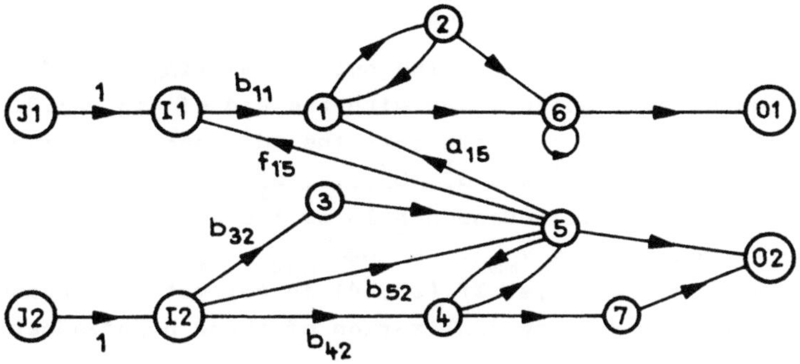

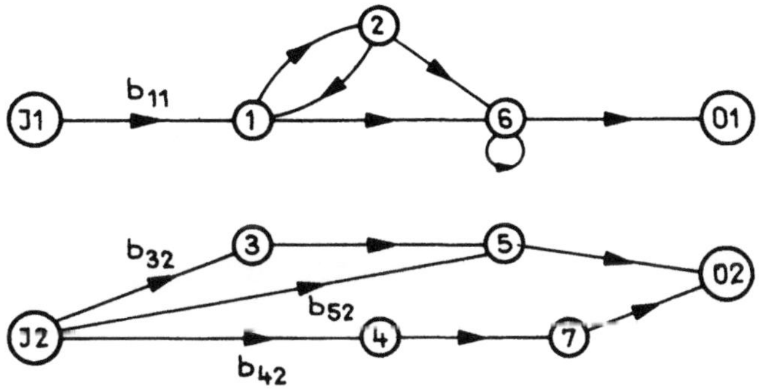

Fig. 23.15

Nevertheless, decoupling is easily possible, see Fig. 23.15. Introducing a new input vector

$$\begin{pmatrix} w_1 \\ w_2 \end{pmatrix} = \begin{pmatrix} u_1 \\ u_2 \end{pmatrix} + \begin{pmatrix} -f_{15}x_5 \\ 0 \end{pmatrix} = \begin{pmatrix} u_1 \\ u_2 \end{pmatrix} + \begin{pmatrix} \frac{a_{15}}{b_{11}}\, x_5 \\ 0 \end{pmatrix}$$

one obtains a decoupled system shown in Fig. 23.15.b.

The allowable (i.e. noninteraction controls preserving) feedback matrices F have a structure as

$$F = \begin{matrix} & 1 & 2 & 3 & 4 & 5 & 6 & 7 \\ 1 & L & L & 0 & 0 & f_{15} & L & 0 \\ 2 & 0 & 0 & L & L & L & 0 & L \end{matrix} \quad \text{where} \quad f_{15} = -\frac{a_{15}}{b_{11}}$$

The seven F-entries marked by "L" may be arbitrarily chosen.

As a conclusion from the foregoing considerations, we can strengthen Theorem 23.2 as follows:

Theorem 23.3

For decoupling by static state feedback of each of the systems meeting (23.23) it is necessary and sufficient that the rank condition (23.34) holds.

The analogy between Theorem 22.5 and Theorem 23.3 should be noticed.

The meaning of the phrase "each of the systems meeting (23.23)" in Theorem 23.3 can be made quite clear with the aid of Fig. 23.8. The rank condition (23.34) reflects only the adjacency relations between the "hyper-vertices" u^1 and $S_1^1, S_1^2, \ldots, S_1^s$. If (23.34) is met then decoupling is possible independently of the actual realization of all the "hyper-edges" $u^1 \Rightarrow \bar{V}_{max}$, $u^2 \Rightarrow \bar{V}_{max}$, $V_{max}^{\sigma} \Rightarrow \bar{V}_{max}$, $\bar{V}_{max} \Rightarrow S_1^{\sigma}$ for $\sigma = 1,2,\ldots,s$. If (23.34) is not met then decoupling is impossible for at least one realization of these hyper-edges.

Chapter 3. Digraph approach to controller synthesis based on static output feedback

31 Transfer function matrices and closed-loop characteristic polynomials in graph-theoretic terms

In Chapter 3, we consider closed-loop systems obtained by applying static output feedback

$$u(t) = F\ y(t) \tag{31.1}$$

to the plant

$$\dot{x}(t) = A\ x(t) + B\ u(t) \tag{31.2}$$

$$y(t) = C\ x(t) \tag{31.3}$$

where $x(t) \in \mathbb{R}^n$, $u(t) \in \mathbb{R}^m$, $y(t) \in \mathbb{R}^r$, rank $B = m$, rank $C = r$.
From (31.1), (31.2), (31.3) we get, by the Laplace transform,

$$U(s) = F\ Y(s) \tag{31.1'}$$

$$s\ X(s) - x(0) = A\ X(s) + B\ U(s) \tag{31.2'}$$

$$Y(s) = C\ X(s) \tag{31.3'}$$

This set of equations can equally be written as

$$\begin{pmatrix} I_r & -C & 0 \\ 0 & sI_n - A & -B \\ -F & 0 & I_m \end{pmatrix} \begin{pmatrix} Y \\ X \\ U \end{pmatrix} = \begin{pmatrix} 0 \\ x(0) \\ 0 \end{pmatrix} \tag{31.4}$$

where I_r, I_n, and I_m are unit matrices of dimension r, n, and m, respectively.

31.1 Transfer functions and their graphical interpretation

Multiplying the second hyper-row of (31.4) by $(sI_n - A)^{-1}$ from the left and substituting the result into the first hyper-row, we obtain

$$Y(s) = C(sI_n - A)^{-1}B\,U(s) + C(sI_n - A)^{-1}x(0) \qquad (31.5)$$

On the right-hand side of (31.5), the so-called transfer function matrix $T(s)$ occurs,

$$T(s) = C(sI_n - A)^{-1}B \qquad (31.6)$$

The (j,i)-entry of $T(s)$ is the transfer function of the plant from input u_i to output y_j $(1 \leq i \leq m,\ 1 \leq j \leq r)$

$$\begin{aligned} t_{ji}(s) &= c_j'(sI_n - A)^{-1}b_i \\ &= [\det(sI_n - A)]^{-1}\, c_j'(sI_n - A)_{adj}\, b_i \\ &= \frac{z_{ji}(s)}{\det(sI_n - A)} \end{aligned}$$

The denominator is the open-loop characteristic polynomial introduced in (21.15),

$$\det(sI_n - A) = s^n + p_1^o s^{n-1} + \dots + p_{n-1}^o s + p_n^o \qquad (31.7)$$

The calculation of $\det(sI_n - A)$ has been discussed in Section 21 and in Section A2.3. In Theorem A2.5, the rules how to get the polynomial coefficients p_i^o $(1 \leq i \leq n)$ from the digraph $G(A)$ have been derived. Here, we merely add some comments concerning large-scale systems.

If the number of state variables is not too small (say, $n > 10$) then the determinant (31.7) of order n splits up into a product of t (> 1) subdeterminants of order $n_1, n_2, \dots, n_t$ in most cases:

$$\det(sI_n - A) = \prod_{i=1}^{t} \det(sI_{n_i} - A_{(ii)}), \quad \sum_{i=1}^{t} n_i = n \qquad (31.7')$$

Each submatrix $A_{(ii)}$ is associated with an equivalence class of strongly connected vertices within the digraph $G(A)$, comp. Section 13.

Corollary 31.1

If the square matrix A is not irreducible, i.e., the digraph $G(A)$ is not strongly connected, then there exist $t > 1$ equivalence classes of states which correspond to square submatrices $A_{(ii)}$ and associated strongly connected subgraphs $G(A_{(ii)})$.

Before applying Theorem A2.5, the digraph G(A) should be reduced to the t disjoint subgraphs $G(A_{(ii)})$. Then Eq. (31.7) be profitably used to determine the determinant $\det(sI_n - A)$.

Example 31.1: Consider the 16×16 matrix A explained in Fig. 13.1. The corresponding reduced graph has been drawn in Fig. 31.1.

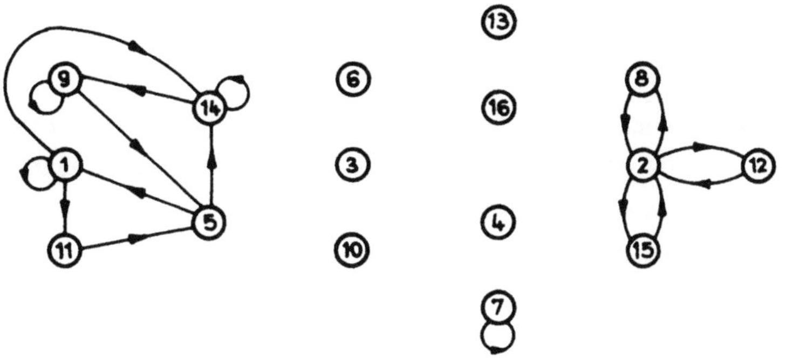

Fig. 31.1

The subgraph associated with the equivalence class K_5 in Fig. 13.1 supplies a subdeterminant

$$s^5 + p_1^{(5)} s^4 + p_2^{(5)} s^3 + p_3^{(5)} s^2 + p_4^{(5)} s + p_5^{(5)} = \det(sI_5 - A_{(55)})$$

The determination of the coefficients $p_i^{(5)}$ for $i = 1,2,3,4,5$ with the aid of Theorem A2.5 is left to the reader.
The subgraph associated with the equivalence class K_6 in Fig. 13.1 supplies a subdeterminant

$$s^4 + p_2^{(6)} s^2 = \det(sI_4 - A_{(66)})$$

where/as the subclass K_7 supplies a subdeterminant $s - a_{77}$.
Each of the remaining six "acyclic" subclasses provides a factor s.
Finally, Eq. (31.7') gives

$$\det(sI_{16} - A) = s^6(s-a_{77})\det(sI_4 - A_{(66)})\cdot\det(sI_5 - A_{(55)})$$

The numerator polynomial

$$Z_{ji}(s) = c_j'(sI_n - A)_{adj}\, b_i \qquad (31.8)$$

is also a polynomial in s,

$$Z_{ji}(s) = p_1^{ij} s^{n-1} + p_2^{ij} s^{n-2} + \ldots + p_{n-1}^{ij} s + p_n^{ij} \qquad (31.9)$$

To find an appropriate graph-theoretic interpretation of the coefficients p_k^{ij} $(1 \le k \le n)$ we consider the $(n+2) \times (n+2)$ matrix

$$Q^{ij} = \begin{pmatrix} 0 & c_j' & 0 \\ 0 & A & b_i \\ -1 & 0 & 0 \end{pmatrix} \qquad (31.10)$$

and the associated digraph $G(Q^{ij})$.

It should be noticed that Q^{ij} may be regarded as a special case of the matrix Q_4 introduced in Section 11, Eq. (11.16).

Definition 31.1

In digraphs such as $G(Q^{ij})$ or $G(Q_4)$, the edges associated with entries of the feedback submatrix F are called feedback edges.

Cycle families which contain at least one feedback edge are called feedback cycle families.

Theorem 31.1

The coefficients p_k^{ij} $(1 \leq k \leq n)$ of the numerator polynomial (31.9) are determined by the feedback cycle families of width k within the digraph $G(Q^{ij})$.

If there exists no cycle family of width k within the structure digraph $G([Q^{ij}])$ then the coefficient p_k^{ij} vanishes identically (i.e., $p_k^{ij} = 0$ for all admissible realizations (A, b_i, c_j) within the structure class defined by $[A, b_i, c_j]$, comp. Section 12). Otherwise, each feedback cycle family of width k corresponds to one summand of p_k^{ij}. The numerical value of the summand is equal to the weight of the corresponding feedback cycle family. This value must be multiplied by a sign-factor $(-1)^d$ if the cycle family under consideration consists of d disjoint cycles.

Numerical cancellation of the p_k^{ij}-summands is possible if there are two or more feedback cycle families of width k.

Proof: From well-known properties of determinants - as for the notations, see Section 21.2 - we conclude on the one hand

$$\det \begin{pmatrix} sI_n - A & b_i \\ c_j' & 0 \end{pmatrix} = s^{n-1} \sum_{1 \leq h_1 \leq n} \begin{pmatrix} -A & b_i \\ c_j' & 0 \end{pmatrix}_{h_1 n+1}^{h_1 n+1} + s^{n-2} \sum_{h_1 < h_2} \sum \begin{pmatrix} -A & b_i \\ c_j' & 0 \end{pmatrix}_{h_1 h_2 n+1}^{h_1 h_2 n+1}$$

$$+ s \sum_{h_1 <} \sum_{h_2 < \ldots <} \sum_{h_{n-1}} \begin{pmatrix} -A & b_i \\ c_j' & 0 \end{pmatrix}_{h_1 h_2 \ldots h_{n-1} n+1}^{h_1 h_2 \ldots h_{n-1} n+1}$$

$$+ \det \begin{pmatrix} -A & b_i \\ c_j' & 0 \end{pmatrix}$$

and on the other hand

$$\det\begin{pmatrix} sI_n - A & b_i \\ c_j' & 0 \end{pmatrix} = \det(sI_n - A)\cdot\det(0 - c_j'(sI_n - A)^{-1}b_i)$$

$$= -c_j'(sI_n - A)_{adj}\, b_i \;=\; -\sum_{k=1}^{n} p_k^{ij} s^{n-k}$$

Comparison with respect to the powers of s provides

$$p_k^{ij} = \sum_{1\le h_1<} \cdots \sum_{\dots < h_k \le n} - \begin{pmatrix} -A & b_i \\ c_j' & 0 \end{pmatrix}_{h_1\dots h_k n+1}^{h_1\dots h_k n+1}$$

$$= \sum_{h_1<} \cdots \sum_{\dots < h_k} - \begin{pmatrix} 1 & 0 & 0 \\ 0 & -A & b_i \\ 0 & c_j' & 0 \end{pmatrix}_{1,h_1+1,\dots,h_k+1,n+2}^{1,h_1+1,\dots,h_k+1,n+2}$$

$$= \sum_{h_1<} \cdots \sum_{\dots < h_k} - \begin{pmatrix} 0 & c_j' & 0 \\ 0 & -A & b_i \\ -1 & 0 & 0 \end{pmatrix}_{1,h_1+1,\dots,h_k+1,n+2}^{1,h_1+1,\dots,h_k+1,n+2}$$

$$p_k^{ij} = \sum_{h_1<} \cdots \sum_{\dots < h_k} (-1)^k \begin{pmatrix} 0 & c_j' & 0 \\ 0 & A & b_i \\ -1 & 0 & 0 \end{pmatrix}_{1,h_1+1,\dots,h_k+1,n+2}^{1,h_1+1,\dots,h_k+1,n+2} \qquad (31.11)$$

Now we apply Theorem A2.1. The minor

$$\begin{pmatrix} 0 & c_j' & 0 \\ 0 & A & b_i \\ -1 & 0 & 0 \end{pmatrix}_{1,h_1+1,\dots,h_k+1,n+2}^{1,h_1+1,\dots,h_k+1,n+2} \qquad (31.12)$$

is a determinant of order $k+2$. Its value is given by the weights of all the feedback cycle families in $G(Q^{ij})$ that involve the k state vertices $h_1,\dots,h_k$. If such a cycle family consists of d disjoint cycles then the sign factor is $(-1)^{k+2-d}$.

Taking into account the sign factor $(-1)^k$ in Eq. (31.11), the total sign factor becomes

$$(-1)^k(-1)^{k+2-d} = (-1)^d.$$

This completes the proof.

▲

Example 31.2: Fig. 31.2.a shows the digraph $G(Q^{ij})$ for an example system with $m = r = 1$, $n = 4$, comp. Section 15.
The open-loop characteristic polynomial can be immediately seen from the subgraph G(A) encircled by a dashed line,

$$\det(sI - A) = s^4 - a_{43}s^2.$$

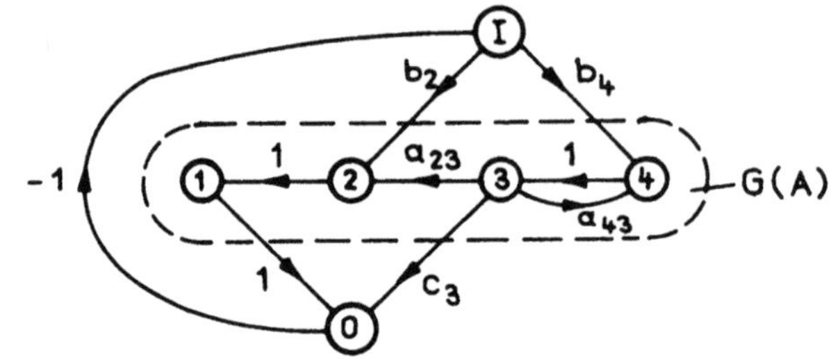

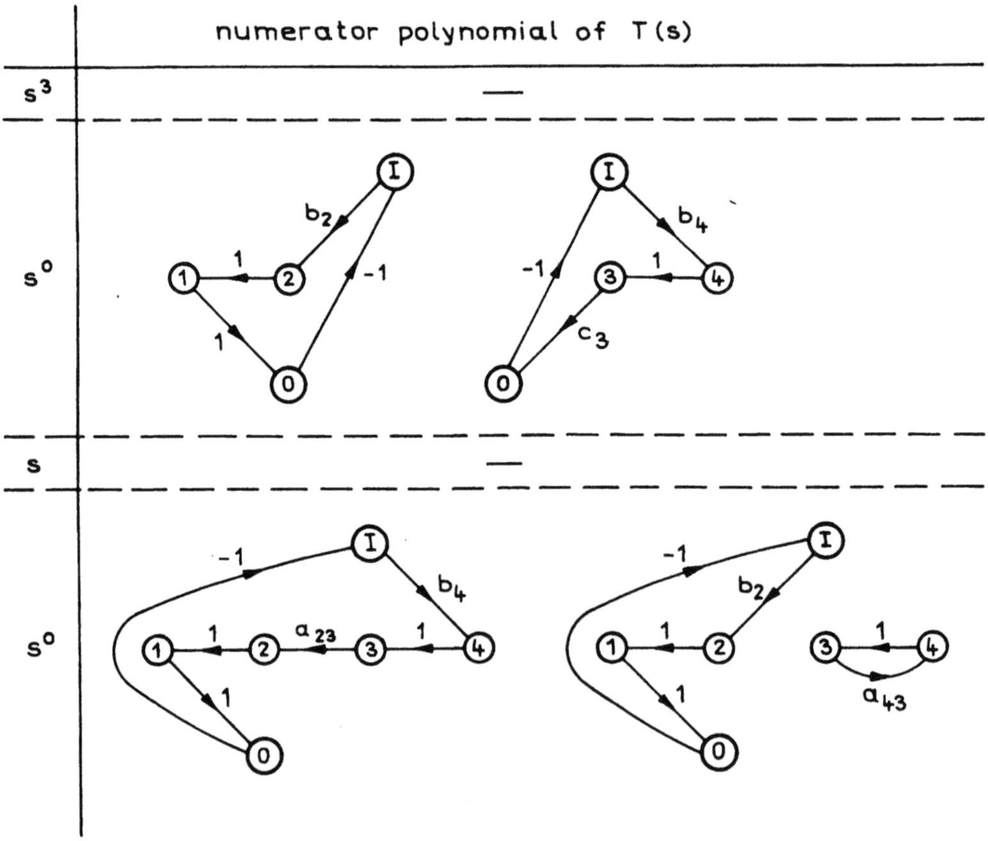

Fig. 31.2

The subgraphs that provide the coefficients of the numerator polynomial of the transfer function have been depicted in Fig. 32.2.b. For this example, the transfer function is

$$T(s) = \frac{(b_2 + b_4 c_3)s^2 - b_2 a_{43} + b_4 a_{23}}{s^4 - a_{43} s^2} .$$

The numerator polynomial and the denominator polynomial of a transfer function $T_{ji}(s)$ have sometimes <u>common divisors in the structural sense</u>. This means there are common zeros of both polynomials for each numerical realization $(A, b_i, c_j) \in [A, b_i, c_j]$. Such a situation can be easily recognized from the digraph $G(Q^{ij})$.

<u>Theorem 31.2</u>

If both the digraphs $G(A)$ and $G(Q^{ij})$ do not contain cycle families and feedback cycle families of width $n, n-1, \ldots, n'$ then $s = 0$ is a common structural zero of multiplicity $n-n'+1$ of the numerator and the denominator of the transfer function $T_{ji}(s)$.

If there is a subgraph $G(A^{ij}_{sub})$ of $G(A)$ not both input and output connected in $G(Q^{ij})$, then the polynomial

$$\det(sI_{sub} - A^{ij}_{sub}) \qquad (31.13)$$

is a common divisor (in the structural sense) of the numerator and the denominator of the transfer function $T_{ji}(s)$.

<u>Proof</u>: The assumption of the first sentence ensures that both the coefficients

$p^o_n, p^o_{n-1}, \ldots, p^o_{n'}$ in (31.7)

and

$p^{ij}_n, p^{ij}_{n-1}, \ldots, p^{ij}_{n'}$ in (31.9)

vanish.

This implies immediately the statement of the first sentence.

To verify the second statement consider the equivalence classes of strongly connected vertices within $G(A)$. Those equivalence classes which are both input and output connected within $G(Q^{ij})$ form a single equivalence class in $G(Q^{ij})$. Any other equivalence class of $G(A)$ (including single vertices that form an "acyclic"equivalence class in the sense of Section 13) constitutes also an equivalence class in $G(Q^{ij})$.

Now we can carry out a reordering of the state variables as explained in detail in Section 13. First we enumerate all the state vertices associated with the submatrix A^{ij}_{sub} and then the remaining state vertices being both input and output connected in $G(Q^{ij})$. Thus we obtain a matrix partitioning

$$\begin{pmatrix} A & b_i \\ c_j' & 0 \end{pmatrix} = \left(\begin{array}{cc|c} A_{sub}^{ij} & X & b_{i,sub} \\ 0 & A_{rem}^{ij} & b_{i,rem} \\ \hline 0 & c_{j,rem}' & 0 \end{array}\right)$$

or

$$\begin{pmatrix} A & b_i \\ c_j' & 0 \end{pmatrix} = \left(\begin{array}{cc|c} A_{sub}^{ij} & 0 & 0 \\ X & A_{rem}^{ij} & b_{i,rem} \\ \hline c_{j,s}' & c_{j,r}' & 0 \end{array}\right)$$

where X symbolizes a not necessarily vanishing submatrix.

The denominator polynomial $\det(sI_n - A)$ splits up into

$$\det(sI_n - A) = \det(sI_{sub} - A_{sub}^{ij}) \det(sI_{rem} - A_{rem}^{ij})$$

and the numerator polynomial becomes

$$z_{ji}(s) = \det \begin{pmatrix} sI_n - A & b_i \\ c_j' & 0 \end{pmatrix} = \det(sI_{sub} - A_{sub}^{ij}) \det \begin{pmatrix} sI_{rem} - A_{rem}^{ij} & b_{i,rem} \\ c_{j,rem}' & 0 \end{pmatrix}$$

The common structural divisor is $\det(sI_{sub} - A_{sub}^{ij})$ as claimed in Theorem 31.2.

▲

Example 31.3: Fig. 31.3 shows a structure digraph of a system with $n = 6$ states, $r = 3$ outputs and $m = 1$ input. For clearness, the subgraphs $G(A)$, $G(Q^{12})$ and $G(A_{sub}^{12})$ have been encircled by dashed lines.

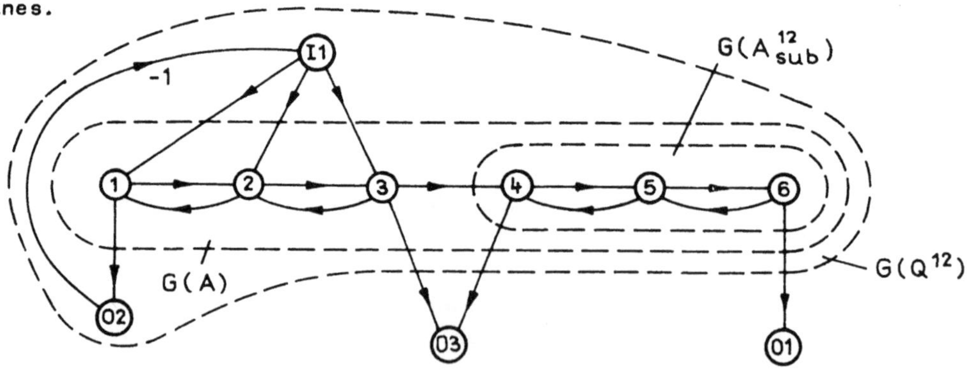

Fig. 31.3

The structure of all interesting polynomials can be immediately seen from Fig. 31.3.

Open-loop characteristic polynomial:

$$\det(sI_6 - A) = s^6 + p_2 s^4 + p_4 s^2.$$

Numerator polynomials:

$$z_{11}(s) = p_4^{11} s^2 + p_5^{11} s + p_6^{11} ,$$
$$z_{21}(s) = p_1^{21} s^5 + p_2^{21} s^4 + p_3^{21} s^3 + p_4^{21} s^2 + p_5^{21} s ,$$
$$z_{31}(s) = p_1^{31} s^5 + p_2^{31} s^4 + p_3^{31} s^3 + p_4^{31} s^2 + p_5^{31} s + p_6^{31} .$$

The state vertices 4, 5, and 6 are not connected with the output O2. The submatrix A_{sub}^{12} becomes

$$A_{sub}^{12} = \begin{pmatrix} 0 & a_{45} & 0 \\ a_{54} & 0 & a_{56} \\ 0 & a_{65} & 0 \end{pmatrix}$$

The denominator and the numerator polynomials of the transfer function $T_{21}(s)$ have the common divisor

$$\det(sI_3 - A_{sub}^{12}) = s^3 + (a_{45}a_{56} + a_{56}a_{65})s$$

independently of the numerical realization of the non-zero entries of A.

31.2 Feedback dependencies of the closed-loop characteristic-polynomial coefficients

The coefficients matrix of the compound system of equations (31.4) is denoted by $S(s)$ and called closed-loop system matrix in the sequel,

$$S(s) = \begin{pmatrix} I_r & -C & 0 \\ 0 & sI_n - A & -B \\ -F & 0 & I_m \end{pmatrix} \qquad (31.15)$$

Its determinant is called the closed-loop characteristic polynomial,

$$\det S(s) = s^n + p_1 s^{n-1} + \dots + p_{n-1} s + p_n \qquad (31.16)$$

Traditionally, the closed-loop characteristic polynomial is introduced as $\det(sI_n - A - B\,F\,C)$. In this context, the reader should be aware of the relation

$$\det S(s) = \det(sI_n - A - B\,F\,C). \qquad (31.17)$$

The identity (31.17) follows from the known rules for evaluating the determinants of block matrices, comp. Eq. (21.8),

$$\det \begin{pmatrix} I_r & -C & 0 \\ 0 & sI_n - A & -B \\ -F & 0 & I_m \end{pmatrix} = \det I_r \det\left(\begin{pmatrix} sI_n - A & -B \\ 0 & I_m \end{pmatrix} + \begin{pmatrix} 0 \\ F \end{pmatrix} I_r (-C \quad 0)\right)$$

$$= \det \begin{pmatrix} sI_n - A & -B \\ -FC & I_m \end{pmatrix} = \det I_m \det(sI_n - A - B\,F\,C) \qquad (31.18)$$

The left-hand side of Eq. (31.18) may also be transformed in the following manner:

$$\det \begin{pmatrix} sI_n - A & -B \\ -FC & I_m \end{pmatrix} = \det(sI_n - A) \det(I_m - F\cdot C(sI_n - A)^{-1}B)$$

Taking into account (31.6) and (31.18) we get

$$\det(sI_n - A - B\,F\,C) = \det(sI_n - A) \det(I_m - F\cdot T(s)) \qquad (31.19)$$

The $m \times m$ matrix $(I_m - F\cdot T(s))$ occurring on the right-hand side of (31.19) is sometimes called return-difference matrix, comp. Section 32.

Now, the submatrices A, B, C of the closed-loop system matrix S(s) are assumed to be numerically fixed. Then the coefficients p_w, $1 \leq w \leq n$, of the closed-loop characteristic polynomial (31.18) depend on feedback gains reflected by the indeterminate entries of the submatrix F.
In general, the characteristic polynomial depends non-linearly upon the feedback gains. Fortunately, the type o non-linear dependenc occurring may easily be seen from the special structure of the closed-loop system matrix (31.17).

Lemma 31.1 gives the key to seeing how the closed-loop characteristic polynomial depends on the feedback gains.

<u>Lemma 31.1</u>

The closed-loop characteristic polynomial det S(s) shows the following F-dependence:
det S(s) appears as a sum where

(a) the F-independent summands give together

$$\det(sI_n - A) = s^n + \sum_{w=1}^{n} p_w^o \, s^{n-w} \tag{31.20}$$

(b) the summands involving one feedback gain are

$$f_{ij} \det \begin{pmatrix} sI_n - A & b_i \\ c_j' & 0 \end{pmatrix} = f_{ij} \sum_{w=1}^{n} p_w^{ij} \, s^{n-w} \tag{31.21}$$

for $i = 1,2,\ldots,m$, $j = 1,2,\ldots,r$;

(c) the summands involving two feedback gains are

$$f_{ij} f_{kl} \det \begin{pmatrix} sI_n - A & b_i & b_k \\ c_j' & 0 & 0 \\ c_l' & 0 & 0 \end{pmatrix} = f_{ij} f_{kl} \sum_{w=2}^{n} p_w^{ij \cdot kl} \, s^{n-w} \tag{31.22}$$

for both $i \neq k$ and $j \neq l$;

(d) the summands involving three feedback gains are

$$f_{ij} f_{kl} f_{gh} \det \begin{pmatrix} sI_n - A & b_i & b_k & b_g \\ c_j' & 0 & 0 & 0 \\ c_l' & 0 & 0 & 0 \\ c_h' & 0 & 0 & 0 \end{pmatrix} = f_{ij} f_{kl} f_{gh} \sum_{w=3}^{n} p^{ij \cdot kl \cdot gh} \, s^{n-w} \tag{31.23}$$

for both $i \neq k \neq g$ and $j \neq l \neq k$;

etc.

<u>Proof</u>: Because of the algebraic Def.(A2.1) of determinants, each summand of det S(s) is a product of r+n+m elements of S(s) placed both

in different rows and in different columns.

For $0 \leq q \leq \min(m,r)$, choose q elements

$$f_{i_1 j_1}, \ldots, f_{i_q j_q}$$

of F situated in different rows as well as in different columns.
Replace the remaining elements of the involved rows and columns of S(s) by zeros.
Replace all F-entries apart from $f_{i_1 j_1}, \ldots, f_{i_q j_q}$ by zeros.
In this way, one obtains a modified system matrix $\widetilde{S}(s)$ whose determinant consists exactly of those summands of $\det S(s)$ that have

$$f_{i_1 j_1}, \ldots, f_{i_q j_q}$$

as common factor of involved feedback gains.
As every column (row) of $\widetilde{S}(s)$ contributes to $\det \widetilde{S}(s)$, the $r-q(m-q)$ diagonal elements of the unit matrix I_r (I_m) not replaced by zeros must be involved in $\det \widetilde{S}(s)$. Developing $\det \widetilde{S}(s)$ with respect to these columns of I_r, and rows of I_m, we obtain

$$\det \widetilde{S}(s) = \det \begin{pmatrix} 0 & \ldots & 0 & -c'_{j_1} & 0 & \ldots & 0 \\ \vdots & & \vdots & \vdots & \vdots & & \vdots \\ 0 & \ldots & 0 & -c'_{j_q} & 0 & \ldots & 0 \\ 0 & \ldots & 0 & sI_n-A & -b_{i_1} & \ldots & -b_{i_q} \\ -f_{i_1 j_1} & & & 0' & 0 & \ldots & 0 \\ & \ddots & & \vdots & \vdots & & \vdots \\ & & -f_{i_q j_q} & 0' & 0 & \ldots & 0 \end{pmatrix}$$

$$= f_{i_1 j_1} \cdots f_{i_q j_q} (-1)^{3q} \det \begin{pmatrix} 0 & c'_{j_1} & 0 & \ldots & 0 \\ \vdots & \vdots & \vdots & & \vdots \\ 0 & c'_{j_q} & 0 & \ldots & 0 \\ 0 & sI_n-A & b_{i_1} & \ldots & b_{i_q} \\ I_q & 0 & 0 & \ldots & 0 \end{pmatrix}$$

$$= f_{i_1 j_1} \cdots f_{i_q j_q} \det \begin{pmatrix} sI_n-A & b_{i_1} & \ldots & b_{i_q} \\ c'_{j_1} & 0 & \ldots & 0 \\ \vdots & \vdots & & \vdots \\ c'_{j_q} & 0 & \ldots & 0 \end{pmatrix}$$

In particular, for $q = 0, 1, 2, 3$ we have (31.20) to (31.23). This completes the proof. ▲

Using Rosenbrock's system matrix

$$P(s) = \begin{pmatrix} sI_n - A & B \\ C & 0 \end{pmatrix} \qquad (31.24)$$

the contents of Lemma 31.1 may be written more condensely,

$$\det S(s) = P(s)^{1\ldots n}_{1\ldots n} - \sum_{i_1=1}^{m} \sum_{j_1=1}^{r} f_{i_1j_1} P(s)^{1\ldots n,n+j_1}_{1\ldots n,n+i_1} \qquad (31.25)$$

$$+ \sum_{i_1<} \sum_{i_2} \sum_{j_1 \neq} \sum_{j_2} f_{i_1j_1} f_{i_2j_2} P(s)^{1\ldots n,n+j_1,n+j_2}_{1\ldots n,n+i_1,n+i_2} - + \ldots$$

$$+ (-1)^q \sum_{i_1<} \cdots \sum_{<i_q} \sum_{j_1 \neq} \cdots \sum_{\neq j_q} f_{i_1j_1} \cdots f_{i_qj_q} P(s)^{1..n,n+j_1\ldots n+j_q}_{1..n,n+i_1\ldots n+i_q}$$

where $q \leq \min(m,r)$.

After this preparation, we are able to formulate explicit relations between the closed-loop characteristic-polynomial coefficients p_w, $1 \leq w \leq n$, and the elements of the output feedback matrix.

<u>Theorem 31.3</u>

Let $p = (p_1, p_2, \ldots, p_n)'$ be a column vector that encompasses the coefficients of the closed-loop characteristic polynomial (31.16). Then there holds

$$p = p^o + \sum_{i_1=1}^{m} \sum_{j_1=1}^{r} p^{i_1j_1} f_{i_1j_1} + \sum_{i_1<} \sum_{i_2} \sum_{j_1 \neq} \sum_{j_2} p^{i_1j_1 \cdot i_2j_2} f_{i_1j_1} f_{i_2j_2} \qquad (31.26)$$

$$+ \ldots + \sum_{i_1<} \sum_{i_2<} \cdots \sum_{<i_q} \sum_{j_1 \neq} \sum_{j_2 \neq} \cdots \sum_{\neq j_q} p^{i_1j_1 \cdot i_2j_2 \cdot\cdot i_qj_q} f_{i_1j_1} f_{i_2j_2} \cdot\cdot f_{i_qj_q}$$

with $q \leq \min(m,r)$,

where, for $1 \leq h_1 < h_2 < \ldots < h_w \leq n$ and $1 \leq w \leq n$

$$p_w^o = \sum \sum \cdots \sum_{1 \leq h_1 < h_2 < \ldots < h_w \leq n} P(0)^{h_1h_2\ldots h_w}_{h_1h_2\ldots h_w} \qquad (31.27)$$

$$p_w^{i_1j_1} = \sum \sum \cdots \sum_{1 \leq h_1 < h_2 < \ldots < h_w \leq n} P(0)^{h_1h_2\ldots h_w n+j_1}_{h_1h_2\ldots h_w n+i_1} \qquad (31.28)$$

and, generally, for $2 \le q \le w \le n$, (31.29)

$$p_w^{i_1j_1 \cdot i_2j_2 \cdots i_qj_q} = \sum_{h_1<h_2<\ldots<h_w} \sum \ldots \sum P(0)^{h_1h_2 \ldots h_w n+j_1, n+j_2, \ldots, n+j_q}_{h_1h_2 \ldots h_w n+i_1, n+i_2, \ldots, n+i_q}$$

Further,

$$p_w^{i_1j_1 \cdot i_2j_2 \cdots i_qj_q} = 0 \quad \text{for} \quad w = 1,2,\ldots,q-1. \tag{31.30}$$

Proof: According to Lemma 31.1, the summands of the closed-loop characteristic polynomial that involve the q feedback gains

$$f_{i_1j_1},\ f_{i_2j_2}, \ldots,\ f_{i_qj_q}$$

are given by

$$f_{i_1j_1} \cdots f_{i_qj_q} \det \begin{pmatrix} sI_n - A & b_{i_1} \ldots b_{i_q} \\ c'_{j_1} & 0 \ldots 0 \\ \vdots & \vdots \quad \vdots \\ c'_{j_q} & 0 \ldots 0 \end{pmatrix}$$

$$= f_{i_1j_1} \cdots f_{i_qj_q} P(s)^{1 \ldots n, n+j_1 \ldots n+j_q}_{1 \ldots n, n+i_1 \ldots n+i_q}$$

$$= f_{i_1j_1} \cdots f_{i_qj_q} \Big\{ s^{n-q} \sum_{h_1< \ldots <h_q} \ldots \sum P(0)^{h_1 \ldots h_q n+j_1 \ldots n+j_q}_{h_1 \ldots h_q n+i_1 \ldots n+i_q}$$

$$+ s^{n-q-1} \sum_{h_1< \ldots <h_q<h_{q+1}} \ldots \sum \sum P(0)^{h_1 \ldots h_q h_{q+1} n+j_1 \ldots n+j_q}_{h_1 \ldots h_q h_{q+1} n+i_1 \ldots n+i_q}$$

$$+ \ldots + s \sum_{h_1<h_2< \ldots <h_{n-1}} \sum \ldots \sum P(0)^{h_1h_2 \ldots h_{n-1} n+j_1 \ldots n+j_q}_{h_1h_2 \ldots h_{n-1} n+i_1 \ldots n+i_q}$$

$$+ P(0)^{1\ 2\ \ldots\ n, n+j_1 \ldots n+j_q}_{1\ 2\ \ldots\ n, n+i_1 \ldots n+i_q} \Big\}$$

This representation proves the validity of the statements of Theorem 31.3.

▲

Example 31.4: Consider a plant with $n = 6$ state variables, $m = 2$ inputs and $r = 3$ outputs that is described mathematically by the matrices A, B, C as follows:

$$A = \begin{pmatrix} 0 & 0 & A^1_3 & 0 & 0 & 0 \\ A^2_1 & 0 & 0 & A^2_4 & A^2_5 & 0 \\ 0 & 0 & A^3_3 & 0 & 0 & A^3_6 \\ 0 & A^4_2 & 0 & 0 & 0 & 0 \\ 0 & 0 & 0 & A^5_4 & A^5_5 & 0 \\ 0 & 0 & 0 & 0 & 0 & 0 \end{pmatrix} \qquad B = (b_1, b_2) = \begin{pmatrix} 0 & 0 \\ 0 & 0 \\ 0 & 0 \\ 0 & 0 \\ B^5_1 & 0 \\ 0 & B^6_2 \end{pmatrix} \qquad (31.31)$$

$$C = \begin{pmatrix} c'_1 \\ c'_2 \\ c'_3 \end{pmatrix} = \begin{pmatrix} 0 & C^1_2 & 0 & 0 & C^1_5 & 0 \\ C^2_1 & 0 & 0 & 0 & 0 & 0 \\ C^3_1 & 0 & C^3_3 & 0 & 0 & 0 \end{pmatrix}$$

The non-vanishing entries of the matrices are characterized by an upper index indicating the row and a lower index indicating the column. We use this rather unusual kind of notation here for two reasons. First: in order to save space; and second:in order to make thorough use of the representation (21.18) for minors of arbitrary order. (Matrix elements may be regarded as minors of order 1.) Let us determine the closed-loop characteristic polynomial with the aid of Theorem 31.3.

The coefficients of the open-loop characteristic polynomial

$$\det(sI_n - A) = P(s)^{1\,2\,\ldots\,n}_{1\,2\,\ldots\,n}$$

become

$$p^o_1 = \sum_{h=1}^{6} (-A)^h_h = -A^3_3 - A^5_5$$

$$p^o_2 = \sum_{h_1 < h_2} \sum (-A)^{h_1 h_2}_{h_1 h_2} = (-A)^{3\,5}_{3\,5} + (-A)^{2\,4}_{2\,4} = A^3_3 A^5_5 - A^2_4 A^4_2$$

$$p^o_3 = \sum_{h_1 < h_2 < h_3} \sum \sum (-A)^{h_1 h_2 h_3}_{h_1 h_2 h_3} = A^2_4 A^3_3 A^4_2 - A^2_5 A^4_2 A^5_4 + A^2_4 A^4_2 A^5_5 \qquad (31.32)$$

$$p^o_4 = \sum_{h_1 < h_2 < h_3 < h_4} \sum \sum \sum (-A)^{h_1 h_2 h_3 h_4}_{h_1 h_2 h_3 h_4} = -A^2_4 A^3_3 A^4_2 A^5_5 + A^2_5 A^3_3 A^4_2 A^5_4$$

$$p^o_5 = \sum_{h_1 < h_2 < h_3 < h_4 < h_5} \sum \sum \sum \sum (-A)^{h_1 h_2 h_3 h_4 h_5}_{h_1 h_2 h_3 h_4 h_5} = 0$$

$$p^o_6 = (-A)^{1\,2\,3\,4\,5\,6}_{1\,2\,3\,4\,5\,6} = 0$$

All summands of the closed-loop characteristic polynomial that depend on the feedback gain f_{11} (and no other feedback gain) are comprised of the vector p^{11}, whose components are

$$
\begin{aligned}
p_1^{11} &= \sum_{h=1}^{6} P(0)^{h\;7}_{h\;7} = -\,B_1^5 C_5^1 \\
p_2^{11} &= \sum_{h_1<h_2}\sum P(0)^{h_1h_2 7}_{h_1h_2 7} = A_3^3 B_1^5 C_1^5 - A_5^2 B_1^5 C_2^1 \\
p_3^{11} &= \sum_{h_1<h_2<h_3}\sum\sum P(0)^{h_1h_2h_3 7}_{h_1h_2h_3 7} = A_4^2 A_2^4 B_1^5 C_5^1 + A_5^2 A_3^3 B_1^5 C_2^1 \\
p_4^{11} &= \sum_{h_1<h_2<h_3<h_4}\sum\sum\sum P(0)^{h_1h_2h_3h_4 7}_{h_1h_2h_3h_4 7} = -A_4^2 A_3^3 A_2^4 B_1^5 C_5^1 \\
p_5^{11} &= \sum_{h_1<h_2<h_3<h_4<h_5}\sum\sum\sum\sum P(0)^{h_1h_2h_3h_3h_5 7}_{h_1h_2h_3h_4h_5 7} = 0 \\
p_6^{11} &= P(0)^{1\;2\;3\;4\;5\;6\;7}_{1\;2\;3\;4\;5\;6\;7} = 0
\end{aligned}
\qquad (31.33a)
$$

Similarly, one finds the other summands of the closed-loop characteristic polynomial depending on exactly one feedback gain:

$$
p^{21} = \begin{pmatrix} 0 \\ 0 \\ 0 \\ -A_3^1 A_1^2 A_6^3 B_2^6 C_2^1 \\ A_3^1 A_1^2 A_6^3 A_5^5 B_2^6 C_2^1 \\ -A_3^1 A_1^2 A_6^3 A_2^4 A_4^5 B_2^6 C_2^1 \end{pmatrix}, \quad p^{12} = 0, \quad p^{22} = \begin{pmatrix} 0 \\ 0 \\ -A_2^1 A_6^3 B_2^6 C_2^1 \\ A_3^1 A_6^3 A_5^5 B_2^6 C_2^1 \\ A_3^1 A_4^2 A_6^3 A_2^4 B_2^6 C_2^1 \\ A_3^1 A_4^2 A_6^3 A_5^4 A_2^5 B_2^6 C_2^1 - A_2^1 A_4^2 A_6^3 A_2^4 A_5^5 B_2^6 C_2^1 \end{pmatrix}
$$

$$
p^{13} = 0, \quad p^{23} = \begin{pmatrix} 0 \\ -A_6^3 B_2^6 C_3^3 \\ -A_3^1 A_6^3 B_2^6 C_1^3 + A_6^3 A_5^5 B_2^6 C_3^3 \\ A_4^2 A_6^3 A_2^4 B_2^6 C_3^3 + A_3^1 A_6^3 A_5^5 B_2^6 C_1^3 \\ A_3^1 A_4^2 A_6^3 A_2^4 B_2^6 C_1^3 - A_4^2 A_6^3 A_2^4 A_5^5 B_2^6 C_3^3 + A_5^2 A_6^3 A_2^4 A_4^5 B_2^6 C_3^3 \\ A_3^1 A_5^2 A_6^3 A_2^4 A_4^5 B_2^6 C_1^3 - A_3^1 A_4^2 A_6^3 A_2^4 A_5^5 B_2^6 C_1^3 \end{pmatrix} \qquad (31.33b)
$$

Next, the summands of the closed-loop characteristic polynomial that depend on two feedback gains are determined. For example, the components of the vector $p^{13\cdot 21}$ which is associated with the product $f_{13}\cdot f_{21}$, may be obtained as follows:

$$p_1^{13\cdot 21} = 0$$

$$p_2^{13\cdot 21} = \sum_{1 \le h_1 <} \sum_{h_2 \le 6} P(0)\begin{matrix} h_1 h_2 78 \\ h_1 h_2 78 \end{matrix} = 0$$

$$p_3^{13\cdot 21} = \sum \sum \sum_{h_1<h_2<h_3} P(0)\begin{matrix} h_1 h_2 h_3 78 \\ h_1 h_2 h_3 78 \end{matrix} = -A_6^3 B_1^5 C_3^3 B_2^6 C_5^1 \qquad (31.34a)$$

$$p_4^{13\cdot 21} = \sum \sum \sum \sum_{h_1<h_2<h_3<h_4} P(0)\begin{matrix} h_1 h_2 h_3 h_4 78 \\ h_1 h_2 h_3 h_4 78 \end{matrix} = -A_5^2 A_6^3 B_1^5 B_2^6 C_3^3 C_2^1 - A_3^1 A_6^3 B_1^5 B_2^6 C_1^3 C_5^1$$

$$p_5^{13\cdot 21} = A_4^2 A_6^3 A_2^4 B_1^5 B_2^6 C_3^3 C_5^1 - A_3^1 A_5^2 A_6^3 B_1^5 B_2^6 C_1^3 C_2^1$$

$$p_6^{13\cdot 21} = A_3^1 A_4^2 A_6^3 A_2^4 B_1^5 B_2^6 C_1^3 C_5^1$$

The other parts of the closed-loop characteristic polynomial involving two feedback gains are associated with the vectors

$$p^{11\cdot 22} = \begin{pmatrix} 0 \\ 0 \\ 0 \\ A_3^1 A_6^3 B_1^5 B_2^6 C_5^1 C_1^2 \\ A_3^1 A_5^2 A_6^3 B_1^5 B_2^6 C_2^1 C_1^2 \\ -A_3^1 A_4^2 A_6^3 A_2^4 B_1^5 B_2^6 C_5^1 C_1^2 \end{pmatrix}, \quad p^{12\cdot 21} = \begin{pmatrix} 0 \\ 0 \\ 0 \\ -A_3^1 A_6^3 B_1^5 B_2^6 C_5^1 C_1^2 \\ -A_3^1 A_5^2 A_6^3 B_1^5 B_2^6 C_1^2 C_2^1 \\ A_3^1 A_4^2 A_6^3 A_2^4 B_1^5 B_2^6 C_1^2 C_5^1 \end{pmatrix}, \quad p^{12\cdot 23} = 0,$$

$$p^{11\cdot 23} = \begin{pmatrix} 0 \\ 0 \\ A_6^3 B_1^5 B_2^6 C_5^1 C_3^3 \\ A_3^1 A_6^3 B_1^5 B_2^6 C_5^1 C_1^3 + A_5^2 A_6^3 B_1^5 B_2^6 C_2^1 C_3^3 \\ -A_4^2 A_6^3 A_2^4 B_1^5 B_2^6 C_5^1 C_3^3 + A_5^2 A_3^1 A_6^3 B_1^5 B_2^6 C_2^1 C_1^3 \\ -A_3^1 A_4^2 A_6^3 A_2^4 B_1^5 B_2^6 C_5^1 C_1^3 \end{pmatrix}, \quad p^{13.22} = 0 . \qquad (31.34b)$$

Together, the coefficients

$$(p_1, p_2, p_3, p_4, p_5, p_6)' = p$$

of the closed-loop characteristic polynomial are subject to the vector equation

$$p = p^{o} + f_{11}\, p^{11} + f_{21}\, p^{21} + f_{22}\, p^{22} + f_{23}\, p^{23}$$

$$+ f_{11} f_{22}\, p^{11.22} + f_{12} f_{21}\, p^{12.21} + f_{11} f_{23}\, p^{11.23} + f_{13} f_{21}\, p^{13.21}$$

or, in equivalent component-wise representation and indicating the non-vanishing entries by "L",

$$\begin{pmatrix} p_1 \\ p_2 \\ p_3 \\ p_4 \\ p_5 \\ p_6 \end{pmatrix} = \begin{pmatrix} L \\ L \\ L \\ L \\ 0 \\ 0 \end{pmatrix} + f_{11} \begin{pmatrix} L \\ L \\ L \\ L \\ 0 \\ 0 \end{pmatrix} + f_{21} \begin{pmatrix} 0 \\ 0 \\ 0 \\ L \\ L \\ L \end{pmatrix} + f_{22} \begin{pmatrix} 0 \\ 0 \\ L \\ L \\ L \\ L \end{pmatrix} + f_{23} \begin{pmatrix} 0 \\ L \\ L \\ L \\ L \\ L \end{pmatrix}$$

$$+ f_{11} f_{22} \begin{pmatrix} 0 \\ 0 \\ 0 \\ L \\ L \\ L \end{pmatrix} + f_{12} f_{21} \begin{pmatrix} 0 \\ 0 \\ 0 \\ L \\ L \\ L \end{pmatrix} + f_{11} f_{23} \begin{pmatrix} 0 \\ 0 \\ L \\ L \\ L \\ L \end{pmatrix} + f_{13} f_{12} \begin{pmatrix} 0 \\ 0 \\ L \\ L \\ L \\ L \end{pmatrix}$$

31.3 Graph-theoretic interpretation of the closed-loop characteristic-polynomial coefficients

Equation (31.26) yields the desired explicit dependencies between the closed-loop characteristic polynomial (31.18) and the output feedback gains. The determination of the occurring coefficients

$$p_w^{i_1 j_1 \ldots i_q j_q}$$

requires the evaluation of determinants, see (31.27) to (31.30). As this is a known standard task of linear algebra, until now nothing has been said of how to compute determinants practically. There are different procedures. The present author would like to recommend the following approach.

Consider determinants as combinatorial products either of their row vectors or of their column vectors and use the rules of combinatorial multiplication published by H.G. Grassmann in 1844. It lies beyond the scope of this monograph to report Grassmann's penetrating work, which found little acceptance by other mathematicians during his lifetime and, apparently, is not fully understood by the scientific community even now. As for the formal rules of Grassmann's combinatorial multiplication, the reader may find them in an article published in English recently (Chen and Ahmad 1984). Here we shall add a graph-theoretic interpretation of Theorem 31.3. This interpretation reflects Grassmann's combinatorial multiplication rules in graph-theoretic terms.

The closed-loop system (31.1), (31.2), (31.3) may be represented by a digraph $G(Q_4)$ where

$$Q_4 = \begin{pmatrix} 0 & C & 0 \\ 0 & A & B \\ F & 0 & 0 \end{pmatrix} \qquad (31.35)$$

comp. (11.16).

Theorem 31.4

The coefficients occurring in (31.26) allow the following interpretation within the digraph $G(Q_4)$ defined by (31.35):

p_w^0 results from the sum of the weights of all cycle families of width w that do not contain any feedback edge;

p_w^{ij} results from the sum of the weights of all cycle families of width w that contain exactly one feedback edge (with the weight f_{ij});

⋮

$p_w^{i_1j_1\ldots i_qj_q}$ results from the sum of the weights of all cycle families of width w that contain exactly q feedback edges (with the weights $f_{i_1j_1}, \ldots f_{i_qj_q}$).

Theorem 31.4 is an immediate consequence of the graph-theoretic characterization of determinants by cycle families. A formal proof was published by the author in 1984. It seems to be unnecessary to repeat it here. (The proof of Theorem 21.1 given above exploits the same ideas.) Instead, the contents of Theorem 31.4 will be explained with the aid of the example system (31.31).

<u>Example 31.5</u>: The digraph associated with the system treated algebraically as Example 31.4 has been depicted in Fig. 12.1, see above. Fig. 31.4 shows the cycle families that provide the components of the vector p^o, comp. (31.32).

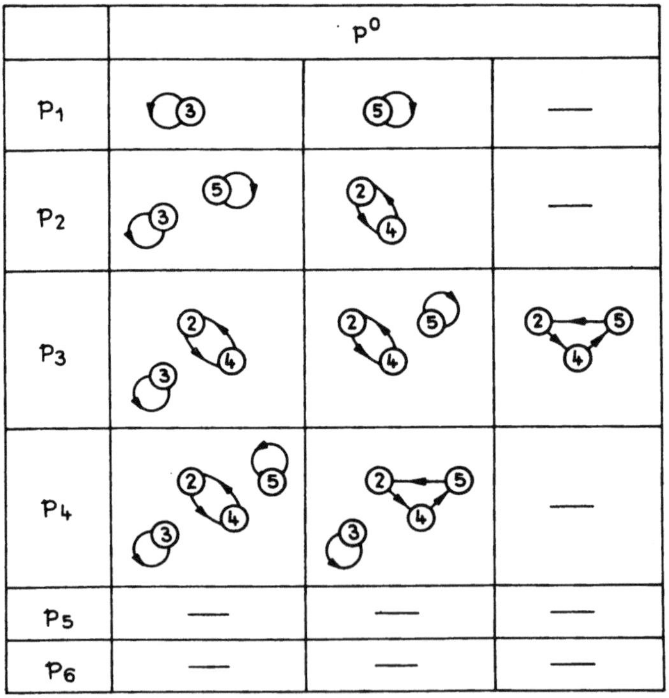

Fig. 31.4

Fig. 31.4 may be regarded as a direct application of Theorem A2.5. Taking into consideration Corollary 31.1 it can be seen that the open-loop characteristic polynomial splits up into three subdeterminants,

Fig. 31.5 →

	$f_{11}\,p^{11}$		$f_{22}\,p^{22}$	
P_1		—	—	—
P_2	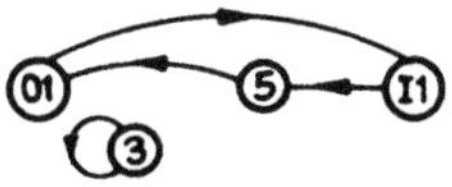		—	—
P_3	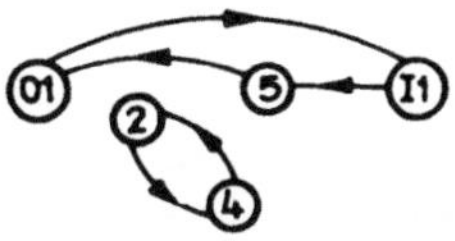	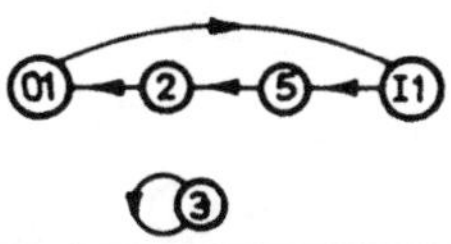	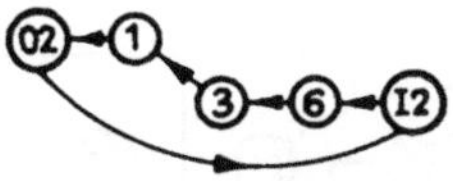	—
P_4	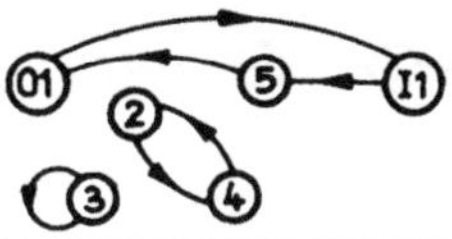	—	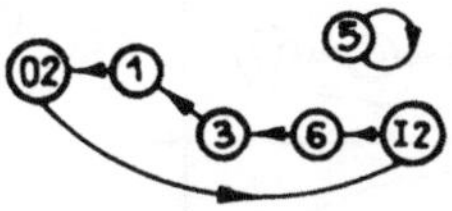	—
P_5	—	—	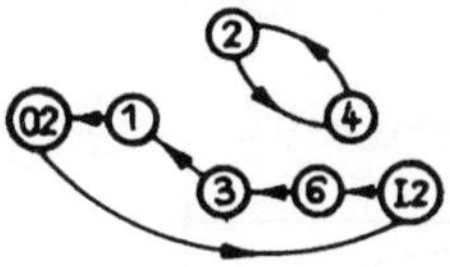	—
P_6	—	—	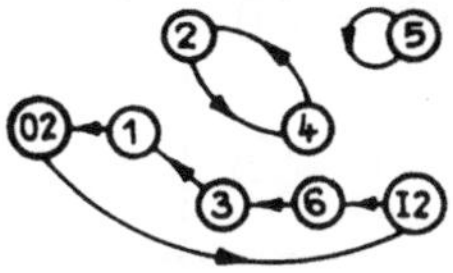	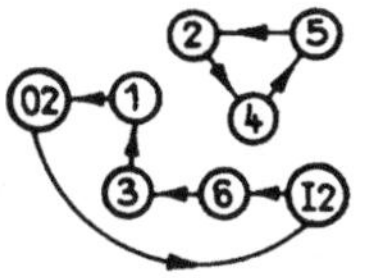

	$f_{21}\,p^{21}$	$f_{23}\,p^{23}$		
P_1	—	—	—	—
P_2	—		—	—
P_3	—		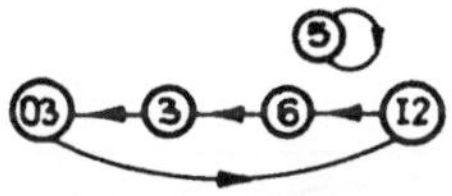	—
P_4	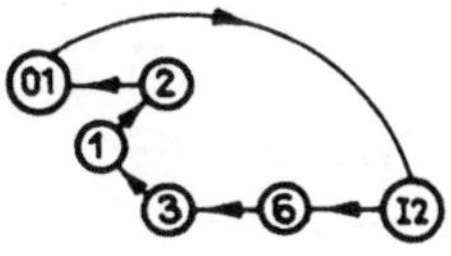	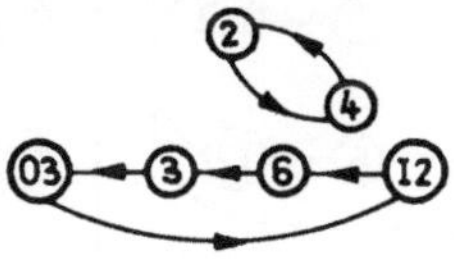	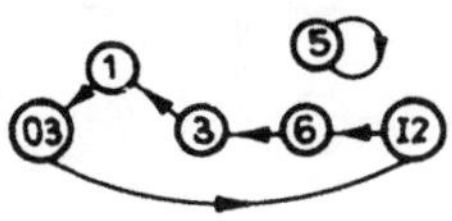	—
P_5	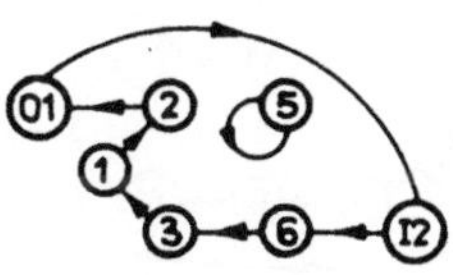	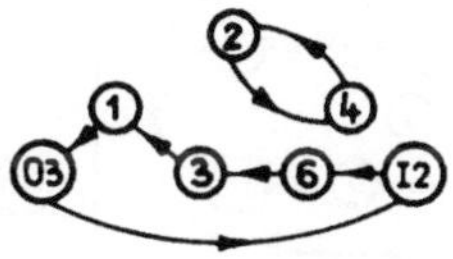	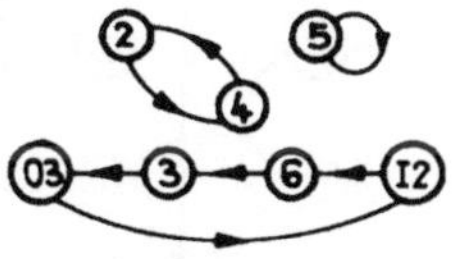	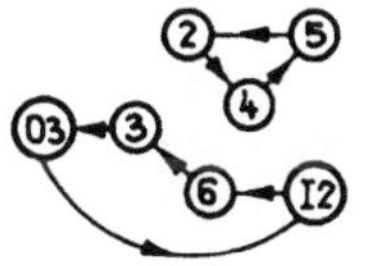
P_6	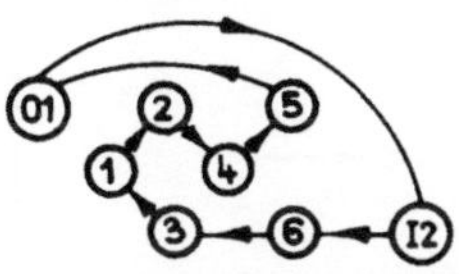	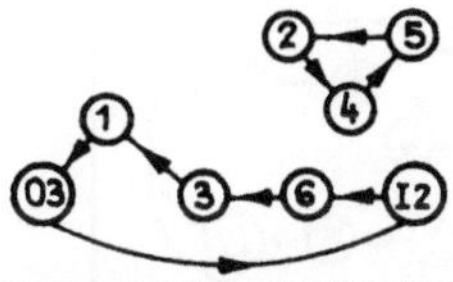	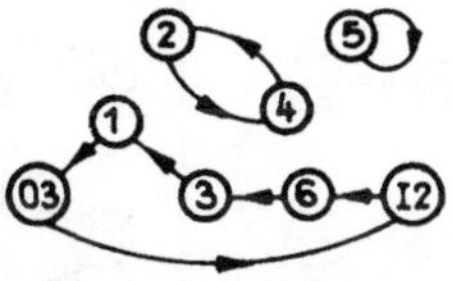	—

	$f_{11} f_{22} p^{11\cdot 22}$	$f_{11} f_{23} p^{11\cdot 23}$	
P_1	—	—	—
P_2	—	—	—
P_3	—	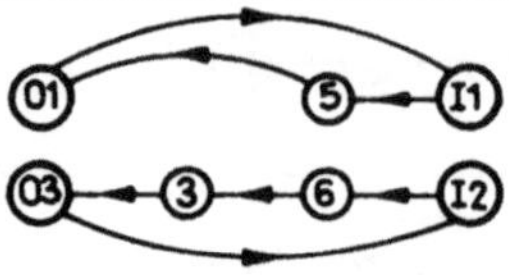	—
P_4	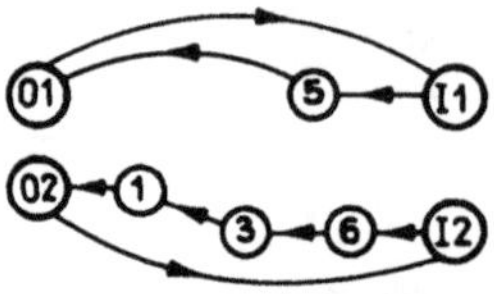	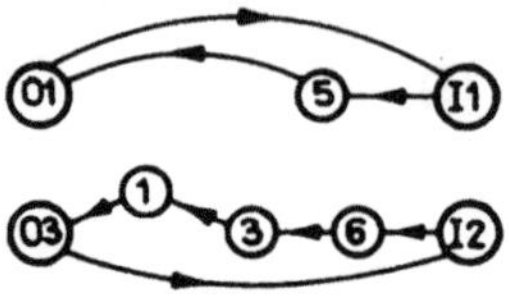	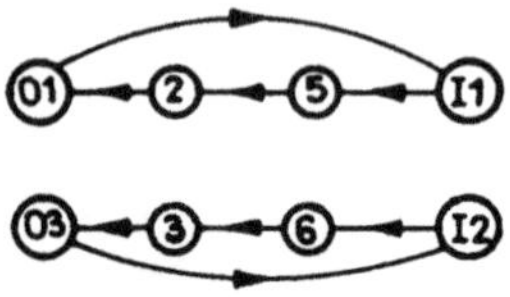
P_5	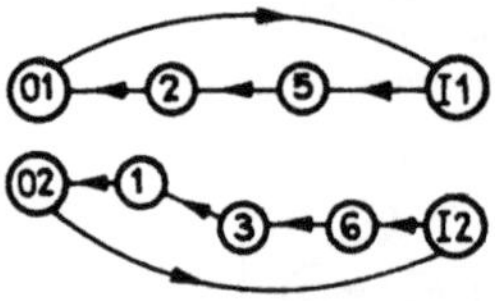	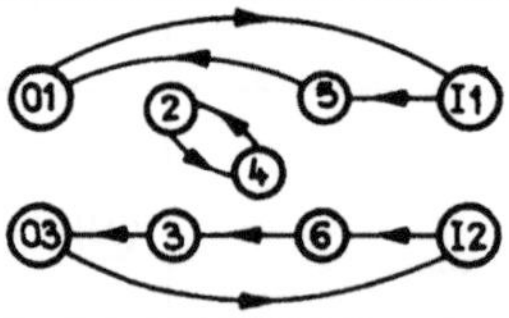	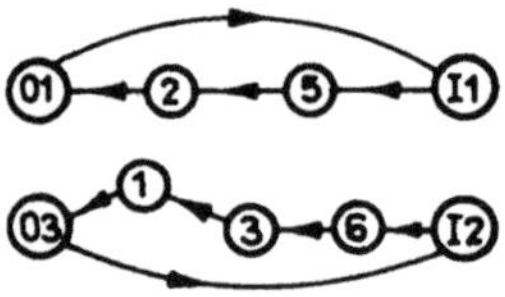
P_6	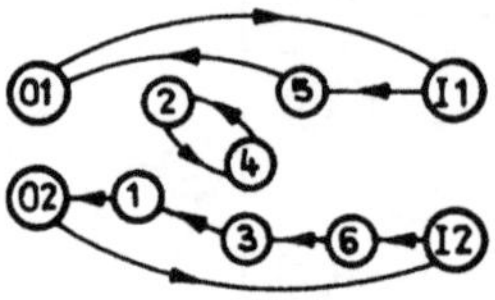	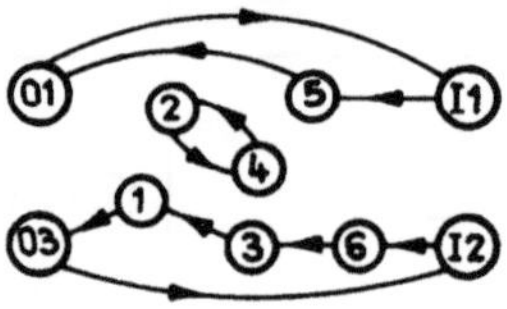	—

	$f_{12} f_{21} p^{12\cdot 21}$	$f_{13} f_{21} p^{13\cdot 21}$	
P_1	—	—	—
P_2	—	—	—
P_3	—	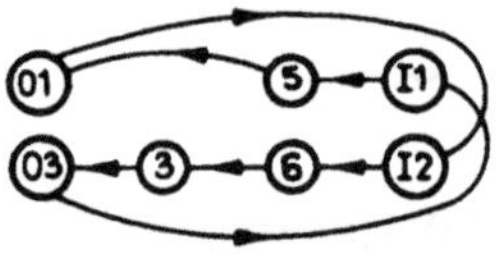	—
P_4	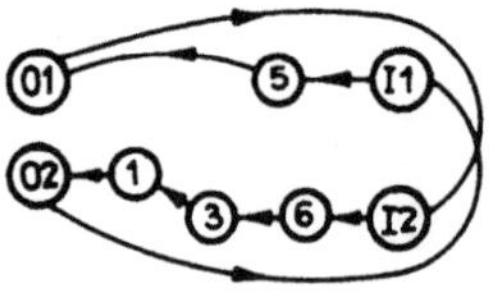	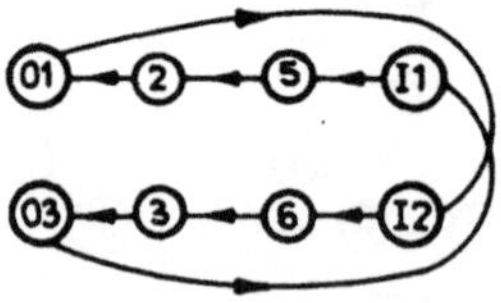	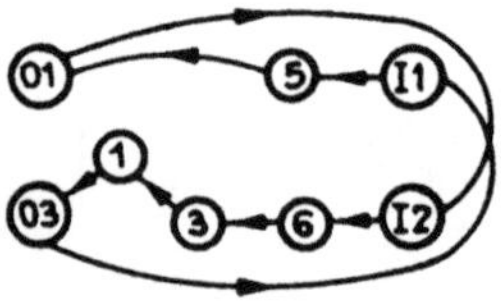
P_5	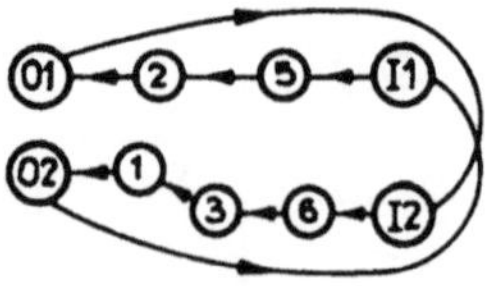	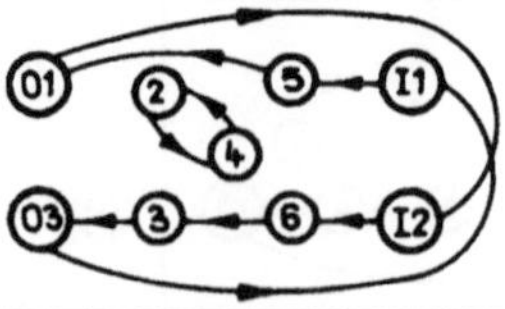	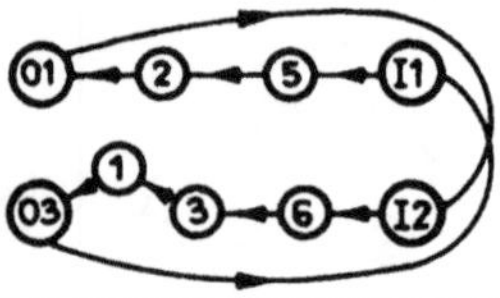
P_6	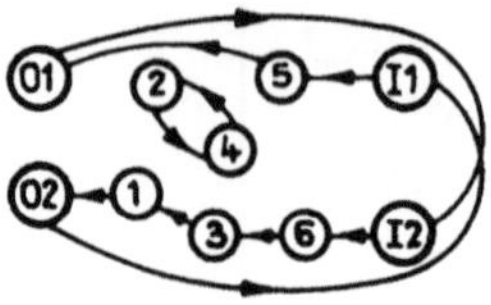	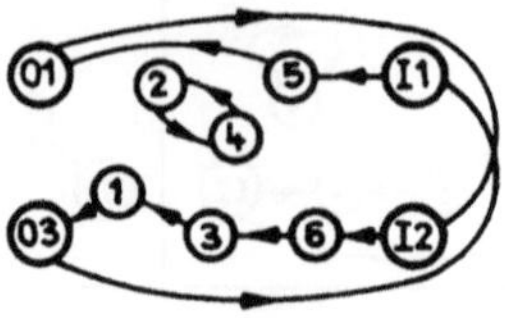	—

$$\det(sI_6 - A) = s^2\,(s - a_{33})\,\det\begin{pmatrix} s - a_{55} & -a_{54} & 0 \\ 0 & s & -a_{42} \\ -a_{25} & -a_{24} & s \end{pmatrix}$$

$$= s^2(s-a_{33})(s^3-a_{55}s^2-a_{24}a_{42}s-a_{54}a_{42}a_{25}+a_{55}a_{42}a_{24})$$

$$= s^6-(a_{33}+a_{55})s^5+(a_{33}a_{55}-a_{42}a_{24})s^4$$

$$+((a_{33}+a_{55})(a_{24}a_{42})+a_{25}a_{54}a_{42})s^3$$

$$+\ a_{33}(a_{25}a_{54}a_{42}-a_{24}a_{42}a_{55})s^2$$

In Fig. 31.5, the cycle families associated with the components $f_{ij}p^{ij}$ for $i = 1,2$ and $j = 1,2,3$ have been drawn. The identically-vanishing vectors

$f_{12}p^{12}$ and $f_{13}p^{13}$

have been omitted, comp. (31.33).

In Fig. 31.6 all cycle families belonging to the vectors

$$f_{i_1j_1}f_{i_2j_2}\,p^{i_1j_1\cdot i_2j_2}$$

have been composed, comp. (31.34).

←Fig. 31.6

32 Poles and zeros of multivariable systems

32.1 Poles and zeros of single-input single-output systems

For convenience, let us first recall the concepts of poles and zeros for systems with $m = 1$ input and $r = 1$ output. In case of such single-input single-output systems, for short, SISO systems, there is a scalar transfer function

$$T(s) = \frac{c'(sI_n - A)_{adj}\, b}{\det(sI_n - A)} = \frac{Z(s)}{\det(sI_n - A)} = \frac{N(s)}{D(s)} \tag{32.1}$$

where the numerator $N(s)$ and the denominator $D(s)$ are assumed to be relatively prime polynomials in the complex frequency s.

A complex number p is called a pole of T(s) if the denominator polynomial $D(s)$ vanishes at $s = p$.
More exactly, p is said to be a pole of multiplicity h_p of $T(s)$ if $(s - p)^{h_p}$ is a divisor of $D(s)$.

A complex number z is said to be a (finite) zero of multiplicity h_z if $(s - z)^{h_z}$ is a divisor of $N(s)$.

The transfer function $T(s)$ has an infinite zero of multiplicity h_{inf} if $T(1/w)$ has a finite zero of multiplicity h_{inf} at $w = 0$.
The last definition implies a simple possibility to obtain h_{inf},

$$h_{inf} = \deg \det(sI_n - a) - \deg Z(s) = \deg D(s) - \deg N(s) \tag{32.2}$$

Usually, the eigenvalues of the square matrix A are called open-loop system poles or zeros of the open-loop characteristic polynomial.

Eigenvalues of A not contained in the set of poles of $T(s)$ are called non-minimal poles or decoupling poles.

Zeros of $Z(s) = c'(sI_n - A)_{adj}\, b$ not contained in the set of zeros of $T(s)$ are called non-minimal zeros or decoupling zeros.

The set of decoupling poles and zeros reflects the "cancellable" pole-zero terms in the transfer function.

With these definitions, the following relationships can be established:

$$\{\text{open-loop system poles}\} = \{\text{zeros of } \det(sI_n - A)\} \tag{32.3}$$

$$= \{\text{poles of } T(s)\} \cup \{\text{decoupling poles}\} \tag{32.4}$$

$$\{\text{open-loop system zeros}\} = \{\text{zeros of } c'(sI_n - A)_{adj}\, b\} \tag{32.5}$$

$$= \{\text{zeros of } T(s)\} \cup \{\text{decoupling zeros}\} \tag{32.6}$$

$$\{\text{decoupling zeros}\} = \{\text{decoupling poles}\} \tag{32.7}$$

If output feedback is applied through a real feedback gain f, then the system matrix A is replaced by

$$A^c = A + f\, b\, c'. \tag{32.8}$$

The poles and zeros of the closed-loop system are defined by

$$\{\text{closed-loop system poles}\} = \{\text{zeros of } \det(sI_n - A - f\, b\, c')\} \tag{32.9}$$

$$\{\text{closed-loop system zeros}\} = \{\text{zeros of } c'(sI_n - A - fbc')_{adj}\, b\} \tag{32.10}$$

There holds the relationship

$$\{\text{closed-loop system zeros}\} = \{\text{open-loop system zeros}\} \tag{32.11}$$

This can be easily seen using elementary properties of determinants,

$$c'(sI_n - A - fbc')_{adj}\, b = -\det\begin{pmatrix} sI_n - A - fbc' & b \\ c' & 0 \end{pmatrix} = -\det\begin{pmatrix} sI_n - A & b \\ c' & 0 \end{pmatrix}$$

$$= c'(sI_n - A)_{adj}\, b$$

A further interesting relationship exists between the open-loop characteristic polynomial (OLCP) and the closed-loop characteristic polynomial (CLCP),

$$\frac{\text{CLCP}(s)}{\text{OLCP}(s)} = \frac{\det(sI_n - A - fbc')}{\det(sI_n - A)} = \frac{\det(sI_n - A) - fc'(sI_n - A)_{adj}\, b}{\det(sI_n - A)}$$

$$= 1 - f\, T(s), \tag{32.12}$$

comp. (31.19).

The quantity $1 - f\,T(s)$ occurring on the right-hand side of (32.12) is called the <u>return difference</u>, for the following reason.

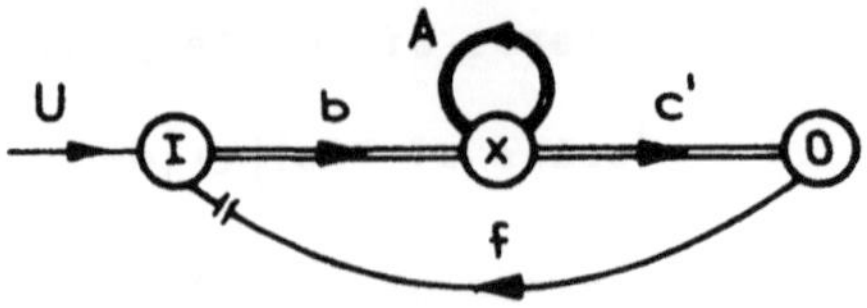

Fig. 32.1

Suppose the feedback edge in Fig. 32.1 to be broken just before the input vertex I and a signal U injected into I. Then the signal which returns to the other side of the broken feedback edge is $f \cdot T(s) \cdot U$. The difference between the injected and the returned signal is $(1 - f \cdot T(s)) \cdot U$.

From (32.12) we draw an important conclusion:
If the feedback gain f tends to infinity, then those closed-loop system poles which remain finite tend to the zeros of the open-loop system. More exactly,

$$\lim_{f \to \infty} \frac{CLCP(s)}{f} = - c'(sI_n - A)_{adj}\, b \qquad (32.13)$$

32.2 Poles and finite zeros of multivariable systems

In the case of multi-input multi-output systems, for short, MIMO systems, the scalar transfer function (32.1) is replaced by the $r \times m$ transfer function matrix

$$T(s) = C(sI_n - A)^{-1}B = [\det(sI_n - A)]^{-1} C(sI_n - A)_{adj} B , \qquad (32.14)$$

comp. (31.6).

It is not trivial to extend the notions of the zeros and poles of a scalar rational function to the zeros and poles of a rational transfer function matrix. This problem has had much attention for many years (see Rosenbrock 1970, Morse 1973, Kaufman 1973, Desoer and Schulman 1974, Davison and Wang 1974, Patel 1975, MacFarlane and Karcanias 1976, Kouvaritakis and MacFarlane 1976, Wolovich 1977, Ferreira and Bhattacharyya 1977, Verghese 1978, Pugh and Ratcliffe 1979, Söte 1980, Vardulakis 1980, Kailath 1980, Commault and Dion 1982, Van der Weiden 1983, Dion 1983, Suda and Umahashi 1984, Svaricek 1985, Patel 1986 and many others).

Most of these works use similarity and equivalence transformations to get canonical forms of rational matrices. Following this way, Smith-McMillan or Wiener-Hopf factorizations appear as the key tools to reveal structural invariants such as finite zeros, poles and zeros at infinity. Unfortunately, such powerful algebraic tools are far from the elementary concept of the digraph approach to control problems. Within the digraph approach, the transformations just mentioned cannot be carried out. The given system (A, B, C) remains practically unchanged in the framework of digraph modelling. Only permutation transformations are allowed at the very most. Nevertheless, the graph-theoretic approach has proved to be useful even for investigating poles and zeros of multivariable systems. Of course, a definition of poles, finite and infinite zeros in terms of the given matrices A, B, and C is a prerequisite for their graph-theoretic interpretation.

Definition 32.1

A complex number p is said to be a *pole of multiplicity* h_p *of the transfer function matrix* $T(s)$ if some element of $T(s)$ has a pole of multiplicity h_p at p, and no element has a pole of larger multiplicity than h_p at p.
Usually, the poles of T(s) are denoted as *transmission poles*.

In agreement with the SISO-case, comp. (32.3), the *open-loop system poles* are defined as the eigenvalues of the matrix A,

$$\{\text{open-loop system poles}\} = \{\text{zeros of } \det(sI_n - A)\} = \sigma(A) \qquad (32.15)$$

The open-loop system poles are either transmission poles or they are simultaneously zeros of all numerator polynomials

$$Z_{ji}(s) = c_j'(sI_n - A)_{adj}\, b_i \qquad (i = 1,\dots,m;\quad j = 1,\dots,r), \tag{32.16}$$

what leads to mutual cancellation of zeros of each numerator polynomial and the common denominator polynomial $\det(sI_n - A)$.

Instead of the zeros of the rational scalar transfer function

$$c'(sI_n - A)^{-1}b \tag{32.17}$$

in the SISO-case, we have to investigate a rational $m \times r$ transfer matrix

$$T(s) = C(sI_n - A)^{-1}B \tag{32.14}$$

in the multivariable case.

A zero of a scalar transfer function is simply a complex value for which this function vanishes. The main problem which the first investigators of the MIMO-case had to solve was to define a "zero" of the $m \times r$ transfer matrix (32.14). An appropriate generalization is reflected by the following

<u>Definition 32.2</u>

The rank of a rational matrix $T(s)$ is a function of s which is computed at a complex value s' by determining the numerical rank of the (complex) matrix $T(s')$.

For almost all values $s' \in \mathbf{C}$ this will be a certain quantity which is called the <u>normal rank</u> of $T(s)$,

$$\text{normal rank } T(s) = \max_{s \in \mathbf{C}} \text{ rank } T(s) \tag{32.18}$$

For a finite number of specific complex values s' the rank of $T(s')$ may be less than the normal rank. Such an exceptional quantity is called the <u>local rank</u> of $T(s)$ for that value $s = s'$.

A "zero" of $T(s)$ is a value of s at which the rank of $T(s)$ is locally reduced. Usually, the zeros of $T(s)$ are denoted as <u>transmission zeros</u>.

In the SISO-case there is only one alternative,

$T(s) = 0$, i.e. rank $T(s) = 1$ or

$T(s) = 0$, i.e. rank $T(s) = 0$.

In the MIMO-case, however, there are many possibilities.

Let q_{max} denote the normal rank of $T(s)$,

$$q_{max} = \text{normal rank } T(s) \tag{32.19}$$

Obviously,

$$q_{max} \leq \min (m, r) \tag{32.20}$$

Systems (A, B, C) with the property

$$q_{max} < \min (m, r)$$

have been called <u>degenerate systems</u> by most of the authors cited at the beginning of Section 32.2. In the sequel, it will be demonstrated that such a subdivision into non-degenerate systems (if $q_{max} = \min (m, r)$) and degenerate systems (if $q_{max} < \min (m, r)$) does not play an essential role. Rather, in the context of system zeros of MIMO-systems, that usage of the attribute "degenerate" seems to be more or less misleading. (See Example 32.3 discussed below.)

<u>Definition 32.3</u>

A complex number z is said to be a (finite) <u>transmission zero</u> if

$$\text{rank } T(z) < q_{max} \tag{32.21}$$

A transmission zero has the <u>multiplicity</u> h if all the $q_{max} \times q_{max}$ minors

$$T(s)^{j_1 j_2 \ldots j_{q_{max}}}_{i_1 i_2 \ldots i_{q_{max}}} \quad \text{with } 1 \leq i_1 < i_2 < \ldots < i_{q_{max}} \leq m;\ 1 \leq j_1 < \ldots < j_{q_{max}} \leq r$$

have a zero of multiplicity $\geq$ h at z, and at least one $q_{max} \times q_{max}$ minor has a zero of multiplicity not larger than h at z.

A transmission zero z is said to be a <u>transmission zero of rank</u> k if

$$\text{rank } T(z) = k \tag{32.22}$$

where $k \in \{0, 1, \ldots, q_{max} - 1\}$

In particular, for $k = 0$, $\text{rank } T(z) = 0$ means $T(z) = 0$.

Each transmission zero of rank k is characterized by $q_{max} - k$ different multiplicities $h_{k+1}, \ldots, h_{q_{max}}$ where h_1 is the minimal multiplicity of z regarded as common zero of all 1×1 minors of $T(s)$.

From the foregoing definition it is seen that the minors of different order of the rational transfer matrix T(s) play a crucial role. At first glance, one could believe that, for minors of different order, different denominator polynomials must be investigated. Fortunately, this is not the case. As early as in 1970 H.H. Rosenbrock used the relation

$$T(s)^{j_1 j_2 \dots j_q}_{i_1 i_2 \dots i_q} = (-1)^q [\det(sI_n - A)]^{-1} \, P(s)^{1\dots n, n+j_1, \dots, n+j_q}_{1\dots n, n+i_1, \dots, n+i_q} \tag{32.23}$$

$$\text{for} \quad q = 1, 2, \dots, \min(m, r)$$

where P(s) is Rosenbrock's system matrix, comp. (31.24),

$$P(s) = \begin{pmatrix} sI_n - A & B \\ C & 0 \end{pmatrix} \tag{32.24}$$

Eq. (32.23) follows immediately from I. Schur's well-known rule for evaluating the determinant of block matrices. In the case of $q = 2$, for example, there holds

$$P(s)^{1\dots n, n+j_1, n+j_2}_{1\dots n, n+i_1, n+i_2} = \det \begin{pmatrix} sI_n - A & b_{i_1} & b_{i_2} \\ c'_{j_1} & 0 & 0 \\ c'_{j_2} & 0 & 0 \end{pmatrix}$$

$$= \det(sI_n - A) \det\left(-\begin{pmatrix} c'_{j_1} \\ c'_{j_2} \end{pmatrix} (sI_n - A)^{-1} (b_{i_1}, b_{i_2})\right) \tag{32.25}$$

$$= \det(sI_n - A) \det \begin{pmatrix} -T_{j_1 i_1} & -T_{j_1 i_2} \\ -T_{j_2 i_1} & -T_{j_2 i_2} \end{pmatrix} = (-1)^2 \det(sI_n - A) \, T(s)^{j_1 j_2}_{i_1 i_2}$$

<u>Lemma 32.1</u>

The normal rank of T(s) can be determined as follows:

$$q_{max} = \text{normal rank } T(s)$$

$$= \max_{1 \le q \le \min(m,r)} \left\{ q: \; P(s)^{1\dots n, n+j_1 \dots n+j_q}_{1\dots n, n+i_1 \dots n+i_q} \neq 0 \text{ for some } s \in \mathbb{C} \right\} \tag{32.26}$$

<u>**Proof**</u>: Consider the finite set S' of complex values defined by

$$S' = \left\{ s' \in \mathbb{C}: \; [\det(s' I_n - A) = 0] \vee [\; \text{rank } T(s') < q_{max}] \right\}$$

For every $s \in \mathbb{C} \setminus S'$, due to Definition 32.2 and Eq. (32. 23), there is a minor

$$P(s)^{1 \ldots n, n+j_1 \ldots n+j_{q_{max}}}_{1 \ldots n, n+i_1 \ldots n+i_{q_{max}}} = (-1)^{q_{max}} \det(sI_n - A)\; T(s)^{j_1 \ldots j_{q_{max}}}_{i_1 \ldots i_{q_{max}}} \neq 0$$

while each $(n+q_{max}+1) \times (n+q_{max}+1)$ minor vanishes.

▲

At this point, let us recall single-input/output systems. Investigating system zeros of SISO-systems we have only to consider the polynomial

$$\begin{pmatrix} sI_n - A & b \\ c' & 0 \end{pmatrix}^{1 \ldots n, n+1}_{1 \ldots n, n+1} = \det \begin{pmatrix} sI_n - A & b \\ c' & 0 \end{pmatrix} = -c'(sI_n - A)_{adj}\, b, \qquad (32.27)$$

comp.(32.1) and (32.5).

For MIMO-systems, however, the complete collection of polynomials

$$\begin{pmatrix} sI_n - A & B \\ C & 0 \end{pmatrix}^{1 \ldots n, n+j_1 \ldots n+j_q}_{1 \ldots n, n+i_1 \ldots n+i_q} \qquad (32.28)$$

where $1 \leq q \leq \min(m, r)$, $1 \leq j_1 < \ldots < j_q \leq r$, $1 \leq i_1 < \ldots < i_q \leq m$

must be taken into consideration.

The transition from (32.27) to (32.28) is a quite natural generalization, indeed.

<u>Theorem 32.1</u>

A complex number z that is assumed not to be an eigenvalue of A, i.e. $z \in \mathbb{C} \setminus \sigma(A)$
is a transmission zero of multiplicity h if all the $(n+q_{max}) \times (n+q_{max})$ minors

$$P(z)^{1 \ldots n, n+j_1 \ldots n+j_{q_{max}}}_{1 \ldots n, n+i_1 \ldots n+i_{q_{max}}}$$

with $1 \leq i_1 < \ldots < i_{q_{max}} \leq m$; $1 \leq j_1 < \ldots < j_{q_{max}} \leq r$

have a zero of multiplicity $\geq$ h at z , and one of these minors has a zero of multiplicity not larger than h at z.

A complex number $z \in \mathbb{C} \setminus \sigma(A)$ is a transmission zero of rank k if, for $q > k$, all $(n+q) \times (n+q)$ minors

$$P(z)\begin{matrix}1\ldots n,\ n+j_1\ldots n+j_q\\ 1\ldots n, n+i_1\ldots n+i_q\end{matrix}$$

vanish while at least one $(n+k)\times(n+k)$ minor does not vanish. In particular, transmission zeros of rank $k = 0$ are characterized by

$$P(z)\begin{matrix}1\ldots n,n+j\\ 1,\ldots n,n+i\end{matrix} = -c_j'(zI_n - A)_{adj}\, b_i = 0 \quad \text{for all } j = 1,\ldots,r,\ i = 1,\ldots,m$$

and

$\det(zI_n - A) \neq 0.$

For each transmission zero of rank k, the $q_{max} - k$ multiplicities h_l, $k+1 \leq l \leq q_{max}$, are determined by the minimal multiplicity of z regarded as common zero of all $(n+l)\times(n+l)$ minors

$$P(z)\begin{matrix}1\ldots n,n+j_1\ldots n+j_l\\ 1\ldots n,n+i_1\ldots n+i_l\end{matrix}$$

Obviously,

$h_{q_{max}} = h$

in the sense of the first sentence of this Theorem.

<u>Proof</u>: Theorem 32.1 is an immediate consequence of Eq.(32.33) and Definition 32.3.

▲

The subsequent assertions are also easy to verify.

If $k_1 \neq k_2$ then

$$\{\text{transmission zeros of rank } k_1\} \cap \{\text{transmission zeros of rank } k_2\} = \emptyset \tag{32.29}$$

$$\bigcup_{k=0}^{q_{max}-1} \{\text{transmission zeros of rank } k\} = \{\text{transmission zeros}\} \tag{32.30}$$

Let us now consider the common zeros of $\det(sI_n - A)$ and all $(n+q)\times(n+q)$ minors

$$P(s)\begin{matrix}1\ldots n,n+j_1\ldots n+j_q\\ 1\ldots n,n+i_1\ldots n+i_q\end{matrix}$$

where $q \geq 1$, $1 \leq j_1 < \ldots < j_q \leq r$, $1 \leq i_1 < \ldots < i_q \leq m$.

<u>Lemma 32.2</u>

For any complex value z there holds

$$\det(zI_n - A) = P(z)\begin{matrix}1\ldots n\\ 1\ldots n\end{matrix} = 0 \tag{32.31a}$$

and

$$P(z)\begin{matrix}1\ldots n,n+j_1\ldots n+j_q\\1\ldots n,n+i_1\ldots n+i_q\end{matrix} = 0 \qquad (32.31b)$$

for all $1 \le q \le q_{max}$, $1 \le j_1 < \ldots < j_q \le r$, $1 \le i_1 < \ldots < i_q \le m$

if and only if

$$\text{rank}\ (zI_n - A,\ B) < n \qquad (32.32a)$$

or

$$\text{rank}\begin{pmatrix} zI_n - A \\ C \end{pmatrix} < n \qquad (32.32b)$$

Proof: We shall demonstrate first that (32.31) follows from (32.32). Assume (32.32a) to be true. Then rank $(zI_n - A) < n$ or, equivalently, (32.31a) is valid.

Further, the first n rows of the $(n+q)\times(n+q)$ matrix Q whose determinant is

$$P(z)\begin{matrix}1\ldots n,n+j_1\ldots n+j_q\\1\ldots n,n+i_1\ldots n+i_q\end{matrix}$$

are linearly dependent. This implies the n+q rows of Q to be linearly dependent or, equivalently, (32.31b) to be valid.

Assume (32.32b) to be true. Then rank $(zI_n - A) < n$ or, equivalently, (32.31a) is valid. Further, the first n columns of the $(n+q)\times(n+q)$ matrix Q whose determinant is

$$P(z)\begin{matrix}1\ldots n,n+j_1\ldots n+j_q\\1\ldots n,n+i_1\ldots n+i_q\end{matrix}$$

are linearly dependent. This implies the n+q columns of Q to be linearly dependent or, equivalently, (32.31b) to be valid.

Next we shall demonstrate that the non-validity of (32.32) implies the non-validity of (32.31).

Non-validity of (32.32) means

$$\text{rank}\ (zI_n - A,\ B) = n \quad \text{and} \quad \text{rank}\begin{pmatrix} zI_n - A \\ C \end{pmatrix} = n \qquad (32.33)$$

Assume (32.33) to be true.

Case 1: rank $(zI_n - A) = n$. Then (32.31a) cannot be valid.

Case 2: rank $(zI_n - A) = n-d < n$ (32.34)

Comparing (32.34) and (32.33) we conclude that there exists an $n \times d$ submatrix $\bar{B}$ of B and a $d \times n$ submatrix $\bar{C}$ of C such that

$$\text{rank}\ (zI_n - A,\ \bar{B}) = n \quad \text{and} \quad \text{rank}\begin{pmatrix} zI_n - A \\ \bar{C} \end{pmatrix} = n$$

Moreover,

$$\text{rank}\begin{pmatrix} zI_n - A & \bar{B} \\ \bar{C} & 0 \end{pmatrix} = n + d$$

or, equivalently, there is an $(n+d) \times (n+d)$ minor

$$P(z)\begin{matrix} 1\ldots n, n+j_1 \ldots n+j_d \\ 1\ldots n, n+i_1 \ldots n+i_d \end{matrix} \neq 0 ,$$

in contradiction to (32.31b).

Thus, we have shown that (32.33) implies the non-validity of (32.31). This completes the proof.

▲

Using the notations coined by H.H. Rosenbrock in 1970, we formulate

Definition 32.4

A complex value z is said to be an input decoupling zero if

$$\text{rank}\ (zI_n - A,\ B) < n \qquad (32.35a)$$

A complex value z is said to be an output decoupling zero if

$$\text{rank}\begin{pmatrix} zI_n - A \\ C \end{pmatrix} < n \qquad (32.35b)$$

The set of (finite) system zeros is formed by the union of the set of transmission zeros and the set of decoupling zeros, i.e.

$$\{\text{system zeros}\} = \{\text{transmission zeros}\} \cup \{\text{input decoupling zeros}\} \cup \{\text{output decoupling zeros}\}$$

The relationships between finite poles and the various sets of finite zeros can be simply illustrated by an Euler-Venn diagram as sketched in Fig. 32.2.

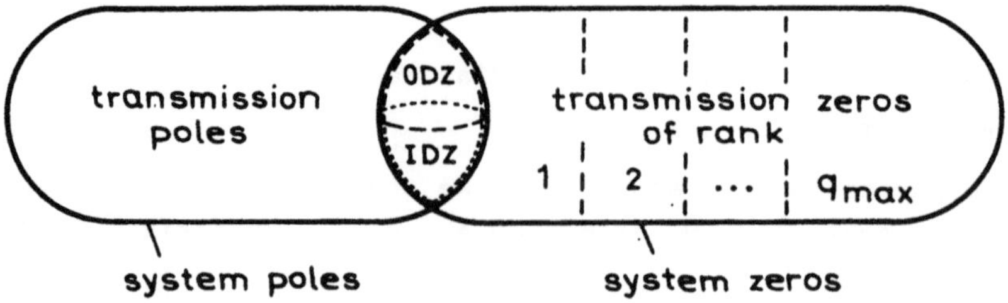

Fig. 32.2

If output feedback is applied through an $m \times r$ real feedback matrix F then the system matrix A of the open-loop system is replaced by

$$A^{c} = A + B\,F\,C \tag{32.36}$$

The <u>poles of the closed-loop system</u> are defined by

$$\{\text{closed-loop system poles}\} = \{\text{zeros of } \det(sI_n - A - BFC)\} \tag{32.37}$$

Generally, the location of the set of closed-loop system poles depends on F.
For $F = 0$, the set of closed-loop system poles coincides with the set of open-loop system poles.
For every $F \neq 0$, the decoupling zeros defined by (32.35) remain zeros of $\det(sI_n - A - BFC)$, i.e., that subset of system poles does not depend on F.
As for the complementary subset of closed-loop transmission poles , we take advantage of Eq.(31.25) and represent the closed-loop characteristic polynomial as follows:

$$\mathrm{CLCP}(s) = P(s)^{1\ldots n}_{1\ldots n} - \sum_{i_1=1}^{m} \sum_{j_1=1}^{r} f_{i_1 j_1} P(s)^{1\ldots n,n+j_1}_{1\ldots n,n+i_1}$$

$$+ \sum_{i_1<i_2} \sum \sum_{j_1 \neq j_2} \sum f_{i_1 j_1} f_{i_2 j_2} P(s)^{1\ldots n,n+j_1,n+j_2}_{1\ldots n,n+i_1,n+i_2} - + \ldots +$$

$$+(-1)^{q_{max}} \sum_{i_1<\ldots<i_{q_{max}}} \sum_{j_1 \neq \ldots \neq j_{q_{max}}} \sum \ldots \sum f_{i_1 j_1} \ldots f_{i_{q_{max}} j_{q_{max}}} P(s)^{1..n,n+j_1..n+j_{q_{max}}}_{1..n,n+i_1..n+i_{q_{max}}}$$

$$= P(s)^{1\ldots n}_{1\ldots n} - \sum_{i_1} \sum_{j_1} F^{i_1}_{j_1} P(s)^{1\ldots n,n+j_1}_{1\ldots n,n+i_1}$$

$$+ \sum_{i_1<i_2} \sum \sum_{j_1<j_2} \sum F^{i_1 i_2}_{j_1 j_2} P(s)^{1\ldots n,n+j_1,n+j_2}_{11..n,n+i_1,n+i_2} - + \ldots + \tag{32.38}$$

$$+(-1)^{q_{max}} \sum_{i_1<\ldots<i_{q_{max}}} \ldots \sum \sum_{j_1<\ldots<j_{q_{max}}} \ldots \sum F^{i_1..i_{q_{max}}}_{j_1..j_{q_{max}}} P(s)^{1..n,n+j_1..n+j_{q_{max}}}_{1..n,n+i_1..n+i_{q_{max}}}$$

Those complex values s at which all the $P(s)$-minors occurring on the right-hand side of (32.38) vanish are nothing else than the decoupling zeros, comp. Lemma 32.2.

If all feedback gains tend to infinity, then those closed-loop transmission poles which remain finite tend to the open-loop transmission zeros. The corresponding relation (32.13) for SISO-systems can be generalized for MIMO-systems in the following way:

Let $\bar{F}$ be a certain fixed feedback matrix all entries of which do not vanish. Consider the closed-loop characteristic polynomial

$$CLCP(s) = \det(sI_n - A - BFC) \quad \text{where} \quad F = c\,\bar{F} \tag{32.39}$$

Then Eq. (32.38) yields

$$\lim_{c \to \infty} c^{-q_{max}} \, CLCP(s) \tag{32.40}$$

$$= (-1)^{q_{max}} \sum_{i_1 < \dots < i_{q_{max}}} \dots \sum \; \sum_{j_1 < \dots < j_{q_{max}}} \dots \sum \; \bar{F}^{\,i_1 \dots i_{q_{max}}}_{\,j_1 \dots j_{q_{max}}} \; P(s)^{\,1..n,n+j_1..n+j_{q_{max}}}_{\,1..n,n+i_1..n+i_{q_{max}}}$$

Provided that s is a transmission zero then the right-hand side of (32.40) vanishes independently of the choice of $\bar{F}$.

Similarly to Definition 32.3, the <u>transmission zeros of the closed-loop system</u> are defined as the zeros of the closed-loop transfer matrix

$$T^c(s) = C(sI_n - A_c)^{-1}B$$

$$= [\det(sI_n - A^c)]^{-1} \, C(sI_n - A^c)_{adj} \, B \tag{32.41}$$

In analogy to (32.33), (32.34) there holds

$$T^c(s)^{\,j_1 \dots j_q}_{\,i_1 \dots i_q} = [\det(sI_n - A^c)]^{-1} \, P^c(s)^{\,1..n,n+j_1..n+j_q}_{\,1..n,n+i_1..n+i_q} \tag{32.42}$$

where

$$P^c(s) = \begin{pmatrix} sI_n - A^c & B \\ C & 0 \end{pmatrix} \tag{32.43}$$

Further,

$$P^c(s)^{\,1..n,n+j_1..n+j_q}_{\,1..n,n+i_1..n+i_q} = \det \begin{pmatrix} sI_n - A - BFC & b_{i_1} \dots b_{i_q} \\ c'_{j_1} & 0 \;\dots\; 0 \\ \vdots & \vdots \;\; \vdots \;\; \vdots \\ c'_{j_q} & 0 \;\dots\; 0 \end{pmatrix}$$

$$= \det \begin{pmatrix} sI_n - A - B_{red}F_{red}C_{red} & b_{i_1} \dots b_{i_q} \\ c'_{j_1} & 0 \dots 0 \\ \vdots & \vdots \; \vdots \vdots \vdots \; \vdots \\ c'_{j_q} & 0 \dots 0 \end{pmatrix} \qquad (32.44)$$

where

the $n \times (m-q)$ matrix B_{red} results from B by deleting the columns b_{i_1}, $\dots, b_{i_q}$,

the $(r-q) \times n$ matrix C_{red} results from C by deleting the rows $c'_{j_1}, \dots, c'_{j_q}$,

and F_{red} is an $(m-q) \times (r-q)$ matrix.

For $q = q_{max}$, we have

$$P^c(s)\begin{matrix} 1..n,n+j_1..n+j_{q_{max}} \\ 1..n,n+i_1..n+i_{q_{max}} \end{matrix} = P(s)\begin{matrix} 1..n,n+j_1..n+j_{q_{max}} \\ 1..n,n+i_1..n+i_{q_{max}} \end{matrix} \qquad (32.45)$$

From Eq. (32.45) and Theorem 32.1 we conclude

<u>Theorem 32.2</u>

The closed-loop transmission zeros (including their multiplicities) coincide with the open-loop transmission zeros.

The closed-loop transmission zeros do not depend on the choice of F.

The subdivision of the set of closed-loop transmission zeros into subsets of transmission zeros of different rank depends on F.

32.3 Graph-theoretic characterization of structural properties of finite zeros and poles

In the foregoing Section 32.2, a summary of the chief points on finite zeros and poles of multivariable systems has been made. Intentionally, all statements have been formulated in terms of the given matrices A, B, and C. We succeeded in expressing all facts with the aid of minors of Rosenbrock's system matrix P(s), see (32.24). Each such minor is the determinant of a square matrix which can be associated with weighted digraphs, comp. Section A2. Therefore, based on the graph-theoretic interpretation of determinants, all algebraic statements of Section 32.2 may be translated into graph-theoretic statements. Here, we do not intend to develop this idea in detail. Instead, the digraph approach will be used for identifying important structural properties of finite zeros and poles which are common to all or almost all actual systems belonging to a given structure class defined by [A], [B], and [C]. For this purpose, we shall consider the (n+m) × (n+m) structure matrix

$$[Q_1] = \begin{pmatrix} [A] & [B] \\ [E] & 0 \end{pmatrix}, \text{ comp. (11.13)},$$

the (r+n) × (r+n) structure matrix

$$[Q_2] = \begin{pmatrix} 0 & [C] \\ [E] & [A] \end{pmatrix}, \text{ comp. (11.14)},$$

the (r+n+m) × (r+n+m) structure matrix

$$[Q_3] = \begin{pmatrix} 0 & [C] & 0 \\ 0 & [A] & [B] \\ [E] & 0 & 0 \end{pmatrix}, \text{ comp. (11.15)},$$

and their associated structure digraphs $G([Q_1])$, $G([Q_2])$, and $G([Q_3])$.

Definition 32.5

A class of systems defined by [A], [B], [C] has <u>structural input decoupling zeros</u> if, for each admissible pair $(A, B) \in [A, B]$, there exists a complex value z such that

$$\text{rank } (zI_n - A, B) < n.$$

Taking into account Lemma 14.2, we are able to formulate the following criterion:

Lemma 32.2

A class of systems defined by [A], [B], [C] has structural input decoupling zeros if and only if each admissible pair is not controllable or, equivalently, if and only if the structure matrix pair [A, B] is not structurally controllable.

Now, all the necessary and sufficient conditions for s-controllability derived in Section 14 may be reformulated as conditions for the existence of structural input decoupling zeros. For example, based on Theorem 14.2 one obtains the following graph-theoretic criterion:

Theorem 32.3

A class of systems defined by [A], [B], [C] has structural input decoupling zeros if and only if the digraph $G([Q_1])$ meets at least one of the following conditions:

(a) At least one state-vertex in $G([Q_1])$ is not input-connectable.

(b) There is no cycle family of width n in $G([Q_1])$.

Analogously, the existence of output decoupling zeros is closely related to non-observability.

Definition 32.6

A class of systems defined by [A], [B], [C] has structural output decoupling zeros if, for each admissible pair

$$\begin{pmatrix} C \\ A \end{pmatrix} \in \begin{bmatrix} C \\ A \end{bmatrix},$$

there exists a complex value z such that

$$\operatorname{rank} \begin{pmatrix} C \\ zI_n - A \end{pmatrix} < n$$

Theorem 32.4

A class of systems defined by [A], [B], [C] has structural output decoupling zeros if and only if the digraph $G([Q_2])$ meets at least one of the following conditions:

(a) At least one state-vertex in $G([Q_2])$ is not output-connectable.

(b) There is no cycle family of width n in $G([Q_2])$.

Definition 32.7

$$\{\text{decoupling zeros}\} = \{\text{input decoupling zeros}\} \cup \{\text{output decoupling zeros}\} \tag{32.46}$$

A class of systems defined by [A], [B], [C] has structural decoupling zeros if each admissible system (A, B, C) ∈ [A, B, C] has decoupling zeros.

Lemma 32.3

A class of systems defined by [A], [B], [C] has structural decoupling zeros if and only if it is not structurally complete in the sense of Definition 14.6.

Theorem 32.5

A class of systems defined by [A], [B], [C] has structural decoupling zeros if and only if the digraph $G([Q_3])$ meets at least one of the following conditions:

(a) At least one state-vertex in $G([Q_3])$ is not input-connectable or not output-connectable.

(b) There is no cycle family of width n in $G([Q_3])$.

Definition 32.8

The number of decoupling zeros typical of the class of systems defined by [A, B, C] is called number of structural decoupling zeros and denoted by n_{sdz}.

Each admissible actual system defined by (A, B, C) ∈ [A, B, C] has a certain number of decoupling zeros. This number is equal to the number of structural decoupling zeros for almost all admissible realizations (A, B, C) ∈ [A, B, C]. Possibly, for specific realizations, the number of actual decoupling zeros may exceed the number of structural decoupling zeros.

Let S be the subset of state-vertices not both input and output connectable,

$$S = \left\{ i: \ i \in \{1,2,\dots,n\}, \ [i \text{ is not input-connectable}] \vee [i \text{ is not output-connectable}] \right\} \tag{32.47}$$

Let A_{red} be the $n \times n$ matrix obtained from A replacing by zero lines those rows and columns that belong to state-vertices contained in S.

Consider the digraph $G([Q_3, red])$ associated with

$$[Q_3, red] = \begin{pmatrix} 0 & [C] & 0 \\ 0 & [A_{red}] & [B] \\ [E] & 0 & 0 \end{pmatrix} \tag{32.48}$$

and look for cycle families of maximal width within $G([Q_3, red])$. This maximal width is denoted by w_{max}. The number of structural decoupling zeros may be obtained as follows:

Theorem 32.6

The number n_{sdz} of structural decoupling zeros (taking into account their multiplicities) of a class of systems defined by [A, B, C] is equal to the difference between n and the maximal width w_{max} of cycle families in $G([Q_3,red])$, see (32.48), i.e.

$$n_{sdz} = n - w_{max} . \tag{32.49}$$

The number of structural decoupling zeros $\neq 0$ is less than or equal to card(S), see (32.47).

Proof: We apply Theorem 31.2.
The state-vertices of S and all the state-edges only incident with state-vertices of S form a subgraph $G(A_{sub})$ of G(A). The polynomial

$$\det(sI_{card(S)} - A_{sub}) \tag{32.50}$$

is a common divisor (in the structural sense) of both the numerator polynomials $c'(sI_n - A)_{adj}\, b_i$ and the denominator polynomial $\det(sI_n - A)$ of all the transfer functions $T_{ji}(s)$ for $i = 1,\dots,m$ and $j = 1,\dots,r$.
Obviously, the card(S) zeros of (32.50) can be $\neq 0$.
In case of

$$w_{max} = n - card(S)$$

nothing remains to prove.
If, however,

$$w_{max} < n - card(S)$$

then both $G(A_{red})$ and $G(Q^{ij}_{red})$ where

$$Q^{ij}_{red} = \begin{pmatrix} 0 & c'_j & 0 \\ 0 & A_{red} & b_i \\ -1 & 0 & 0 \end{pmatrix} \tag{32.51}$$

for $i = 1,\dots,m,\ j = 1,\dots,r$
do not contain cycle families and feedback cycle families of width $n\text{-}card(S),\dots,w_{max}+1$.
Consequently, $s = 0$ is a common structural zero of multiplicity $n - card(S) - w_{max}$ of the numerator and the denominator of all transfer functions
$T_{ji}(s)$ for $i = 1,\dots,m,\ j = 1,\dots,r$.
This completes the proof.

▲

Example 32.1: Consider a class of systems with $n = 6$, $m = r = 2$ defined by the structure matrices

$$[A] = \begin{pmatrix} L & L & 0 & 0 & 0 & 0 \\ L & 0 & 0 & 0 & 0 & 0 \\ 0 & 0 & L & 0 & L & 0 \\ 0 & 0 & L & 0 & L & 0 \\ 0 & 0 & 0 & 0 & L & 0 \\ L & 0 & 0 & 0 & 0 & L \end{pmatrix}, \quad [B] = \begin{pmatrix} 0 & 0 \\ 0 & 0 \\ L & L \\ 0 & L \\ L & 0 \\ 0 & 0 \end{pmatrix}, \quad [C] = \begin{pmatrix} 0 & 0 & 0 & 0 & 0 & L \\ 0 & 0 & 0 & L & L & 0 \end{pmatrix}$$

The associated digraph $G([Q_3])$ has been drawn in Fig. 32.3.

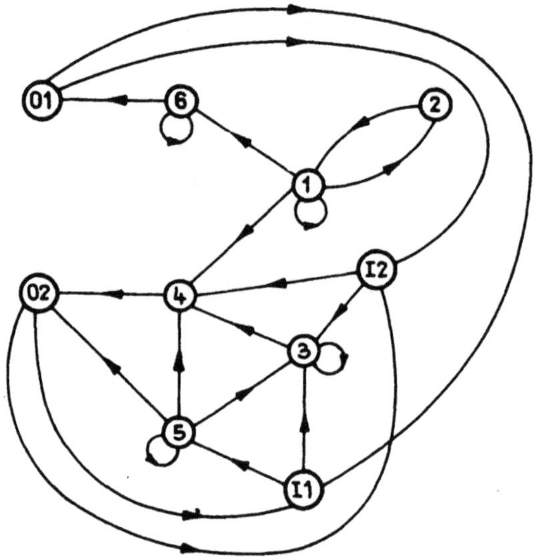

Fig. 32.3

It is easy to verify that

$$\text{s-rank } [A, B] = 6 = n, \quad \text{s-rank} \begin{bmatrix} C \\ A \end{bmatrix} = 6 = n$$

That is, the value $s = 0$ cannot be a structural decoupling zero.
Notwithstanding, there are structural decoupling zeros.
The subset S of state-vertices defined by (32.47) becomes

$$S = \{1, 2, 6\}.$$

Hence,

$$[A_{red}] = \begin{pmatrix} 0 & 0 & 0 & 0 & 0 & 0 \\ 0 & 0 & 0 & 0 & 0 & 0 \\ 0 & 0 & L & 0 & L & 0 \\ 0 & 0 & L & 0 & L & 0 \\ 0 & 0 & 0 & 0 & L & 0 \\ 0 & 0 & 0 & 0 & 0 & 0 \end{pmatrix}$$

The digraph $G([Q_3, red])$ has been drawn in Fig. 32.4.

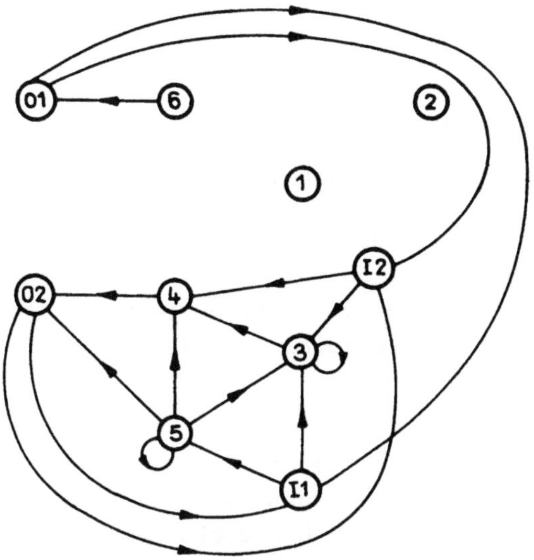

Fig. 32.4

In $G([Q_3, red])$ there are cycle families of width 1, 2, and 3. Becauee of $w_{max} = 3$ there are

$$n_{sdz} = n - w_{max} = 6 - 3 = 3$$

structural decoupling zeros.
For this example, we have

$$[A_{sub}] = \begin{matrix} & \begin{matrix} 1 & 2 & 6 \end{matrix} \\ \begin{matrix} 1 \\ 2 \\ 3 \end{matrix} & \begin{pmatrix} L & L & 0 \\ L & 0 & 0 \\ L & 0 & L \end{pmatrix} \end{matrix} \quad \text{and the polynomial det} \begin{pmatrix} s-a_{11} & -a_{12} & 0 \\ -a_{21} & s & 0 \\ -a_{61} & 0 & s-a_{66} \end{pmatrix}$$

appears as a cancellable factor in both the denominator and the numerator of all transfer functions $T_{ji}(s)$ for $i = 1, 2$; $j = 1, 2$. •

Example 32.2: Fig. 32.5 represents a class of systems with $n = 4$, $m = 2$, and $r = 1$. All state-vertices are both input- and output-connectable.
The set S defined by (32.47) is empty, i.e. $[Q_3, red] = [Q_3]$. The cycle families in $G([Q_3])$ have a maximal width $w_{max} = 3$. Thus we have

$n_{sdz} = 4 - 3 = 1$

structural decoupling zero at $s = 0$.

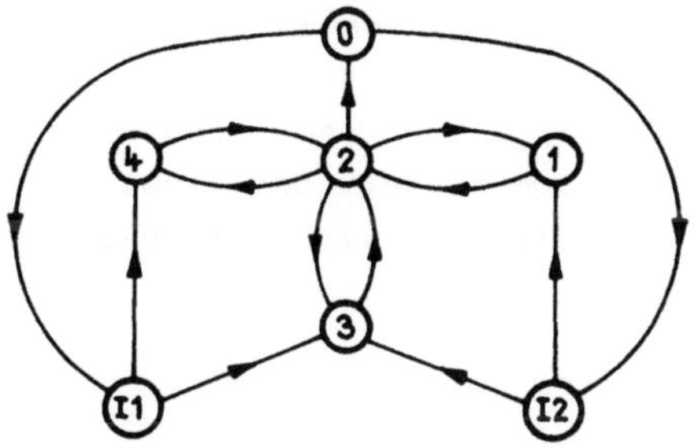

Fig. 32.5

Now, we turn to the graph-theoretic interpretation of the normal rank of T(s), comp. Definition 32.2.

Definition 32.9

The structural normal rank of the transfer matrices T(s) for a class of systems characterized by [A, B, C] is defined as

$$\max_{(A,B,C)\in[A,B,C]} \ \max_{s\in C} \ \operatorname{rank} T(s)$$

The normal rank of almost all actual systems (A,B,C) ∈ [A,B,C] is equal to the structural normal rank.
For specific admissible realizations (A,B,C), the normal rank may be smaller than the structural normal rank.

Theorem 32.7

Consider the digraph $G([Q_3])$ and look for cycle families with as many as possible feedback edges. The maximal number of feedback edges contained in at least one cycle family is equal to the structural normal rank.

Proof: We start from Lemma 32.1.

$$P(s)^{1\ldots n,n+j_1\ldots n+j_q}_{1\ldots n,n+i_1\ldots n+i_q} \neq 0 \quad \text{for some} \quad s \in C$$

is equivalent to

$$\sum_{w=q}^{n} p_w^{i_1 j_1 \ldots i_q j_q} s^{n-w} \neq 0 \quad \text{for some} \quad s \in C,$$

comp. (31.25) and (31.26).
This is true if and only if at least one of the n-q coefficients $p_w^{i_1 j_1 \ldots i_q j_q}$ does not vanish. From Theorem 31.4 it is known that this coefficient results from the sum of the weights of all cycle families of width w in $G([Q_3])$ which contain exactly q feedback edges.
This implies the assertion of Theorem 32.7.

▲

Example 32.3: Consider the plant with n = 6 state-variables, m = 2 inputs and r = 3 outputs discussed above as Example 12.1 and Example 31.4. Fig. 12.1 shows the associated digraph $G([Q_3])$.
It is immediately seen that the structural normal rank is two. All the cycle families with two feedback edges have been depicted in Fig. 31.6.

For the most example systems, the structural normal rank is equal to min(m, r). However, there are structurally complete systems whose structural normal rank is less than min(m, r).

<u>Example 32.4</u>: Fig. 32.6a shows the structure digraph $G([Q_3])$ for a system with $n = 5$ states, $m = 2$ inputs and $r = 2$ outputs.

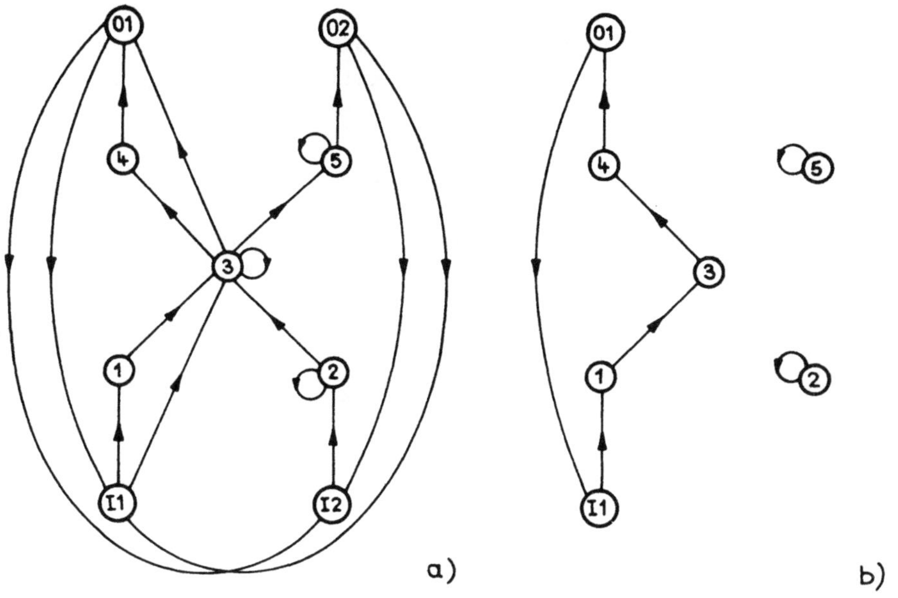

Fig. 32.6

It is easily seen that this example system is structurally complete (comp. Theorem 14.5):

(a) All state-vertices are both input- and output-connectable.

(b) There is a cycle family of width $n = 5$ in $G([Q_3])$, see Fig. 32.6b.

On the other hand, there is no cycle family in $G([Q_3])$ containing more than one feedback edge. Hence, the structural normal rank is equal to one, i.e., it is less than $\min(m, r) = 2$.

<u>Definition 32.10</u>

The number of finite transmission zeros typical of the class of systems defined by $[A,B,C]$ is called <u>number of structural transmission zeros</u>, abbreviated by n_{stz}.

For almost all admissible realizations $(A,B,C) \in [A,B,C]$ whose normal rank agrees with the structural normal rank, there holds that their number of transmission zeros is equal to the number of structural transmission zeros.

Theorem 32.8

Let q_{max} denote the structural normal rank of the transfer matrices of the class of systems defined by [A,B,C].
Consider the subset of cycle families in $G([Q_3])$ which contain exactly q_{max} feedback edges. Denote this subset by S_M.
Let w_{min} be the minimal width of all cycle families contained in the subset S_M.
Upper boundaries for the number n_{stz} of structural transmission zeros (taking into account their multiplicities) are given by the inequalities

$$n_{stz} \leq n - w_{min} - n_{sdz} \leq n - q_{max} - n_{sdz} \tag{32.49}$$

If all cycle families of S_M involve the same set both of input-vertices $Ii_1, Ii_2, \ldots, Ii_{q_{max}}$ and of output-vertices $Oj_1, \ldots, Oj_{q_{max}}$ then there holds

$$n_{stz} = n - w_{min} - n_{sdz} \leq n - q_{max} - n_{sdz} \tag{32.50}$$

If there are cycle families in S_M involving different sets of input-vertices or different sets of output-vertices, then there holds $n_{stz} = 0$ in most applications.

The exceptional cases may be characterized graph-theoretically as follows:
All the cycle families of S_M have a width $< n$ while G(A) contains a cycle family of width n
or
there is a subgraph $G(A_{sub})$ not touched by feedback cycles of S_M while some vertices of $G(A_{sub})$ are strongly connected with state-vertices not contained in $G(A_{sub})$.

In the former case there are structural transmission zeros at $s = 0$.
In the latter case, structural transmission zeros $\neq 0$ are possible.

Proof: For each admissible realization $(A,B,C) \in [A,B,C]$ whose normal rank of T(s) agrees with the structural normal rank, there holds (comp. (32.23))

$$\{\text{finite transmission zeros}\}$$

$$= \left\{\text{common zeros of } P(s)^{1\ldots n, n+j_1 \ldots n+j_{q_{max}}}_{1\ldots n, n+i_1 \ldots n+i_{q_{max}}} \text{ for all } 1 \leq j_1 < \ldots < j_{q_{max}} \leq r,\; 1 \leq i_1 < \ldots < i_{q_{max}} \leq m\right\}$$

$$\setminus \{\text{decoupling zeros}\}$$

The number of zeros of a polynomial (taking int account their multiplicities) is determined by its degree. For almost all systems (A,B,C) [A,B,C] there holds

$$\max_{\substack{1\leq j_1<\ldots<j_{q_{max}}\leq r \\ 1=i_1<\ldots<i_{q_{max}}\leq m}} \deg P(s)^{1\ldots n,n+j_1\ldots n+j_{q_{max}}}_{1\ldots n,n+i_1\ldots n+i_{q_{max}}}$$

$$= \max \deg \sum_{w=q_{max}}^{n} p_w^{i_1 j_1 \ldots i_{q_{max}} j_{q_{max}}} s^{n-w} = n - w_{min} \leq n - q_{max} \qquad (32.52)$$

and

$$\text{number of decoupling zeros} = n_{sdz}. \qquad (32.53)$$

From (32.51), (32.53), and (32.52) follows (32.49).

Provided that both the input set $\{Ii_1,\ldots,Ii_{q_{max}}\}$ and the output set $\{Oj_1,\ldots,Oj_{q_{max}}\}$ are uniquely determined,
then there is only one $(n+q_{max})\times(n+q_{max})$ minor
$P(s)^{1\ldots n,n+j_1\ldots n+j_{q_{max}}}_{1\ldots n,n+i_1\ldots n+i_{q_{max}}}$. It has exactly $n - w_{min}$ zeros.
This implies (32.50).

If there are two or more $(n+q_{max})\times(n+q_{max})$ minors of P(s) to be taken into consideration in (32.51), then the decoupling zeros are common zeros of those minors in any case, see Lemma 32.2.
Further common zeros, apart from the decoupling zeros, presuppose common divisors of all the $(n+q_{max})\times(n+q_{max})$ minors under consideration which, however, must not divide $\det(sI_n - A)$.

Common structural zeros require the existence of common divisors in the structural sense, comp. Theorem 32.6. They can be seen from $G([Q_3])$ as formulated in the last sentences of Theorem 32.8.
For illustration, see Example 32.8.
This completes the proof.

▲

Corollary 32.1

In many applications, there is valid

$q_{max} = m = r.$

Then the relationship (32.50) holds true.

Example 32.7: In Fig. 32.7 the structure digraph $G([Q_3])$ for a class of structurally complete systems with $n = 6$, $m = r = 2$ has been shown.

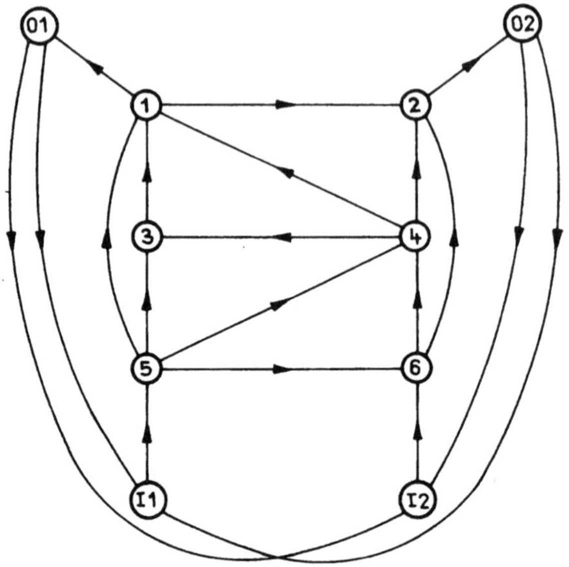

Fig. 32.7

The property of structural completeness means

$$n_{sdz} = 0.$$

The cycle families in $G([Q_3])$ contain $q_{max} = 2$ feedback edges at the most.

The width of cycle families with two feedback edges is at least four, i.e.

$$w_{min} = 4.$$

Applying (32.50) one obtains the number of structural transmission zeros as

$$n_{stz} = 6 - 4 - 0 = 2.$$

If the entries a_{15} and a_{26} are fixed at zero, then the modified system structure class has no structural transmission zeros (because of $w_{min} = 6$).

Example 32.8: Consider the digraph $G([Q_3])$ of a class of structurally complete systems with $n = 3$, $m = 2$, $r = 1$ drawn in Fig. 32.8a.

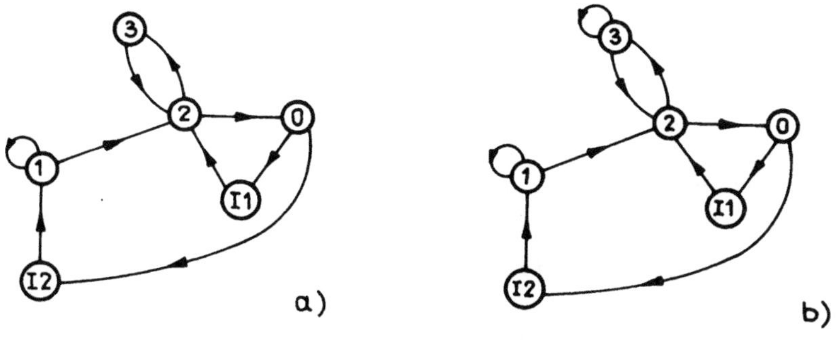

Fig. 32.8

There holds $n_{sdz} = 0$, $q_{max} = 1$, $w_{min} = 1$.
Hence,

$$n_{sdz} = 3 - 1 - 0 = 2$$

This is a rough estimation since we have to take into consideration more than one $(n+q_{max}) \times (n+q_{max}) = 4 \times 4$ minors, namely

$$P(s)^{1\,2\,3\,4}_{1\,2\,3\,4} = -c_{12}b_{21}s^2 + c_{12}b_{21}a_{11}s = -(s-a_{11})c_{12}b_{21}s$$

and

$$P(s)^{1\,2\,3\,4}_{1\,2\,3\,5} = -c_{12}a_{21}b_{12}s$$

The common structural divisor is s. Since

$$\det(sI_3 - A) = s^3 - a_{11}s^2 - a_{23}a_{32}s + a_{11}a_{23}a_{32}$$

is not divided by s, the class of systems exhibited in Fig. 32.8a has $n_{stz} = 1$ structural transmission zero at $s = 0$.

Now, let us consider the slightly modified system structure depicted in Fig.32.8b. Again, we have $n_{sdz} = 0$, $q_{max} = 1$, $w_{min} = 1$.
Further,

$$P(s)^{1\,2\,3\,4}_{1\,2\,3\,4} = (s-a_{11})(s-a_{33})c_{12}b_{21}, \quad P(s)^{1\,2\,3\,4}_{1\,2\,3\,5} = -(s-a_{33})c_{12}a_{21}b_{12}s$$

The common structural divisor is $s-a_{33}$. Since

$$\det(sI_3 - A) = s^3 - (a_{11}+a_{33})s^2 + (a_{11}a_{33}-a_{23}a_{32})s + a_{11}a_{23}a_{32}$$

is not divided by $s-a_{33}$, the class of systems exhibited in Fig. 32.8b has $n_{stz} = 1$ structural transmission zero at $s = a_{33}$.

For this example, the subgraph $G(A_{sub})$ introduced in the last part of Theorem 32.8 is constituted by the state-vertex 3 together with the self-cycle associated with a_{33}.

If the entries a_{23} or a_{32} are fixed at zero then the modified system

class has no structural transmission zeros.
Rather, $s = a_{33}$
becomes a structural decoupling zero.
The detailed verification of this assertion is left to the reader.

Example 32.9: Consider again the class of systems with
$n = 6$ states, $m = 2$ inputs, and $r = 3$ outputs
discussed above as Example 32.3.
It is structurally complete and its structural rank is $q_{max} = 2$.
The set S_M of cycle families with two feedback edges has been drawn in Fig. 31.6.
These cycle families have a minimal width of $w_{min} = 3$.
According to (32.49), there are

$n_{stz} \leq 6 - 3 - 0 = 3$ structural transmission zeros.

On the right-hand side of (32.52) we have to take into consideration two $(6+2) \times (6+2)$ minors, namely

$$P(s)\begin{matrix} 1\,2\,3\,4\,5\,6\,7\,8 \\ \\ 1\,2\,3\,4\,5\,6\,7\,8 \end{matrix} \quad \text{and} \quad P(s)\begin{matrix} 1\,2\,3\,4\,5\,6\,7\,9 \\ \\ 1\,2\,3\,4\,5\,6\,7\,8 \end{matrix}$$

They have no common structural divisor as it can be directly seen from Fig. 31.6.
Hence, $n_{stz} = 0$.

32.4 Infinite zeros of multivariable systems and their graph-theoretic characterization

For multivariable systems, the concept of a zero at infinity was first investigated by Rosenbrock in 1970. He wrote that the transfer matrix

$$T(s) = C(sI_n - A)^{-1}B \tag{32.14}$$

if each minor of order q tends to zero as s tends to infinity, i.e.

$$\lim_{s\to\infty} T(s)^{i_1 \ldots i_q}_{j_1 \ldots j_q} = 0 \tag{32.54}$$

Incorporating the results of further research (Rosenbrock 1974, Hautus 1976, Suda and Mutsuyoshi 1978, Pugh and Ratcliffe 1979, Malabre 1982, Vardulakis and Karcanias 1983, Van der Weiden 1983 and others) we turn over to a definition which seems to be the most natural one.

Definition 32.11

The multivariable system

$$\dot{x} = A\,x + B\,u, \qquad y = C\,x \tag{32.55,32.56}$$

is said to have q_{max} zeros at infinity if

$$\text{normal rank } T(s) = q_{max} \tag{32.57}$$

For each integer value of q, $1 \leq q \leq q_{max}$, consider

$$d_q = \max_{\substack{1\leq i_1<\ldots<i_q\leq m \\ 1\leq j_1<\ldots<j_q\leq r}} \deg P(s)^{1..n,n+i_1..n+i_q}_{1..n,n+j_1..n+j_q} \tag{32.58}$$

Then the q_{max} zeros at infinity are said to have the multiplicities

$$\begin{aligned} h_1 &= n - d_1, \\ h_2 &= d_1 - d_2, \\ &\vdots \\ h_{q_{max}} &= d_{q_{max}-1} - d_{q_{max}} \end{aligned} \tag{32.59}$$

It has been proved in the literature that the number of zeros at infinity and their multiplicities have interesting properties. In particular, they remain invariant under certain types of transformations:

(T1) state-coordinate transformation

$$(A,B,C) \longrightarrow (M^{-1}AM, M^{-1}B, CM) \tag{32.60}$$

with a non-singular $n \times n$ matrix M

(T2) state feedback

$$(A,B,C) \longrightarrow (A + BF,\ B,\ C) \tag{32.61}$$

with an $m \times n$ matrix F

(T3) output feedback

$$(A,B,C) \longrightarrow (A + BFC,\ B,\ C) \tag{32.62}$$

with an $m \times r$ matrix F

(T4) output injection

$$(A,B,C) \longrightarrow (A + FC,\ B,\ C) \tag{32.63}$$

with an $n \times r$ matrix F

(T5) input-coordinate transformation

$$(A,B,C) \longrightarrow (A, BG,\ C) \tag{32.64}$$

with a non-singular $m \times m$ matrix G

(T6) output-coordinate transformation

$$(A,B,C) \longrightarrow (A,\ B, GC) \tag{32.65}$$

with a non-singular $r \times r$ matrix G

To substantiate the invariance properties it is sufficient to show that the integers d_q introduced in (32.58) remain invariant under each of these six types of transformations. A detailed proof is omitted here.

Because of those invariance properties, the so-called "infinite zero structure" plays an essential role for many problems of the analysis and synthesis of multivariable linear control systems such as

left- and right-invertibility (see, e.g., Silverman 1969),
disturbance decoupling (see, e.g., Commault, Dion and Perez 1984),

non-interaction control (see, e.g., Descusse, Lafay and Malabre 1983), exact model matching (see, e.g., Malabre and Kucera 1984).

Based on the geometric interpretation of the infinite zero structure of linear systems (see Malabre 1982, Commault and Dion 1982), zeros at infinity for affine non-linear control systems have been introduced by Nijmeijer and Schumacher 1983, DiBenedetto and Isidori 1984.
Making use of the "infinite zero structure", these authors have successfully tackled all the synthesis problems just mentioned for affine non-linear control systems. We shall not consider such questions here because non-linear systems are beyond the scope of this monograph.

As for the details of the applications of the "infinite zero structure" in case of linear systems, the reader is also referred to the papers cited above. Here, we shall merely add the graph-theoretic characterization of the "infinite zero structure".

For this purpose, we shall neglect the actual numerical values of the entries of the matrices A, B, C which define a given actual system (32.55, 32.56) and ask for the "infinite zero structure" typical of the class of systems defined by the structure matrices [A], [B], [C]. This means we try to extract the information on infinite zeros, which is contained in the structure digraph $G([Q_3])$ introduced by (11.15).

Just at the beginning, it should be realized that we have used the notation "structure" in two different meanings in this Section.
On the one hand, the "infinite zero structure" reflects the degree of certain polynomials where the values of the non-vanishing polynomial coefficients are not taken into account. That is, "structure" is here referred to one actual physical system whose mathematical description may be altered by coordinate transformations and applying feedback. The "structure parameters" remain invariant under the transformations (32.60) to (32.65).
On the other hand, a "structure matrix" reflects the pattern of the zero entries of a matrix where the values of the remaining entries are assumed to be indeterminate. That is, "structure" is here referred to a class of systems characterized by [A], [B], [C].

In order to avoid confusion and strange phrases such as "structural infinite zero structure" we shall speak in the sequel of a "generic infinite zero structure" to express properties of the "infinite zero structure" typical of a whole class of systems defined by its structure matrices [A], [B], [C].

Definition 32.12

The "generic infinite zero structure" provides the "infinite zero structure" of almost all admissible systems given by $(A,B,C) \in [A,B,C]$.

The "generic infinite zero structure" may be obtained from the digraph $G([Q_3])$, see (11.15).

For each actual system $(A,B,C) \in [A,B,C]$, the actual infinite zero structure may be obtained from the weighted digraph $G(Q_3)$.

The generic number of zeros at infinity is defined by

$$q^g_{max} = \max_{(A,B,C) \in [A,B,C]} \ \max_{s \in C} \ \text{rank } C(sI_n - A)^{-1}B \tag{32.66}$$

For each $q = 1,2,\dots,q^g_{max}$ consider

$$d^g_q = \max_{(A,B,C) \in [A,B,C]} \ \max_{\substack{1 \le i_1 < \dots < i_q \le m \\ 1 \le j_1 < \dots < j_q \le r}} \deg P(s)^{1..n,n+j_1..n+j_q}_{1..n,n+i_1..n+i_q} \tag{32.67}$$

The zeros at infinity have the generic multiplicities

$$\begin{aligned} h^g_1 &= n - d^g_1 \\ h^g_2 &= d^g_1 - d^g_2 \\ &\vdots \\ h^g_{q^g_{max}} &= d^g_{q^g_{max}-1} - d^g_{q^g_{max}} \end{aligned} \tag{32.68}$$

Theorem 32.9

The generic number q^g_{max} of zeros at infinity is equal to the maximal number of feedback cycles within at least one cycle family of $G([Q_3])$.

The parameters d^g_q $(1 \le q \le q^g_{max})$ are determined by the minimal $w_{q,min}$ of a cycle family with exactly q feedback edges in $G([Q_3])$.

There holds

$$d^g_q = n - w_{q,min} \le n - q \tag{32.69}$$

$$n-1 \ge d^g_1 > d^g_2 > \dots > d^g_{q^g_{max}} \ge 0 \tag{32.70}$$

$$h^g_1 = w_{1,min}, \quad h^g_q = w_{q,min} - w_{q-1,min} \quad \text{for } q = 2,\dots,q^g_{max} \tag{32.71}$$

Proof: The generic number of zeros at infinity, q^g_{max}, agrees with the structural normal rank of the transfer matrix T(s), comp. Definition 32.12 and Definition 32.9. Therefore, the first sentence of Theorem 32.9 may be regarded as an immediate consequence of Theorem 32.7.

From (31.25) and (31.26) it is known that

$$P(s)^{1..n,n+j_1..n+j_q}_{1..n,n+i_1..n+i_q} = \sum_{w=q}^{n} p_w^{i_1j_1...i_qj_q} s^{n-w} \tag{32.72}$$

There holds

$$p_w^{i_1j_1...i_qj_q} = 0 \quad \text{for some} \quad (A,B,C) \in [A,B,C]$$

if and only if there exists at least one cycle family of width w and exactly q feedback edges in $G([Q_3])$.

Due to (32.67), the parameter d^g_q is equal to $n - w_{q,min}$ where $w_{q,min}$ is the minimal width of a cycle family containing exactly q feedback edges. Each feedback cycle involves at least one state-vertex. Each cycle family with q feedback edges involves at least q state vertices. In other words, the width of a cycle family with q feedback edges cannot be smaller than q. This implies (32.69).

Furthermore, the minimal width within the set of cycle families with q feedback edges must be smaller than the minimal width within the set of cycle families with q+1 feedback edges. This implies (32.70). Finally, (32.71) results from (32.68) and (32.69).

▲

By specialization, a lot of further interesting statements can be derived. Let us mention only one special case.

Corollary 32.2

If $G([Q_3])$ contains a cycle family of width q^g_{max} with q^g_{max} feedback edges, then almost all systems $(A,B,C) \in [A,B,C]$ have q^g_{max} zeros at infinity, each of multiplicity 1.

Proof: By assumption, $d^g_{q^g_{max}} = n - q^g_{max}$.

Taking into account (32.70) one gets

$d^g_1 = n-1, \quad d^g_2 = n-2, \quad \ldots, \quad d^g_{q^g_{max}} = n - q^g_{max}$.

Hence, $h^g_1 = h^g_2 = \ldots = h^g_{q^g_{max}} = 1$, q.e.d.

▲

Example 32.10: Consider the class of "structurally degenerated" systems introduced above as Example 32.4. There are $n = 5$ states, $m = 2$ inputs and $r = 2$ outputs, see Fig. 32.6a for $G([Q_3])$.
Cycle families with two feedback edges do not exist, $q^g_{max} = 1$.
The minimal width of cycle families with one feedback edge is $w_{1,min} = 1$.
(Choose the cycle I1 → 3 → O1 → I1.)
This means the generic number of zeros at infinity is 1. This zero has a generic multiplicity 1.
Now, let us modify this Example by fixing the entry c_{13} at zero.
Then the minmal width of cycle families with one feedback edge becomes $w_{1,min} = 2$. (Choose the cycles I1 → 3 → 5 → O2 → I1 or I1 → 3 → 4 → O1 → I1.)
Thus, the modified class of systems has also one zero at infinity. Its generic multiplicity, however, is 2.

Example 32.11: Consider the digraph $G([Q_3])$ drawn in Fig. 32.7.
There holds $n = 6$, $m = r = 2$.
It is immediately seen that the generic number of zeros at infinity is $q^g_{max} = 2$.
The minimal width of cycles with 1 feedback edge is $w_{1,min} = 2$.
(Choose the cycles I1 → 5 → 1 → O1 → I1 or I2 → 6 → 2 → O2 → I2.)
The minimal width of cycle families with two feedback edges is $w_{2,min} = 4$.
(Choose the cycle family formed by the cycles I1 → 5 → 1 → O1 → I1 and I2 → 6 → 2 → O2 → I2.)

This means that the class of systems under consideration has generically two zeros at infinity where each of them has a generic multiplicity of two.

33 Pole placement by static output feedback

33.1 Problem formulation and preliminary considerations

Consider a linear plant modelled by the Eqs.

$$\dot{x} = A x + B u \tag{33.1}$$

$$y = C x \tag{33.2}$$

see (31.2) and (31.3) for details.

The problem of arbitrary pole assignability by static output feedback is to find a static output feedback law

$$u = F y \tag{33.3}$$

where F is a real $m \times r$ matrix
such that the zeros of the closed-loop characteristic polynomial

$$\det(sI_n - A - BFC) = s^n + p_1 s^{n-1} + \ldots + p_{n-1} s + p_n \tag{33.4}$$

may be assigned to arbitrarily chosen places
where any feasible complex zeros must of course appear in complex conjugate pairs.

It is known that the coefficients $p_1, p_2, \ldots, p_{n-1}, p_n$ of the closed-loop characteristic polynomial depend non-linearly on the feedback gains. In Section 31.2, the kind of non-linearity occurring was studied in detail.

The problem of arbitrary pole placement is equivalent to the assignment of the coefficients $p_1, p_2, \ldots, p_{n-1}, p_n$ to arbitrarily chosen real values.

Definition 33.1

A plant (33.1), (33.2) is said to have the property of (global) pole assignability under static output feedback (33.3) if given any real polynomial $p(s) = s^n + p_1 s^{n-1} + \ldots + p_{n-1} s + p_n$ there exists a real $m \times r$ matrix F such that

$p(s) = \det(sI_n - A - BFC)$.

Let

$p = (p_1, p_2, \ldots, p_{n-1}, p_n)'$

be a column vector encompassing the coefficients of the closed-loop characteristic polynomial (33.4). If A, B, and C are numerically fixed then p may be regarded as a function of F. Eq. (33.4) yields a vector-valued relation

$$p = g(F) \tag{33.5}$$

Each real $m \times r$ matrix F can be interpreted as an element of a real Euclidean vector space $\mathbb{R}^{m \cdot r}$ where $m \cdot r$ is the number of real feedback gains that may be chosen independently of each other.
Relation (33.5) defines a smooth mapping

$$g: R^{m \cdot r} \longrightarrow N \subset \mathbb{R}^n \tag{33.6}$$

where N is a smooth manifold of the n-dimensional vector space R^n. This mapping has been investigated in Section 33.1, see Theorem 31.3 and Theorem 31.4.

Lemma 33.1

The non-linear mapping (33.6) may be represented as follows:

$$p = g(F) = p^o + \sum_{i_1=1}^{m} \sum_{j_1=1}^{r} F_{j_1}^{i_1} \, p^{i_1 j_1} + \sum\sum_{i_1<i_2} \sum\sum_{j_1<j_2} F_{j_1 j_2}^{i_1 i_2} \, p^{i_1 j_1 \cdot i_2 j_2}$$
$$+ \dots + \sum \dots \sum_{i_1<\dots<i_q} \sum \dots \sum_{j_1<\dots<j_q} F_{j_1 \dots j_q}^{i_1 \dots i_q} \, p^{i_1 j_1 \dots i_q j_q} \tag{33.7}$$

with $q \leq \min(m, r)$

Proof: Eq.(33.7) follows immediately from Theorem 31.3, comp. (31.25), (31.26) and (32.38).

▲

In the sequel, we use $T_F \mathbb{R}^{m \cdot r}$ and $T_{g(F)} N$ to denote, respectively, the tangent space to $\mathbb{R}^{m \cdot r}$ at F and the tangent space to N at g(F). The elements of $T_F \mathbb{R}^{m \cdot r}$ and $T_{g(F)} N$ are denoted by dF and dp, respectively. The differential of g at the point $F \in \mathbb{R}^{m \cdot r}$ is a linear map from $T_F \mathbb{R}^{m \cdot r}$ to $T_{g(F)} N$,

$$g_{*F}: T_F \mathbb{R}^{m \cdot r} \longrightarrow T_{g(F)} N \tag{33.8}$$

or, expressed in more traditional terms,

$$dp = g_{*F}(dF). \tag{33.9}$$

As for the notations taken from Differential Topology, see, for example, Bröcker and Jänicke 1973, Hartshorne 1977 or Miscenko and Fomenko 1980.

<u>Lemma 33.2</u>

The linear map (33.8) may be represented as follows:

$$dp = g_{*F}(dF) = \sum_{i_1=1}^{m} \sum_{j_1=1}^{r} \Big(p^{i_1 j_1} + \sum_{\substack{i_2 \\ > i_1}} \sum_{\substack{j_2 \\ > j_1}} F^{i_2}_{j_2} \, p^{i_1 j_1 \cdot i_2 j_2} + \ldots \tag{33.10}$$

$$+ \underbrace{\sum \ldots \sum}_{\substack{i_2 < \ldots < i_q \\ > i_1}} \underbrace{\sum \ldots \sum}_{\substack{j_2 < \ldots < j_q \\ > j_1}} F^{i_2 \ldots i_q}_{j_2 \ldots j_q} \, p^{i_1 j_1 \cdot i_2 j_2 \ldots i_q j_q} \Big) df_{i_1 j_1}$$

<u>Proof</u>: Lemma 33.2 is an immediate consequence of Lemma 33.1. The differential g_{*F} is nothing else than the Jacobian matrix of g at a point F.

▲

As for examples of $g(F)$ and its differential g_{*F}, see p.183. For certain given numerically fixed feedback gain matrices F and the corresponding vector p of closed-loop characteristic-polynomial coefficients, many pole assignable systems show a lack of regularity in the following sense: To some given small variation dp it is impossible to find a variation dF that causes dp.

<u>Definition 33.2</u>

A plant (33.1), (33.2) controlled by static output feedback (33.3) is said to have the property of <u>local pole assignability</u> at $F_o \in \mathbb{R}^{m \cdot r}$, $p_o = g(F_o)$ if given any $dp \in \mathbb{R}^n$ there exists a real $m \times r$ matrix dF such that

$$dp = g_{*F}(dF) \tag{33.11}$$

<u>Definition 33.3</u>

A point $F \in \mathbb{R}^{m \cdot r}$ is said to be a <u>regular point</u> of the mapping (33.6) if

$$\text{rank } g_{*F} = n. \tag{33.12}$$

<u>Lemma 33.3</u>

If there exists at least one regular point $F \in \mathbb{R}^{m \cdot r}$ then the set of regular F-points is open and everywhere dense in $\mathbb{R}^{m \cdot r}$. More exactly, the set M_1 of non-regular F-points forms a real algebraic submanifold in $\mathbb{R}^{m \cdot r}$, i.e..

M_1 is the intersection of hypersurfaces each defined by an algebraic equation.

<u>Proof</u>: The set M_1 of non-regular F-points may be explicitly computed,

$$M_1 = \left\{ F \in \mathbb{R}^{m\cdot r} : \ \text{rank}\ g_{*F} < n \right\} \tag{33.13}$$

$$= \bigcap_{(j)} \left\{ F \in \mathbb{R}^{m\cdot r} : (g_{*F})^{1\ 2\ \dots\ n}_{j_1 j_2 \dots j_n} = 0 \right\} \tag{33.14}$$

where $\bigcap_{(j)}$ symbolizes the intersection of subsets $\left\{ F \in \mathbb{R}^{m\cdot r} : \dots \right\}$ each defined by a sequence $1 \leq j_1 < j_2 < \dots < j_n \leq m\cdot r$.
Eq.(33.10) implies that the $n \times n$ determinants

$$(g_{*F})^{1\ 2\ \dots\ n}_{j_1 j_2 \dots j_n}$$

are polynomials with the feedback gains as arguments. Consequently, M_1 is an algebraic submanifold of $R^{m\cdot r}$.

▲

<u>Definition 33.4</u>

An element $p \in \mathbb{R}^n$ is said to be a <u>regular value</u> of g if $g^{-1}(p)$ (= inverse image of p) is not empty and each $F \in g^{-1}(p)$ is a regular point in $\mathbb{R}^{m\cdot r}$.
An element $p \in \mathbb{R}^n$ is said to be a <u>critical value</u> of g if $g^{-1}(p)$ contains a non-regular point $F \in \mathbb{R}^{m\cdot r}$.

<u>Lemma 33.4</u>

Assume $\max_{F \in \mathbb{R}^{m\cdot r}} \text{rank}\ g_{*F} = n.$

The set N_1 of critical values $p \in \mathbb{R}^n$ is a closed and nowhere dense subset of $\mathbb{R}^n$. N_1 is the image of the set M_1 of non-regular F-points,

$$N_1 = g(M_1). \tag{33.15}$$

<u>Proof</u>: Lemma 33.4 is an immediate consequence of Sard's Lemma proved in almost all texts on Differential Geometry and Topology.

<u>Lemma 33.5</u>

Let p_o be a regular value of g. Then the inverse image

$$M = g^{-1}(p_o) \tag{33.16}$$

is a smooth manifold of dimension

$$\dim M = m \cdot r - n. \qquad (33.17)$$

Some $m \cdot r - n$ local coordinates of $\mathbb{R}^{m \cdot r}$ may be used as local coordinates of M.

<u>Proof</u>: Choose a regular point $F_o \in \mathbb{R}^{m \cdot r}$ such that $p_o = g(F_o)$.
Because of Definition 33.3, rank $g_{*F_o} = n$.
Now apply the Implicit Function Theorem:
There exists a neighbourhood $U \ni F_o$ that is locally homeomorph to the Euclidean space $\mathbb{R}^{m \cdot r - n}$.
Some $m \cdot r - n$ coordinates of $\mathbb{R}^{m \cdot r}$, i.e. $m \cdot r - n$ entries of F, may be chosen as local coordinates in U, say $(f_{j_1}, f_{j_2}, \ldots, f_{j_{m \cdot r - n}})$.

The remaining n local coordinates appear as smooth functions of $(f_{j_1}, f_{j_2}, \ldots, f_{j_{m \cdot r - n}})$ on M.
This implies M to be a smooth manifold of dimension $m \cdot r - n$.

▲

33.2 Necessary and sufficient conditions for local pole assignability

Lemma 33.6

A closed-loop system (33.1), (33.2), (33.3) is nowhere locally pole assignable if and only if

$$\text{rank } g_{*F} < n \quad \text{for all} \quad F \in \mathbb{R}^{m \cdot r} \tag{33.18}$$

Proof: Assume (33.18) to be valid.
Then the image $g(\mathbb{R}^{m \cdot r}) = N \subset \mathbb{R}^n$ forms a smooth submanifold of dimension less than n. Hence, the set N of assignable values p is of measure zero. In other words, almost all $p \in \mathbb{R}^n$ cannot be assigned by means of static output feedback. This means the property of local pole assignability (see Definition 33.2) does nowhere hold.

If (33.18) is not fulfilled then

$$\text{rank } g_{*F_o} = n \quad \text{for at least one } F_o \in \mathbb{R}^{m \cdot r},$$

and there exists a (not necessarily unique) solution of Eq.(33.11) for any given dp. Therefore, the closed-loop system (33.1), (33.2), (33.3) is locally pole assignable at F_o, $p_o = g(F_o)$.
This completes the proof.

▲

Now, the following statement is obvious.

Theorem 33.1

A closed-loop system (33.1), (33.2), (33.3) is locally pole assignable at F_o, $p_o = g(F_o)$ if and only if

$$\text{rank } g_{*F_o} = n.$$

Taking into account Lemma 33.3 one obtains

Theorem 33.2

A closed-loop system (33.1), (33.2), (33.3) is locally pole assignable at F, $p = g(F)$ for almost all F if

$$\text{rank } g_{*F_o} = n \quad \text{for at least one} \quad F_o \in \mathbb{R}^{m \cdot r}. \tag{33.19}$$

The set M_1 of non-regular F-points forms an algebraic subset that may be computed as follows:

$$M_1 = \bigcap_{(j)} \left\{ F \in \mathbb{R}^{m \cdot r} : (g_{*F})^{1\ 2\ \dots\ n}_{j_1 j_2 \dots j_n} = 0 \right\} \tag{33.20}$$

where (j) symbolizes a sequence $1 \leq j_1 < j_2 < \dots < j_n \leq m \cdot r$.

An equivalent characterization is

$$M_1 = \left\{F \in \mathbb{R}^{m \cdot r}: \quad \det[(g_{*F})(g_{*F})'] = 0\right\} \tag{33.20'}$$

The associated set of critical p-values can be easily computed as the image of M_1,

$$N_1 = g(M_1). \tag{33.21}$$

In this Section, the condition (33.19) plays a crucial role.
In the literature (see Willems and Hesselink 1978, Kabamba and Longman 1982, Sevaston 1984) another criterion for local pole assignability has been derived and discussed:

<u>Lemma 33.7</u>

A necessary and sufficient condition for local pole assignability at a feedback matrix F_o for which $A + BF_oC$ has distinct eigenvalues is that the matrices

$$CB,\ C(A+BF_oC)B,\ C(A+BF_oC)^2B,\ \dots,\ C(A+BF_oC)^{n-1}B \tag{33.22}$$

be linearly independent.

The link between (33.19) and Lemma 33.7 is provided by

<u>Lemma 33.8</u>

The condition

$$\operatorname{rank} g_{*F_o} = n$$

holds if and only if the matrices

$$CB,\ C(A+BF_oC)B,\ C(A+BF_oC)^2B,\ \dots,\ C(A+BF_oC)^{n-1}B$$

are linearly independent.

<u>Proof</u>: Substituting $A+BFC = \bar{A}$ the closed-loop characteristic polynomial can be written as

$$s^n + \sum_{k=1}^{n} p_k\, s^{n-k} = \det(sI_n - \bar{A}), \quad \text{comp. (33.4).}$$

Differentiating with respect to the feedback gain f_{ij} provides

$$\sum_{k=1}^{n} \frac{\partial p_k}{\partial f_{ij}} s^{n-k} = \lim_{\Delta f_{ij} \to 0} \left[\frac{1}{\Delta f_{ij}} (\det(sI_n - \bar{A} - b_i\, \Delta f_{ij} c_j') - \det(sI_n - \bar{A}) \right]$$

$$= \lim_{f_{ij} \to 0} \frac{1}{\Delta f_{ij}} [- \Delta f_{ij}\, c_j' (sI_n - \bar{A})_{adj}\, b_i] = -c_j'(sI_n - \bar{A})_{adj}\, b_i \tag{33.23}$$

The second transformation follows from the relationship

$$\det(Q + bc') - \det Q = c'\, Q_{adj}\, b \tag{33.24}$$

which holds true for any square matrices Q.

Based on the known formula (see Section 21, Eq.(21.24))

$$\begin{aligned}(sI_n - \bar{A})_{adj} = \;& I_n\, s^{n-1} \\ &+ (\bar{A} + p_1 I_n) s^{n-2} \\ &+ (\bar{A}^2 + p_1\bar{A} + p_2 I_n) s^{n-3} \\ &\;\vdots \\ &+ (\bar{A}^{n-2} + p_1\bar{A}^{n-3} + p_2\bar{A}^{n-4} + \dots + p_{n-3}\bar{A} + p_{n-2}I_n)\, s \\ &+ (\bar{A}^{n-1} + p_1\bar{A}^{n-2} + p_2\bar{A}^{n-3} + \dots + p_{n-3}\bar{A}^2 + p_{n-2}\bar{A} + p_{n-1}I_n)\end{aligned}$$

we obtain the following representation of the partial derivatives of the closed-loop characteristic-polynomial coefficients, $1 \leq i \leq m$, $1 \leq j \leq r$,

$$\begin{aligned}\frac{\partial p_1}{\partial f_{ij}} &= -\, c_j' b_i \\ \frac{\partial p_2}{\partial f_{ij}} &= -(c_j'\bar{A}\, b_i + p_1 c_j' b_i) \\ \frac{\partial p_3}{\partial f_{ij}} &= -(c_j'\bar{A}^2 b_i + p_1 c_j'\bar{A}\, b_i + p_2 c_j' b_i) \\ &\;\vdots \\ \frac{\partial p_n}{\partial f_{ij}} &= -(c_j'\bar{A}^{n-1} b_i + p_1 c_j'\bar{A}^{n-2} b_i + p_2 c_j'\bar{A}^{n-3} b_i + \dots + p_{n-2} c_j'\bar{A} b_i + p_{n-1} c_j' b_i)\end{aligned} \tag{33.25}$$

The system (33.25) of $n\cdot m\cdot r$ scalar equations can be rewritten as one matrix equation

$$\begin{pmatrix} \frac{\partial p_1}{\partial f_{11}} \cdots \frac{\partial p_1}{\partial f_{ij}} \cdots \frac{\partial p_1}{\partial f_{mr}} \\ \frac{\partial p_2}{\partial f_{11}} \cdots \frac{\partial p_2}{\partial f_{ij}} \cdots \frac{\partial p_2}{\partial f_{mr}} \\ \vdots \\ \frac{\partial p_n}{\partial f_{11}} \cdots \frac{\partial p_n}{\partial f_{ij}} \cdots \frac{\partial p_n}{\partial f_{mr}} \end{pmatrix} = - \begin{pmatrix} 1 & 0 & 0 & \dots & 0 \\ p_1 & 1 & 0 & & 0 \\ \vdots & \vdots & \vdots & \ddots & \vdots \\ p_{n-1} & p_{n-2} & p_{n-3} & \dots & 1 \end{pmatrix} \begin{pmatrix} c_1' b_1 & \dots & c_j' b_i & \dots & c_r' b_m \\ c_1'\bar{A} b_1 & \dots & c_j'\bar{A} b_i & \dots & c_r'\bar{A} b_m \\ \vdots & & \vdots & & \vdots \\ c_1'\bar{A}^{n-1} b_1 & \dots & c_j'\bar{A}^{n-1} b_i & \dots & c_r'\bar{A}^{n-1} b_m \end{pmatrix} \tag{33.26}$$

The matrix on the left-hand side of Eq.(33.26) is nothing else than g_{*F}.

Linear independence of the matrices CB, $C\bar{A}B$, ..., $C\bar{A}^{n-1}B$ is equivalent to linear independence of the row vectors

$$(c_1'b_1,\dots,c_j'b_i,\dots,c_r'b_m)$$

$$(c_1'\overline{A}b_1,\dots,c_j'\overline{A}b_i,\dots,c_r'\overline{A}b_m)$$

$$\vdots$$

$$(c_1'\overline{A}^{n-1}b_1,\dots,c_j'\overline{A}^{n-1}b_i,\dots,c_r'\overline{A}^{n-1}b_m)$$

From (33.26) it is evident that

$$\operatorname{rank} g_{*F} = \operatorname{rank}\begin{pmatrix} c_1'b_1 & \dots & c_j'b_i & \dots & c_r'b_m \\ c_1'\overline{A}b_1 & \dots & c_j'\overline{A}b_i & \dots & c_r'\overline{A}b_m \\ \vdots & \vdots\vdots\vdots & \vdots & \vdots\vdots\vdots & \vdots \\ c_1'\overline{A}^{n-1}b_1 & \dots & c_j'\overline{A}^{n-1}b_i & \dots & c_r'\overline{A}^{n-1}b_m \end{pmatrix} \qquad (33.27)$$

This implies the assertion of Lemma 33.8.

▲

<u>Example 33.1</u>: Consider a plant with $n = 3$ state-variables, $m = 2$ inputs, and $r = 2$ outputs,

$$\dot{x} = A\,x + B\,u = \begin{pmatrix} a_{11} & a_{12} & 0 \\ 0 & 0 & 0 \\ a_{31} & a_{32} & 0 \end{pmatrix} x + \begin{pmatrix} b_{11} & 0 \\ 0 & b_{22} \\ 0 & 0 \end{pmatrix} u$$

$$y = C\,x = \begin{pmatrix} c_{11} & 0 & 0 \\ 0 & 0 & c_{23} \end{pmatrix} x$$

controlled by output feedback

$$u = F\,y = \begin{pmatrix} f_1 & f_2 \\ f_3 & f_4 \end{pmatrix} y\,,$$

comp. Example 11.1.
The associated digraph $G([Q_3])$ has been shown in Fig. 11.3.
For convenience, $G([Q_3])$ has been redrawn in Fig. 33.1

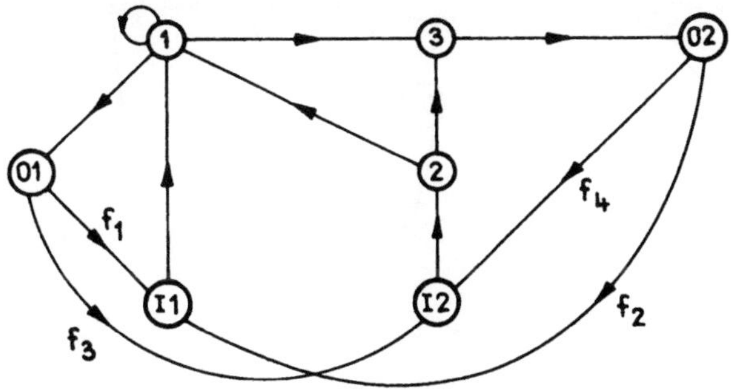

Fig. 33.1

To get the explicit representation of the non-linear mapping $p = g(F)$ according to Lemma 33.1, the cycle families associated with the coefficients p_1, p_2, and p_3 have been composed in Fig. 33.2

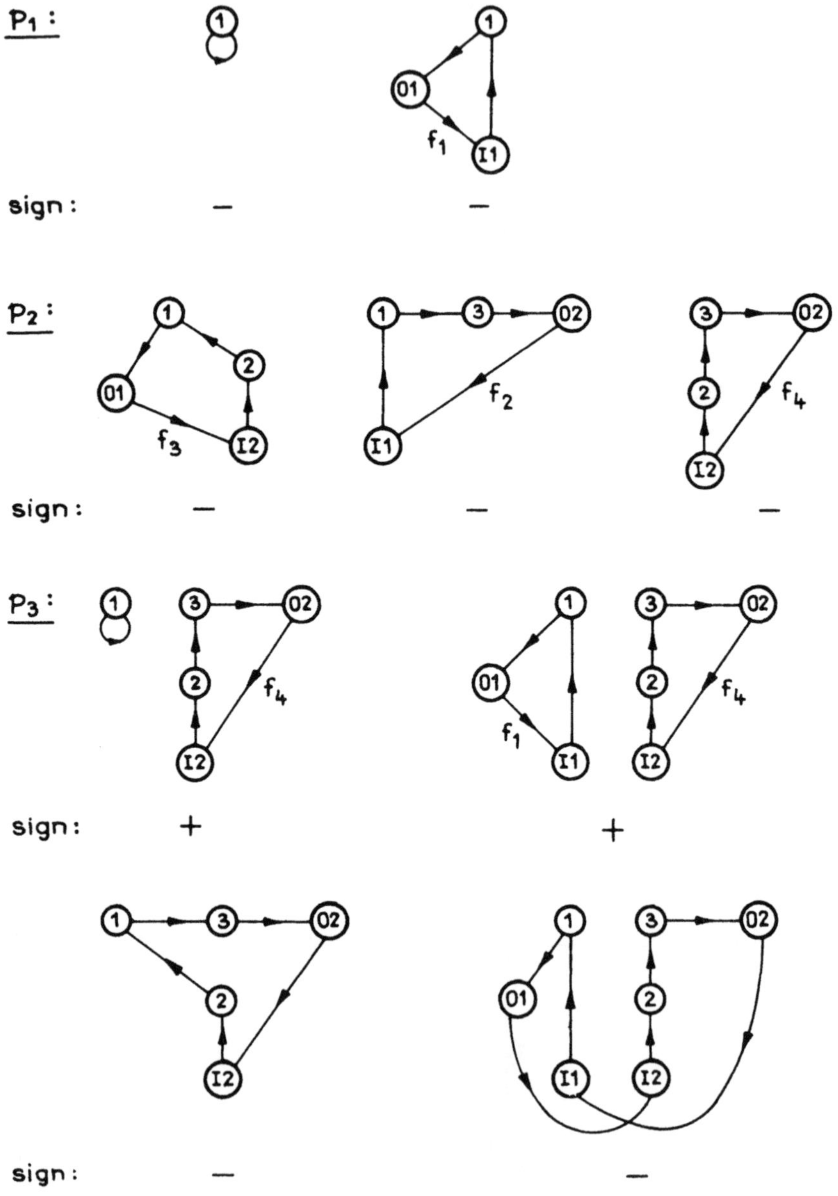

Fig. 33.2

Thus, we obtain

$$p = \begin{pmatrix} a_{11} \\ 0 \\ 0 \end{pmatrix} - \begin{pmatrix} b_{11}c_{11} \\ 0 \\ 0 \end{pmatrix} f_1 - \begin{pmatrix} 0 \\ b_{11}a_{31}c_{23} \\ 0 \end{pmatrix} f_2 - \begin{pmatrix} 0 \\ b_{22}a_{12}c_{11} \\ 0 \end{pmatrix} f_3 +$$

$$+ \begin{pmatrix} 0 \\ -b_{22}a_{32}c_{23} \\ b_{22}c_{23}(a_{11}a_{32}-a_{12}a_{31}) \end{pmatrix} f_4 + \begin{pmatrix} 0 \\ 0 \\ b_{22}a_{32}c_{23}b_{11}c_{11} \end{pmatrix} (f_1f_4-f_2f_3) \tag{33.28}$$

According to Lemma 33.2, the differential g_{*F} follows as

$$dp = \begin{pmatrix} b_{11}c_{11} \\ 0 \\ b_{22}a_{32}c_{23}b_{11}c_{11}\cdot f_4 \end{pmatrix} df_1 - \begin{pmatrix} 0 \\ b_{11}a_{31}c_{23} \\ b_{22}a_{32}c_{23}b_{11}c_{11}\cdot f_3 \end{pmatrix} df_2$$

$$- \begin{pmatrix} 0 \\ b_{22}a_{12}c_{11} \\ b_{22}a_{32}c_{23}b_{11}c_{11}\cdot f_2 \end{pmatrix} df_3 + \begin{pmatrix} 0 \\ -b_{22}a_{32}c_{23} \\ b_{22}c_{23}(a_{11}a_{32}-a_{12}a_{31}+a_{32}b_{11}c_{11}\cdot f_1) \end{pmatrix} df_4$$

Now, let us numerically fix the non-vanishing entries of A, B, and C. It is especially convenient to put all these values equal to 1. Then we get

$$\begin{pmatrix} p_1 \\ p_2 \\ p_3 \end{pmatrix} = g(F) = \begin{pmatrix} -1 \\ 0 \\ 0 \end{pmatrix} - \begin{pmatrix} 1 \\ 0 \\ 0 \end{pmatrix} f_1 - \begin{pmatrix} 0 \\ 1 \\ 0 \end{pmatrix} f_2 - \begin{pmatrix} 0 \\ 1 \\ 0 \end{pmatrix} f_3 - \begin{pmatrix} 0 \\ 1 \\ 0 \end{pmatrix} f_4 + \begin{pmatrix} 0 \\ 0 \\ 1 \end{pmatrix} (f_1f_4-f_2f_3) \tag{33.29}$$

$$\begin{pmatrix} dp_1 \\ dp_2 \\ dp_3 \end{pmatrix} = g_{*F}(dF) = \begin{pmatrix} -1 \\ 0 \\ f_4 \end{pmatrix} df_1 - \begin{pmatrix} 0 \\ 1 \\ f_3 \end{pmatrix} df_2 - \begin{pmatrix} 0 \\ 1 \\ f_2 \end{pmatrix} df_3 + \begin{pmatrix} 0 \\ -1 \\ f_1 \end{pmatrix} df_4$$

In order to decide on local pole assignability by static output feedback we have to investigate the rank of the 3×4 matrix g_{*F},

$$g_{*F} = \begin{pmatrix} -1 & 0 & 0 & 0 \\ 0 & -1 & -1 & -1 \\ f_4 & -f_3 & -f_2 & f_1 \end{pmatrix} \tag{33.30}$$

Because of Theorem 33.1, the system under consideration has the property of local pole assignability at

$$F_o = \begin{pmatrix} 1 & 1 \\ 1 & 1 \end{pmatrix}, \quad p_o = g(F_o) = \begin{pmatrix} -2 \\ -3 \\ 0 \end{pmatrix}$$

For $f_1 = f_2 = f_3 = f_4 = 0$, however, we have rank $g_{*F} = 2 < 3 = n$. Consequently, the given plant cannot be everywhere locally pole assignable.

According to Theorem 33.2, the set M_1 of non-regular F-points can be computed,

$$M_1 = \left\{F \in \mathbb{R}^4: [(g_{*F})^{123}_{123} = 0] \wedge [(g_{*F})^{123}_{124} = 0] \wedge [(g_{*F})^{123}_{134} = 0] \wedge [(g_{*F})^{123}_{234} = 0]\right\}$$

$$= \left\{F \in \mathbb{R}^4: [-f_2+f_3 = 0] \wedge [-f_1-f_3 = 0] \wedge [f_1+f_2 = 0] \wedge [0 = 0]\right\}$$

$$= \left\{F \in \mathbb{R}^4: -f_1 = f_2 = f_3\right\} \tag{33.31}$$

The algebraic submanifold M_1 is a two-dimensional plane in $\mathbb{R}^4$. Its intersection with the three-dimensional (f_1, f_2, f_3)-space is a straight line shown in Fig. 33.3. The coordinate f_4 may be freely chosen, $-\infty < f_4 < \infty$.

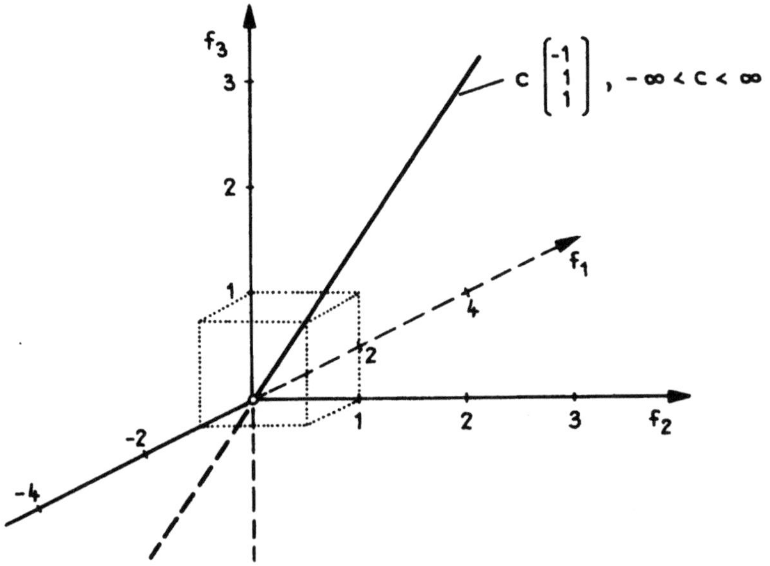

Fig. 33.3

The set N_1 of critical values follows from (33.29) and (33.31) as

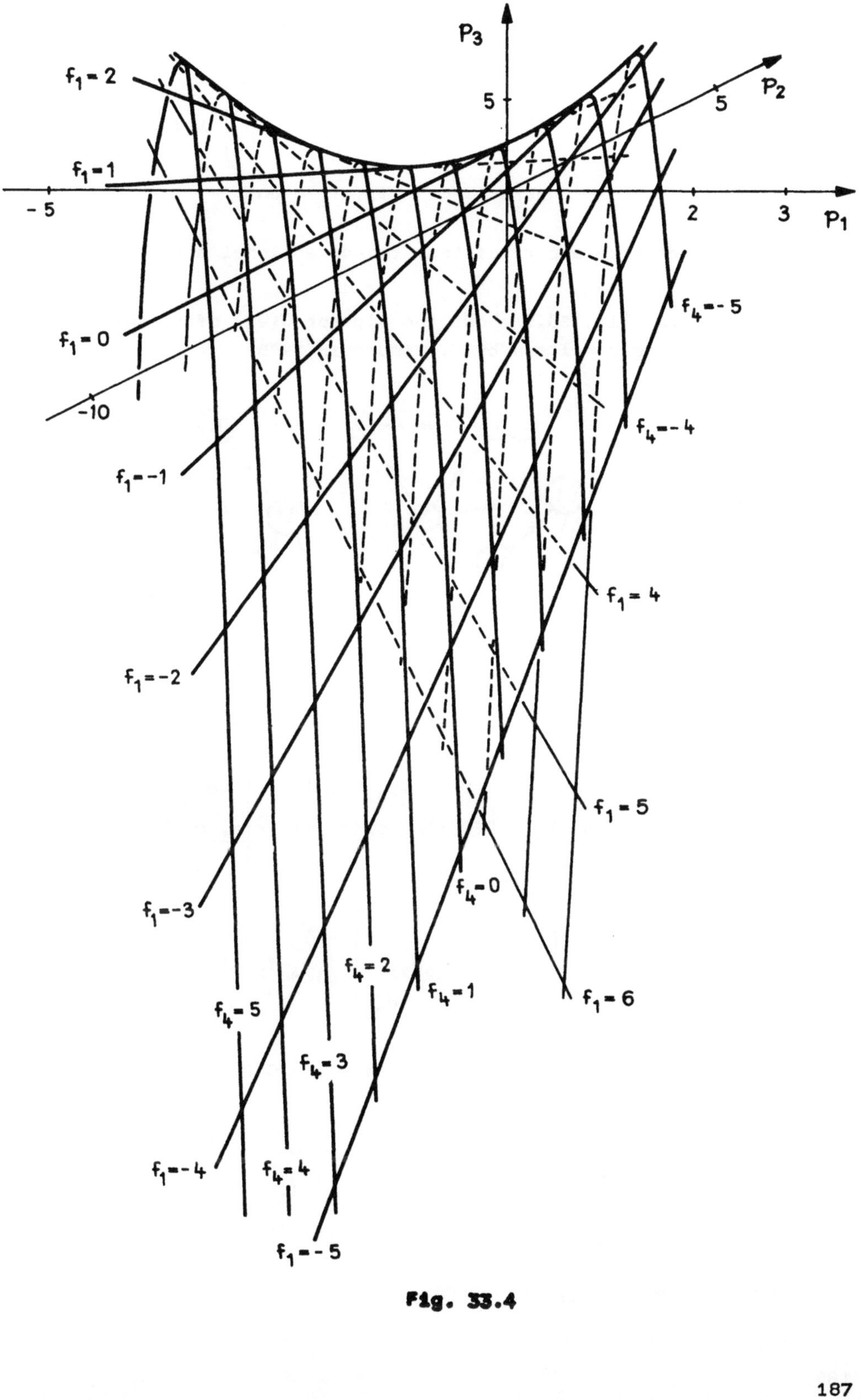

p3
p2
p1
5
5
-5
2
3
-10
f1 = 2
f1 = 1
f1 = 0
f1 = -1
f1 = -2
f1 = -3
f1 = -4
f1 = -5
f1 = 4
f1 = 5
f1 = 6
f4 = -5
f4 = -4
f4 = 0
f4 = 1
f4 = 2
f4 = 3
f4 = 4
f4 = 5

Fig. 33.4

$$N_1 = g(M_1) = \left\{ p \in \mathbb{R}^3 \quad p = g(F) \quad \text{with} \quad F \in M_1 \right\}$$

$$= \left\{ p \in \mathbb{R}^3: \quad p = \begin{pmatrix} -1 \\ 0 \\ 0 \end{pmatrix} + \begin{pmatrix} -1 \\ 2 \\ 0 \end{pmatrix} f_1 - \begin{pmatrix} 0 \\ 1 \\ 0 \end{pmatrix} f_4 - \begin{pmatrix} 0 \\ 0 \\ 1 \end{pmatrix} f_1^2 + \begin{pmatrix} 0 \\ 0 \\ 1 \end{pmatrix} f_1 f_4 \right. \tag{33.32}$$

$$\left. \text{where} \quad -\infty < f_1, \; f_4 < \infty \right\}$$

Fig. 33.4 illustrates a part of this two-dimensional surface.

Example 33.2: Fig. 33.5 shows the digraph $G(Q_3)$ of another example suggested by a paper written by H. Kimura in 1975.

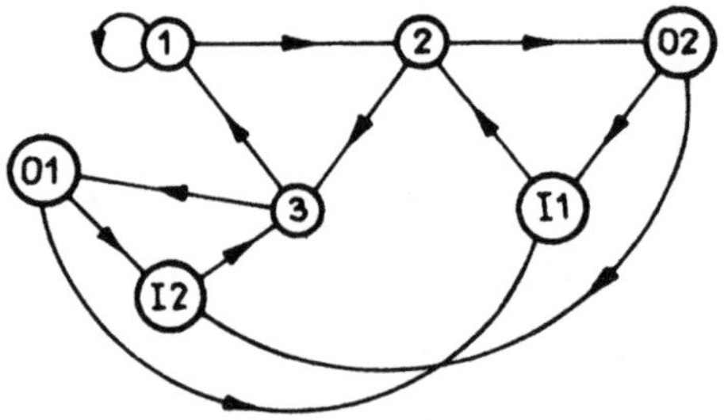

Fig. 33.5

From the cycle families composed in Fig. 33.6 the mapping $p = g(F)$ may be read as

$$p = \begin{pmatrix} -a_{11} \\ 0 \\ -a_{13}a_{32}a_{21} \end{pmatrix} + \begin{pmatrix} 0 \\ -c_{13}a_{32}b_{21} \\ a_{11}c_{13}a_{32}b_{21} \end{pmatrix} f_1 + \begin{pmatrix} -c_{22}b_{21} \\ a_{11}c_{22}b_{21} \\ 0 \end{pmatrix} f_2 + \begin{pmatrix} -c_{13}b_{32} \\ a_{11}c_{13}b_{32} \\ 0 \end{pmatrix} f_3$$

$$- \begin{pmatrix} 0 \\ 0 \\ c_{22}a_{21}a_{13}b_{32} \end{pmatrix} f_4 + \begin{pmatrix} 0 \\ -c_{13}b_{32}c_{22}b_{21} \\ a_{11}c_{13}b_{32}c_{22}b_{21} \end{pmatrix} (f_1 f_4 - f_2 f_3)$$

Putting all the non-vanishing entries of B and C equal to 1, the differential g_{*F} becomes

$$g_{*F} = \begin{pmatrix} 0 & -1 & -1 & 0 \\ -a_{32}-f_4 & a_{11}+f_3 & a_{11}+f_2 & -f_1 \\ a_{11}a_{32}+a_{11}f_2 & -a_{11}f_3 & -a_{11}f_2 & -a_{21}a_{13}+a_{11}f_1 \end{pmatrix} \tag{33.33}$$

The generic rank of g_{*F} is three.

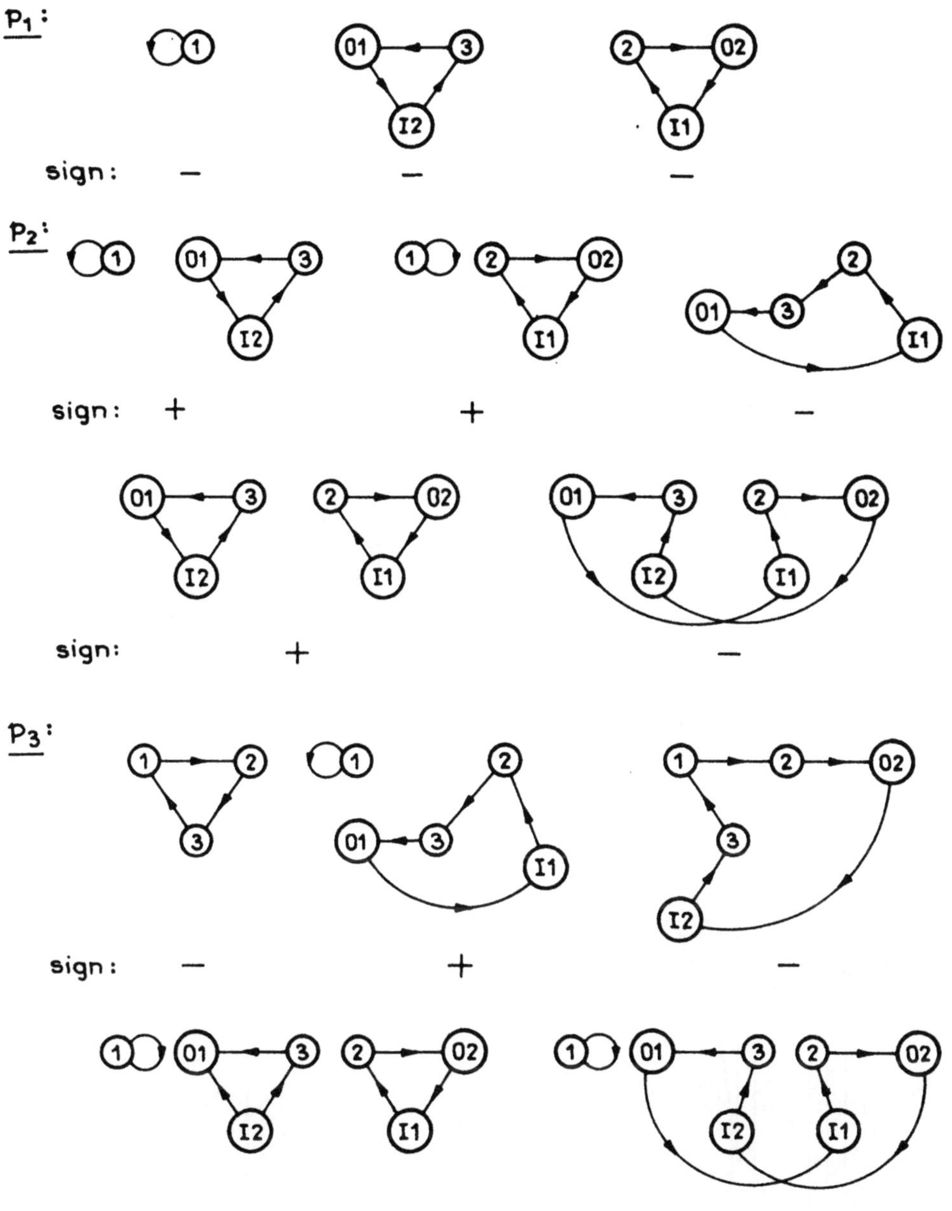

Fig. 33.6

The set of non-regular F-points becomes

$$M_1 = \left\{ F \in \mathbb{R}^4 : \text{rank } g_{*F} < 3 \right\}$$
$$= \left\{ F \in \mathbb{R}^4 : [f_4 = -a_{32}] \wedge [f_2 = f_3] \right\} \tag{33.34}$$

This is a plane in $\mathbb{R}^4$.

The set N_1 of critical p-values follows as

$$N_1 = g(M_1) = \left\{ p \in \mathbb{R}^3 : \begin{pmatrix} -a_{11} \\ 0 \\ 0 \end{pmatrix} + \begin{pmatrix} -2 \\ 2a_{11} \\ 0 \end{pmatrix} f_2 + \begin{pmatrix} 0 \\ 1 \\ -a_{11} \end{pmatrix} f_2^2 \right\} \tag{33.35}$$

This is a plane parabola in $\mathbb{R}^3$, see Fig. 33.7.

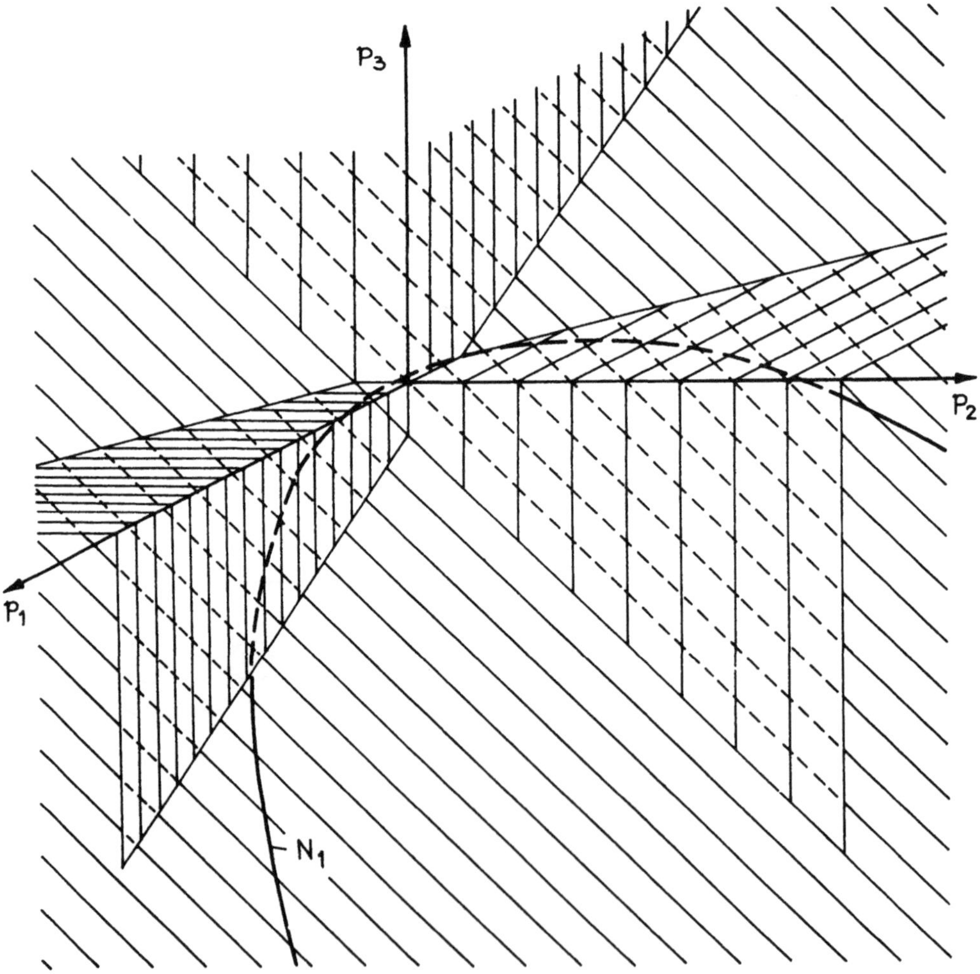

Fig. 33.7

Let us now replace the plane M_1 by a slightly modified plane M_2. For example,

$$M_2 = \left\{ F \in \mathbb{R}^4 : \ [f_4 = -c\ a_{32}] \wedge [f_2 = f_3] \right\} \qquad (33.36)$$

for some real scalar $c = 1$.

The image of M_2 is

$$N_2 = g(M_2) = \left\{ p \quad R^3 : \begin{pmatrix} -a_{11} \\ 0 \\ a_{32}a_{21}a_{13}(c-1) \end{pmatrix} + \begin{pmatrix} 0 \\ a_{32}(c-1) \\ a_{11}a_{32}(1-c) \end{pmatrix} f_1 + \begin{pmatrix} -2 \\ 2a_{11} \\ 0 \end{pmatrix} f_2 + \begin{pmatrix} 0 \\ 1 \\ -a_{11} \end{pmatrix} f_2^2 \right\} \qquad (33.37)$$

This is a two-dimensional surface whose elements may be characterized by the parameter values of f_1 and f_2.
For c — 1, the surface degenerates into a plane curve.

Example 33.3: Let be

$$A = \begin{pmatrix} 0 & 1 & 1 \\ 1 & 0 & 0 \\ 0 & 0 & 1 \end{pmatrix}, \quad B = \begin{pmatrix} 1 & 0 \\ 0 & 0 \\ 0 & 1 \end{pmatrix}, \quad C = \begin{pmatrix} 0 & 1 & 1 \\ 1 & 0 & 0 \end{pmatrix}, \quad F = \begin{pmatrix} f_1 & f_2 \\ f_3 & f_4 \end{pmatrix} \qquad (33.38)$$

The associated digraph $G(Q_3)$ has been drawn in Fig. 33.8.

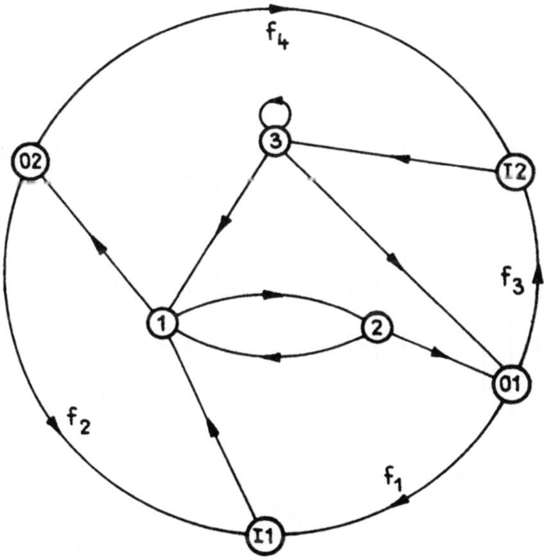

Fig. 33.8

The mapping g and its differential g_{*F} are

$$p = \begin{pmatrix} p_1 \\ p_2 \\ p_3 \end{pmatrix} = g(F) = \begin{pmatrix} -1 \\ 1 \\ 1 \end{pmatrix} + \begin{pmatrix} 0 \\ -1 \\ 1 \end{pmatrix} f_1 + \begin{pmatrix} -1 \\ 0 \\ 0 \end{pmatrix} f_3 + \begin{pmatrix} -1 \\ 1 \\ 0 \end{pmatrix} f_2 - \begin{pmatrix} 0 \\ 1 \\ 0 \end{pmatrix} f_4 - \begin{pmatrix} 0 \\ 1 \\ 0 \end{pmatrix} (f_1 f_4 - f_2 f_3) \qquad (33.39)$$

$$g_{*F} = \begin{pmatrix} 0 & -1 & -1 & 0 \\ -1-f_4 & 1+f_3 & f_2 & -1-f_1 \\ 1 & 0 & 0 & 0 \end{pmatrix} \qquad (33.40)$$

The generic rank of g_{*F} is three.

The set M_1 of non-regular F-points becomes

$$M_1 = \left\{ F \in \mathbb{R}^4 :\ \text{rank } g_{*F} < 3 \right\} = \left\{ F \in \mathbb{R}^4 :\ [f_1 = -1] \wedge [f_3 = f_2 - 1] \right\} \qquad (33.41)$$

The set of critical p-values follows from (33.39) and (33.41) as

$$N_1 = g(M_1) = \left\{ p \in \mathbb{R}^3 :\ p = \begin{pmatrix} -2 \\ 0 \\ 0 \end{pmatrix} f_2 + \begin{pmatrix} 0 \\ 1 \\ 0 \end{pmatrix} f_2^2 \right\} \qquad (33.42)$$

This is a parabola in the p_1, p_2-plane, see Fig. 33.9.

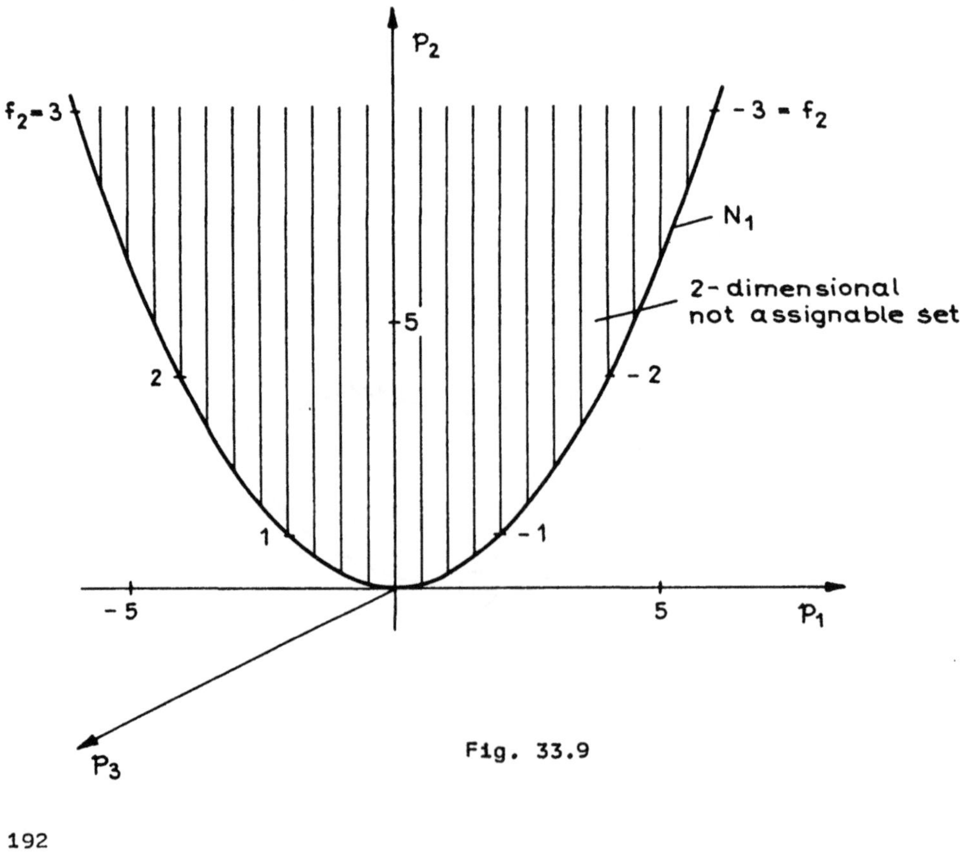

Fig. 33.9

33.3 Conditions for global pole assignability

As for global pole assignability, let us first recall a result published by Willems and Hesselink in 1978 based on a paper of Hermann and Martin which appeared in 1977.

Lemma 33.9

Assume (33.19) to be valid.
Then for any given real polynomial

$$p(s) = s^n + p_1 s^{n-1} + \dots + p_{n-1} s + p_n$$

there is a complex $m \times r$ feedback matrix $F(\mathbb{C})$ such that

$$\det(sI_n - A - B\,F(C)\,C) = p(s).$$

From Theorem 33.2 it is known that (33.19) implies the existence of a subset $N \subset \mathbb{R}^n$ with positive measure each element of which can be achieved by a real feedback matrix F. The subset $\overline{N} = \mathbb{R}^n \setminus N$ of p-values only achievable by complex feedback may be either of positive measure, or of measure zero, or empty.

A given plant (33.1), (33.2) has the property of arbitrary pole assignability in the global sense of Definition 33.1 if and only if the subset $\overline{N}$ is empty or, equivalently, the domain N is the whole space R^n.

We are going to consider the problem of global pole assignability by real static output feedback from two different points of view.
On the one hand, we shall ask for additional conditions on the mapping g(F) that ensure a global version of the Implicit Function Theorem.
On the other hand, we shall investigate the local properties of g(F) in the neighbourhood of non-regular F-points.

In what follows it is assumed that $m \cdot r = n$.
Let F_{red} be a real $m \times r$ matrix with n non-vanishing entries. F_{red} may be interpreted as a submatrix of F. The remaining $(m \cdot r - n)$ entries are the non-vanishing elements of an $m \times r$ matrix F_{rem}. For short,

$$F = F_{red} \cup F_{rem}, \qquad F_{red} \cap F_{rem} = 0 \tag{33.43}$$

and

$$p = g(F) = g(F_{red}, F_{rem}) \tag{33.44}$$

If the non-zero entries of F_{rem} are fixed at particular real values then p depends only on F_{red}.

$$p = \bar{g}(F_{red}) \tag{33.45}$$

This relationship defines a smooth mapping

$$\bar{g}: \mathbb{R}^n \longrightarrow \mathbb{R}^n \tag{33.46}$$

Its differential is denoted by

$$\bar{g}_{*F_{red}}: T_{F_{red}} \mathbb{R}^n \longrightarrow T_{\bar{g}(F_{red})} \mathbb{R}^n \tag{33.47}$$

i.e.,

$$dp = \bar{g}_{*F_{red}}(dF_{red}) \tag{33.48}$$

<u>Theorem 33.3</u>

Assume $m \cdot r \geq n$.
If there exists a partitioning (33.43) of the feedback matrix F such that, after an appropriate fixation of the $m \cdot r - n$ non-zero entries of F_{rem}, the differential

$\bar{g}_{*F_{red}}$ is a regular $n \times n$ matrix with constant entries, (33.49)

then the closed-loop system (33.1), (33.2), (33.3) is globally pole assignable.

<u>Proof</u>: Using the abbreviation

$$\bar{g}_{*F_{red}} = Q \tag{33.50}$$

the assumption (33.49) implies

$$|\det Q| = K_1 = \text{const.} > 0 \tag{33.51}$$

Further, the inverse matrix Q^{-1} exists and has a bounded Euclidean norm,

$$\| Q^{-1} \| = K_2 < \infty \tag{33.52}$$

This may be seen as follows: The existence of Q^{-1} results from (33.51). The Euclidean norm of any regular matrix Q is the square root of the maximal eigenvalue of $Q'Q$,

$$\| Q \| = \sqrt{\lambda_{max}(Q'Q)} \tag{33.53}$$

The determinant is determined by the eigenvalues

$$(\det Q)^2 = \det(Q'Q) = \prod_{i=1}^{n} \lambda_i(Q'Q), \quad \text{all } \lambda_i(Q'Q) > 0 \tag{33.54}$$

This implies

$$\lambda_{min}(Q'Q) \geq K_3 = \text{const.} > 0 \qquad (33.55)$$

Furthermore,

$$(\det Q^{-1})^2 = \det (Q'Q)^{-1} = \prod_{i=1}^{n} \lambda_i((Q'Q)^{-1}) = \prod_{i=1}^{n} (\lambda_i(Q'Q))^{-1}$$

Thus we can write

$$\|(g_{*F_{red}})^{-1}\| = \|Q^{-1}\| = \sqrt{\lambda_{max}((Q'Q^{-1})} = \sqrt{(\lambda_{min}(Q'Q))^{-1}}$$

$$\leq \sqrt{\frac{1}{K_3}} = K_2 < \infty$$

as stated in (33.52).

Now, we are able to apply a Theorem published by Hadamard in 1906 (see, for example, Ortega and Rheinboldt 1970, Ch.5):

From the smoothness of the mapping $\bar{g}$ and from (33.52) it can be concluded that the mapping $\bar{g}$ is a homeomorphism. That is, for each

$p_o \in \mathbb{R}^n$ there is an $F_{o,red}$ such that $\bar{g}(F_{o,red}) = p_o$.

This completes the proof of Theorem 33.3.

▲

For many feedback systems, possibly after reordering of the elements of p and of F_{red}, Eq. (33.44) splits up as follows:

$$^1p = {}^1g(F_{rem}, {}^1F_{red}) \qquad (33.56.1)$$

$$^2p = {}^2g(F_{rem}, {}^1F_{red}, {}^2F_{red}) \qquad (33.56.2)$$

$$\vdots$$

$$^kp = {}^kg(F_{rem}, {}^1F_{red}, \ldots, {}^{k-1}F_{red}, {}^kF_{red}) \qquad (33.56.k)$$

where, for $1 \leq i \leq k$, the number n_i of non-zero entries of $^iF_{red}$ is equal to the number of components of ^ip,

$$\sum_{i=1}^{k} n_i = n,$$

and the entries of the $n_i \times n_i$ Jacobian matrices $\frac{\partial^i p}{\partial^i F_{red}}$ do not depend on $^iF_{red}$,

i.e.

$$\frac{\partial^1 p}{\partial^1 F_{red}} = {}^1J(F_{rem}) \qquad (33.57.1)$$

$$\frac{\partial^2 p}{\partial^2 F_{red}} = {}^2J(F_{rem}, {}^1F_{red}) \qquad (33.57.2)$$

$$\vdots$$

$$\frac{\partial^k p}{\partial^k F_{red}} = {}^kJ(F_{rem}, {}^1F_{red}, \ldots, {}^{k-1}F_{red}) \qquad (33.57.k)$$

By repeated application of Theorem 33.3 we can conclude:

If

$$\det {}^1J(F_{rem}) \neq 0 \qquad (33.58.1)$$

then Eq.(33.56.1) can be solved for ${}^1F_{red}$,

$${}^1F_{red} = {}^1f(F_{rem}, {}^1p). \qquad (33.59.1)$$

If

$$\det {}^2J(F_{rem}, {}^1F_{red}) = \det {}^2J(F_{rem}, {}^1f(F_{rem}, {}^1p)) \neq 0 \qquad (33.58.2)$$

then Eq.(33.56.2) can be solved for ${}^2F_{red}$,

$${}^2F_{red} = {}^2\bar{f}(F_{rem}, {}^1F_{red}, {}^2p) = {}^2\bar{f}(F_{rem}, {}^1f(F_{rem}, {}^1p), {}^2p) \qquad (33.59.2)$$

$$= {}^2f(F_{rem}, {}^1p, {}^2p).$$

$$\vdots$$

If

$$\det {}^kJ(F_{rem}, {}^1F_{red}, \ldots, {}^{k-1}F_{red})$$

$$= \det {}^kJ(F_{rem}, {}^1f(F_{rem}, {}^1p), \ldots, {}^{k-1}f(F_{rem}, {}^1p, \ldots, {}^{k-1}p) \neq 0 \qquad (33.58.k)$$

then Eq.(33.56.k) can be solved for ${}^kF_{red}$,

$${}^kF_{red} = {}^k\bar{f}(F_{rem}, {}^1F_{red}, \ldots, {}^{k-1}F_{red}, {}^kp)$$

$$= {}^k\bar{f}(F_{rem}, {}^1f(F_{rem}, {}^1p), \ldots, {}^{k-1}f(F_{rem}, {}^1p, \ldots, {}^{k-1}p), {}^kp)$$

$$= {}^kf(F_{rem}, {}^1p, {}^2p, \ldots, {}^kp). \qquad (33.59.k)$$

In this way we habe substantiated

Corollary 33.1

Assume the n-vector equation (33.44) to be transformed into k n_i-vector equations (33.56.1 - k) by reordering the rows and F_{red}- entries where $k \geq 1$,

$$\sum_{i=1}^{k} n_i = n,$$

and the $n_i \times n_i$ Jacobian matices ${}^iJ$ (see (33.57.1 - k) are independent of ${}^iF_{red}$.

If the feedback gains gathered in F_{rem} can be fixed in such a way that, for arbitrary choice of ${}^1p, {}^2p, \dots, {}^kp$, each of the k Jacobian determinants (33.58.1 - k) does not vanish, then the closed-loop system (33.1), (33.2), (33.3) is globally pole assignable.

Example 33.4: Consider the example system treated above as Example 33.1. To decide on global pole assignability we try to apply Theorem 33.3. Here, condition (33.49) is equivalent to

$$(g_{*F})^{123}_{123} = 0 \quad \text{for all} \quad f_1,\ f_2,\ f_3 \text{ and a suitably fixed } f_4,$$

or

$$(g_{*F})^{123}_{124} = 0 \quad \text{for all} \quad f_1,\ f_2,\ f_4 \text{ and a suitably fixed } f_3,$$

or

$$(g_{*F})^{123}_{134} = 0 \quad \text{for all} \quad f_1,\ f_3,\ f_4 \text{ and a suitably fixed } f_2,$$

or

$$(g_{*F})^{123}_{234} = 0 \quad \text{for all} \quad f_2,\ f_3,\ f_4 \text{ and a suitably fixed } f_1.$$

The occurring four minors of order 3 have already been computed, see (33.31). It is easily seen that the condition (33.49) cannot be met. Fortunately, Eq. (33.29) splits up into

$$p_1 = {}^1g(f_1) = -1 - f_1$$

$$\begin{pmatrix} p_2 \\ p_3 \end{pmatrix} = {}^2g(f_1, f_2, f_3, f_4) = \begin{pmatrix} -f_2 - f_3 - f_4 \\ f_1 f_4 - f_2 f_3 \end{pmatrix}$$

comp. (33.56.1) and (33.56.2). Application of Corollary 33.1 seems to be promising.

Putting

$${}^1F_{red} = \begin{pmatrix} f_1 & 0 \\ 0 & 0 \end{pmatrix}, \quad {}^2F_{red} = \begin{pmatrix} 0 & 0 \\ f_3 & f_4 \end{pmatrix}, \quad F_{rem} = \begin{pmatrix} 0 & f_2 \\ 0 & 0 \end{pmatrix}$$

one obtains

$$^{1}J(f_2) = \frac{\partial p_1}{\partial f_1} = -1 \neq 0 \quad \text{independently of } f_2,$$

and the first row of the 3-vector equation (33.29) may be solved for f_1,

$$f_1 = -1 - p_1.$$

The second and third row of (33.29) give

$$^{2}J(f_2,f_1) = \frac{\partial(p_2,\ p_3)}{\partial(f_3,\ f_4)} = \det\begin{pmatrix} -1 & -1 \\ -f_2 & f_1 \end{pmatrix} = \det\begin{pmatrix} -1 & -1 \\ -f_2 & -1-p_1 \end{pmatrix} = 1+p_1-f_2$$

$$\neq 0 \quad \text{if} \quad f_2 \neq 1 + p_1$$

and

$$\begin{pmatrix} f_2 \\ f_3 \end{pmatrix} = \begin{pmatrix} -1 & -1 \\ -f_2 & -1-p_1 \end{pmatrix}^{-1} \left[\begin{pmatrix} p_2 \\ p_3 \end{pmatrix} + \begin{pmatrix} 1 \\ 0 \end{pmatrix} f_2\right]$$

This means the example system is pole assignable in the global sense of Definition 33.1 if we fix f_2 at a particular value $\neq 1+p_1$. Then, the other feedback gains f_1, f_2, and f_3 are uniquely determined as functions of any given values p_1, p_2, and p_3.

Example 33.5: Let us turn to the example system introduced as Example 33.3. As for the property of global pole assignability, the situation seems to be similar as in the foregoing Example 33.4. Therefore, we try to apply Corollary 33.1.

Eq. (33.39) splits up into

$$^{1}p = p_3 = {^{1}g}(f_1) = 1 + f_1$$

$$^{2}p = \begin{pmatrix} p_1 \\ p_2 \end{pmatrix} = {^{2}g}(f_1,f_2,f_3,f_4) = \begin{pmatrix} -1 - f_2 - f_3 \\ 1 - f_1 + f_2 - f_4 - f_1 f_4 + f_2 f_3 \end{pmatrix}$$

Putting

$$F_{rem} = \begin{pmatrix} 0 & 0 \\ 0 & f_4 \end{pmatrix}, \quad {^{1}F}_{red} = \begin{pmatrix} f_1 & 0 \\ 0 & 0 \end{pmatrix}, \quad {^{2}F}_{red} = \begin{pmatrix} 0 & f_2 \\ f_3 & 0 \end{pmatrix}$$

one obtains

$$f_1 = p_3 - 1 \quad \text{independently of } f_4.$$

The 2×2 Jacobian matrix

$$\frac{\partial(p_2, p_3)}{\partial(f_2, f_3)} = \begin{pmatrix} -1 & -1 \\ 1+f_3 & f_2 \end{pmatrix}$$

depends on ${}^2F_{red}$, in contradiction to (33.57.2).

Obviously, the choice ${}^1F_{red} = \begin{pmatrix} f_1 & 0 \\ 0 & 0 \end{pmatrix}$ cannot be altered. This implies $f_1 = p_3 - 1$.

Putting

$$F_{rem} = \begin{pmatrix} 0 & 0 \\ f_3 & 0 \end{pmatrix}, \quad {}^2F_{red} = \begin{pmatrix} 0 & f_2 \\ 0 & f_4 \end{pmatrix}$$

the Jacobian matrix

$$\frac{\partial(p_2, p_3)}{\partial(f_2, f_4)} = \begin{pmatrix} -1 & 0 \\ 1+f_3 & -1-f_1 \end{pmatrix}$$

fulfills the condition (33.57.2). The determinant

$$\det \frac{\partial(p_2, p_3)}{\partial(f_2, f_4)} = 1 + f_1 = p_3$$ vanishes for $p_3 = 0$, comp. (33.58.2).

Putting

$$F_{rem} = \begin{pmatrix} 0 & f_2 \\ 0 & 0 \end{pmatrix}, \quad {}^2F_{red} = \begin{pmatrix} 0 & 0 \\ f_3 & f_4 \end{pmatrix}$$

the Jacobian determinant

$$\det \frac{\partial(p_2, p_3)}{\partial(f_3, f_4)} = \det \begin{pmatrix} -1 & 0 \\ f_2 & -1-f_1 \end{pmatrix} = 1 + f_1 = p_3$$

also vanishes for $p_3 = 0$.

Hence, Corollary 33.1 fails to guarantee global pole assignability. However, any desired closed-loop characteristic polynomial p_o with $p_{o,3} \neq 0$ can be achieved by the feedback gains

$f_1 = p_3 - 1$

f_2 = free parameter

$$\begin{pmatrix} f_3 \\ f_4 \end{pmatrix} = \begin{pmatrix} -1 & 0 \\ f_2 & -p_3 \end{pmatrix}^{-1} \begin{pmatrix} p_1 + 1 - f_2 \\ p_2 - p_3 + f_2 \end{pmatrix}$$

Let us now study the behaviour of the solution set $M(p_o)$ for given p_o near the plane $p_3 = 0$.

Eq. (33.39) yields

$$f_1 = p_3 - 1, \; f_2 = -1-f_3-p_1, \; f_4 = 1 - \frac{p_2}{p_3} - \frac{(1+f_3)(1+f_3+p_1)}{p_3} \qquad (33.60)$$

Fig. 33.10 shows the solution sets M(p) in the f_2, f_3, f_4-space for $p_1 = 0$, $p_2 = 1$, and $p_3 = 1, 0.5, 0.2, -0.2, -0.5$, or -1.

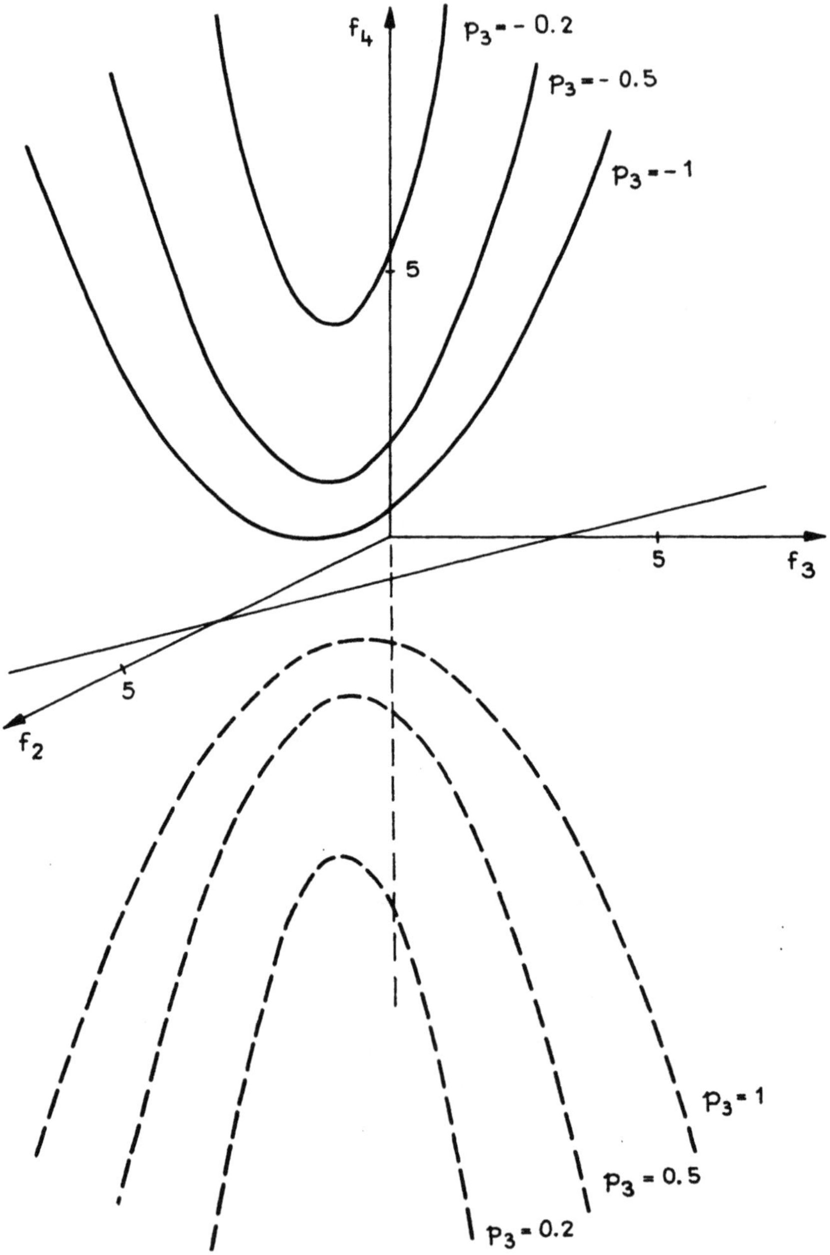

Fig. 33.10

It can be seen that the curves disappear at ∞ ($-\infty$) if p_3 tends to zero from below (from above). The closed-loop characteristic polynomial coefficients $p_o = (0, 1, 0)'$ can never be attained by real output feedback.

Fig. 33.11 shows the solution set M(p) for

$p_1 = 1$, $p_2 = 0$, and $p_3 = 1, 0.5, 0.2, -0.2, -0.5$, or -1.

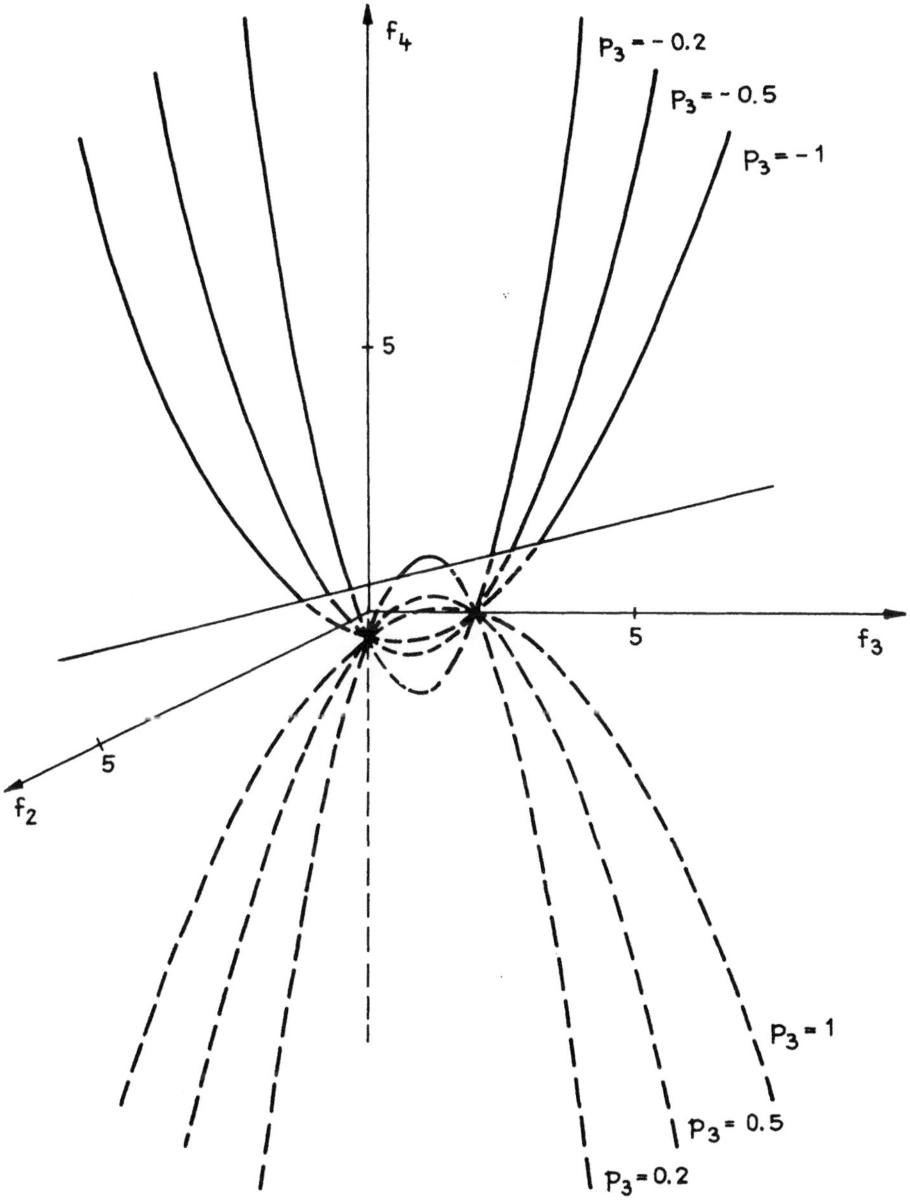

Fig. 33.11

Here, we notice two fixed points that are elements of $M(p)$ for all p_3-values. For $p_o = (1, 0, 0)'$ pole assignment by real output feedback is possible. The one-dimensional smooth curve $M(p)$ degenerates into two straight lines each parallel to the f_4-axis.

Finally, it should be realized that the property of non-assignability of the example system (33.38) is a consequence of the given numerical entries of the matrix A. Consider a slightly modified system of the same structure,

$$A = \begin{pmatrix} 0 & 1 & a_{13} \\ 1 & 0 & 0 \\ 0 & 0 & 1 \end{pmatrix}, \quad B = \begin{pmatrix} 1 & 0 \\ 0 & 0 \\ 0 & 1 \end{pmatrix}, \quad C = \begin{pmatrix} 0 & 1 & 1 \\ 1 & 0 & 0 \end{pmatrix} \quad \text{where } a_{13} = 1. \tag{33.61}$$

For this example system, the mapping g and its differential g_{*F} are

$$p = g(F) = \begin{pmatrix} -1 \\ -1 \\ 1 \end{pmatrix} + \begin{pmatrix} 0 \\ -1 \\ 1 \end{pmatrix} f_1 + \begin{pmatrix} -1 \\ 1 \\ 0 \end{pmatrix} f_2 + \begin{pmatrix} -1 \\ 0 \\ 1-a_{13} \end{pmatrix} f_3 + \begin{pmatrix} 0 \\ -1 \\ 0 \end{pmatrix} f_4 + \begin{pmatrix} 0 \\ 1 \\ 0 \end{pmatrix} (f_2 f_3 - f_1 f_4) \tag{33.62}$$

$$g_{*F} = \begin{pmatrix} 0 & -1 & -1 & 0 \\ -1-f_4 & 1+f_3 & f_2 & -1-f_1 \\ 1 & 0 & 1-a_{13} & 0 \end{pmatrix} \tag{33.63}$$

Placing f_3 at a suitably chosen non-vanishing value one obtains

$$f_1 = p_3 - 1 - (1-a_{13}) f_3,$$

$$f_2 = -p_1 - 1 - f_3,$$

$$f_4 = \frac{1}{1 + f_1} (-1 - f_1 + f_2(1+f_3) - p_2)$$

In terms of Corollary 33.1, there is

$$F_{rem} = \begin{pmatrix} 0 & 0 \\ f_3 & 0 \end{pmatrix}, \quad {}^1F_{red} = \begin{pmatrix} f_1 & 0 \\ 0 & 0 \end{pmatrix}, \quad {}^2F_{red} = \begin{pmatrix} 0 & f_2 \\ 0 & f_4 \end{pmatrix}$$

Eq. (33.56.1) reads as

$${}^1p = p_3 = 1 + (1-a_{13}) f_3 + f_1.$$

Evidently,

$$\frac{\partial p_3}{\partial f_1} = 1 \neq 0$$

Further,

$${}^2p = \begin{pmatrix} p_1 \\ p_2 \end{pmatrix} = \begin{pmatrix} -1-f_2-f_3 \\ -1-f_1+f_2-f_4+f_2 f_3 - f_1 f_4 \end{pmatrix} = \begin{pmatrix} -1-f_3 \\ p_3-(1-a_{13}) f_3 \end{pmatrix} + \begin{pmatrix} -1 & 0 \\ 1+f_3 & -1-f_1 \end{pmatrix} \begin{pmatrix} f_2 \\ f_4 \end{pmatrix}$$

and

$$\det \frac{\partial(p_1, p_2)}{\partial(f_2, f_4)} = \det \begin{pmatrix} -1 & 0 \\ 1+f_3 & -1-f_1 \end{pmatrix} = 1+f_1 = p_3 - (1-a_{13})f_3 \neq 0.$$

This side condition can always be fulfilled by an appropriate choice of the f_3-value. Consequently, the system (33.61) is globally pole assignable by real output feedback.

Now, we are going to investigate the local properties of the mapping (33.6) in the neighbourhood of non-regular F-points.

Let be $F_o \in M_1$. Its image $p_o = g(F_o) \in N_1$ is a critical value. It is candidate for being an element of the boundary set ∂N (= collection of all elements $p \in N$ with the property that every open n-dimensional neighbourhood containing p_o intersects both N and $\overline{N} = \mathbb{R}^n \setminus N$).

The question is whether any closely neighboured value $p_o + \Delta p$ can be achieved by a neighboured feedback matrix $F_o + \Delta F$.

An affirmative answer implies p_o to be an inner point of N.

A negative answer implies p_o to belong to the boundary set ∂N. Thus, we have to decide on the solvability for ΔF of the equation

$$p_o + \Delta p = g(F_o + \Delta F) \qquad (33.64)$$

or, equivalently,

$$\Delta p = g(F_o + \Delta F) - g(F_o) = h_{F_o}(\Delta F) \qquad (33.65)$$

An explicit representation of the mapping

$$h_{F_o} : \mathbb{R}^{m \cdot r} \longrightarrow \mathbb{R}^n \qquad (33.66)$$

may be found, based on Lemma 33.1, using quite elementary determinantal transformations. The arising expressions, however, become rather lengthy. Therefore, the general formulas are not written down here.

The closed-loop system (33.1), (33.2), (33.3) is locally pole assignable at

$$F_o \in M_1, \quad p_o = g(F_o) \in N_1$$

if and only if the solution set of (33.65) contains all inner elements of a small n-dimensional ball

$$U_{p_o}(\varepsilon) = \left\{ p \in \mathbb{R}^n : \varrho(p_o, p) = \sqrt{(p-p_o)'(p-p_o)} = \sqrt{(\Delta p)' \Delta p} = \|\Delta p\| < \varepsilon \right\}$$

provided that $\Delta F = F - F_o$ runs through all the inner points of

$$U_{F_o}(\delta) = \left\{ F \in \mathbb{R}^{m \cdot r}: \quad \varrho(F_o, F) = \sqrt{(\Delta F)' \ \Delta F} = \|\Delta F\| < \delta \right\}$$

Fortunately, we need not solve (33.65) in such a general manner for all $F_o \in M_1$. Due to the smoothness of the mapping (33.6) it is sufficient to execute the subsequent procedure formulated as Theorem 33.4:

<u>Theorem 33.4</u>

The set M_1 of non-regular F-points and its image $N_1 = g(M_1)$ of critical p-values are assumed to be known.

Choose a pair $F_o \in M_1$, $p_o = g(F_o) \in N_1$.

Consider the set of $\mathbb{R}^n$-vectors $\Delta p_\perp$ which meet the orthogonality relation

$$(\Delta p_\perp)' \ g_{*F_o} = 0 \tag{33.67}$$

and have the Euclidean length 1, i.e. $\|\Delta p_\perp\| = 1$.

If, for all these vectors $\Delta p_\perp$ and for each sufficiently small scalar $c > 0$, both

$c \, \Delta p_\perp$ and $-c \, \Delta p_\perp$

fulfil the Eq.(33.65) for some real ΔF, then N_1 does not belong to the boundary set with respect to pole assignability by real static output feedback, for short, $N_1 \not\subset \partial N$.

If, for at least one vector $\Delta p_\perp$, either $c \, \Delta p_\perp$ or $-c \, \Delta p_\perp$ has no real-valued solution of Eq.(33.65), then N_1 is a "local boundary" for pole assignability, i.e., there are p-values neighboured to p_o which cannot be assigned by real F neighboured to F_o.

In case of $m \cdot r = n$, N_1 is also a global boundary beyond of which pole assignability by real static output feedback is impossible, i.e., $N_1 \subset \partial N$.

In case of $m \cdot r > n$, look for other F-points F_1 with $p_o = g(F_1)$. Provided that such F_1 exist and rank $g_{*F_1} = n$, then N_1 cannot belong to the boundary set ∂N.

If rank $g_{*F_1} < n$, the investigation just described may be repeated based on an equation $\Delta p = g(F_1 + \Delta F) - g(F_1) = h_{F_1}(\Delta F)$ made-up in analogy to Eq.(33.56).

Example 33.6: We consider again the example system discussed before as Example 33.1 and Example 33.4.

Investigation of the local properties of the mapping (33.29) in the neighbourhood of the set N_1 (see (33.32)) should also give evidence of global pole assignability. For the example system, Eq. (33.65) becomes at any $\overline{F} \in N_1$:

$$\Delta p = g(\overline{F} + \Delta F) - g(\overline{F}) = h_{\overline{F}}(\Delta F)$$

$$= \begin{pmatrix} -1 \\ 0 \\ \overline{f}_4 \end{pmatrix} \Delta f_1 + \begin{pmatrix} 0 \\ -1 \\ \overline{f}_1 \end{pmatrix} \Delta f_2 + \begin{pmatrix} 0 \\ -1 \\ \overline{f}_1 \end{pmatrix} \Delta f_3 + \begin{pmatrix} 0 \\ -1 \\ \overline{f}_1 \end{pmatrix} \Delta f_4 + \begin{pmatrix} 0 \\ 0 \\ 1 \end{pmatrix} (\Delta f_1 \Delta f_4 - \Delta f_2 \Delta f_3) \qquad (33.68)$$

whereas Eq.(33.67) reads as

$$(\Delta p_{\perp})' \begin{pmatrix} -1 & 0 & 0 & 0 \\ 0 & -1 & -1 & -1 \\ \overline{f}_4 & \overline{f}_1 & \overline{f}_1 & \overline{f}_1 \end{pmatrix} = 0, \; |\Delta p_{\perp}| = 1 \qquad (33.69)$$

There is a unique solution

$$\Delta p_{\perp} = \begin{pmatrix} \overline{f}_4 \\ \overline{f}_1 \\ 1 \end{pmatrix} (\overline{f}_4^2 + \overline{f}_1^2 + 1)^{1/2} \qquad (33.70)$$

According to Theorem 33.4, we look for a solution of (33.68) with the left-hand side $\Delta p = c\,\Delta p_{\perp}$ where $|c|$ is a small value.

There exists a one-dimensional variety of solutions. For example, both the vectors

$$\begin{pmatrix} \Delta f_1 \\ \Delta f_2 \\ \Delta f_3 \\ \Delta f_4 \end{pmatrix} = \begin{pmatrix} -c\,\overline{f}_4(1+\overline{f}_1^2+\overline{f}_4^2)^{-1/2} \\ 0 \\ -c\,\overline{f}_1(1+\overline{f}_1^2+\overline{f}_4^2)^{-1/2} + \overline{f}_4^{-1}(1+\overline{f}_1^2+\overline{f}_4^2) \\ -\overline{f}_4^{-1}(1+\overline{f}_1^2+\overline{f}_4^2) \end{pmatrix} \qquad (33.71.a)$$

$$\begin{pmatrix} \Delta f_1 \\ \Delta f_2 \\ \Delta f_3 \\ \Delta f_4 \end{pmatrix} = \begin{pmatrix} -c\,\overline{f}_4(1+\overline{f}_1^2+\overline{f}_4^2)^{-1/2} \\ -c\,\overline{f}_1(1+\overline{f}_1^2+\overline{f}_4^2)^{-1/2} \\ (\overline{f}_4+\overline{f}_1)^{-1}(1+\overline{f}_1^2+\overline{f}_4^2) \\ -(\overline{f}_4+\overline{f}_1)^{-1}(1+\overline{f}_1^2+\overline{f}_4^2) \end{pmatrix} \qquad (33.71.b)$$

meet the equation

$$c(1+\overline{f}_1^2+\overline{f}_4^2)^{-1/2}\begin{pmatrix}\overline{f}_4\\ \overline{f}_1\\ 1\end{pmatrix}=\begin{pmatrix}-1\\ 0\\ \overline{f}_4\end{pmatrix}\Delta f_1+\begin{pmatrix}0\\ -1\\ \overline{f}_1\end{pmatrix}\Delta f_2+\begin{pmatrix}0\\ -1\\ \overline{f}_1\end{pmatrix}\Delta f_3+\begin{pmatrix}0\\ -1\\ \overline{f}_1\end{pmatrix}\Delta f_4+\begin{pmatrix}0\\ 0\\ 1\end{pmatrix}(\Delta f_1\Delta f_4-\Delta f_2\Delta f_3)$$

(33.72)

Obviously, there exist real solutions both for c negative and for c positive. This means the p-values on N_1 do not belong to a boundary set ∂N. Rather, they are inner elements of the domain N of p-values achievable by real feedback gains.

<u>Example 33.7</u>: We continue to deal with the example system discussed above as Example 33.3 and Example 33.5.
Let us investigate the neighbourhood of critical values $p \in N_1$ (see Lemma 33.10 and Theorem 33.4). We obtain

$$p = g(\overline{F}+\Delta F) - g(\overline{F}) = h_{\overline{F}}(\Delta F)$$

$$=\begin{pmatrix}0\\ -1-\overline{f}_4\\ 1\end{pmatrix}\Delta f_1+\begin{pmatrix}-1\\ \overline{f}_2\\ 0\end{pmatrix}\Delta f_2+\begin{pmatrix}-1\\ \overline{f}_2\\ 0\end{pmatrix}\Delta f_3+\begin{pmatrix}0\\ -1\\ 0\end{pmatrix}(\Delta f_1\Delta f_4-\Delta f_2\Delta f_3) \qquad (33.73)$$

Only vectors p that are located in the plane $p_3 = 0$ are of interest. Then, Eq.(33.67) reads as

$$(\Delta p_\perp)'\, g_{*\overline{F}} = (\Delta p_\perp)'\begin{pmatrix}0 & -1 & -1 & 0\\ 0 & \overline{f}_2 & \overline{f}_2 & 0\\ 0 & 0 & 0 & 0\end{pmatrix} = 0, \qquad \Delta p_\perp = 1, \qquad (33.74)$$

which is uniquely solved by

$$\Delta p_\perp = \begin{pmatrix}\overline{f}_2\\ 1\\ 0\end{pmatrix}(1+\overline{f}_2^2)^{-1/2} \qquad (33.75)$$

The test condition

$$c\,\Delta p_\perp = \begin{pmatrix}-1\\ \overline{f}_2\\ 0\end{pmatrix}(\Delta f_2+\Delta f_3)+\begin{pmatrix}0\\ 1\\ 0\end{pmatrix}\Delta f_2\Delta f_3 \qquad (33.76)$$

gives

$$\Delta f_2 + \Delta f_3 = -c(\overline{f}_2^2+1)^{-1/2}\,\overline{f}_2$$

$$\overline{f}_2(\Delta f_2+\Delta f_3) + \Delta f_2\Delta f_3 = c(1+\overline{f}_2^2)^{-1/2}$$

whence

$$(\Delta f_2)_{1,2} = -\frac{c}{2}\,\overline{f}_2(1+\overline{f}_2^2)^{-1/2} \pm\sqrt{\frac{c^2}{2}\,\overline{f}_2^2(1+\overline{f}_2^2)^{-1} - c(1+\overline{f}_2^2)^{1/2}}$$

(33.77)

This means there exist no real solutions of (33.76) within the open interval

$$c \in (0,\ 4\cdot\overline{f}_2^{-2}(1+\overline{f}_2^2)^{3/2}) \tag{33.78}$$

All the inner elements of the two-dimensional area whose boundary is the parabola

$$p_2 = \left(\frac{p_1}{2}\right)^2 \quad , \quad p_3 = 0 \tag{33.79}$$

cannot be achieved by real output feedback. This not attainable area has been shaded in Fig. 33.9.

33.4 Numerical computation of the smooth submanifold $M \in \mathbb{R}^{m\cdot r}$ belonging to a given regular p-value

Suppose $p_o \in \mathbb{R}^n$ to be a regular value of the mapping (33.6).
In Fig. 33.12 an algorithm has been outlined for the computation of one point of the $(m\cdot r - n)$-dimensional smooth manifold

$$M = \left\{ F \in \mathbb{R}^{m\cdot r} : \ p_o = g(F) \in \mathbb{R}^{m\cdot r} \right\} \tag{33.80}$$

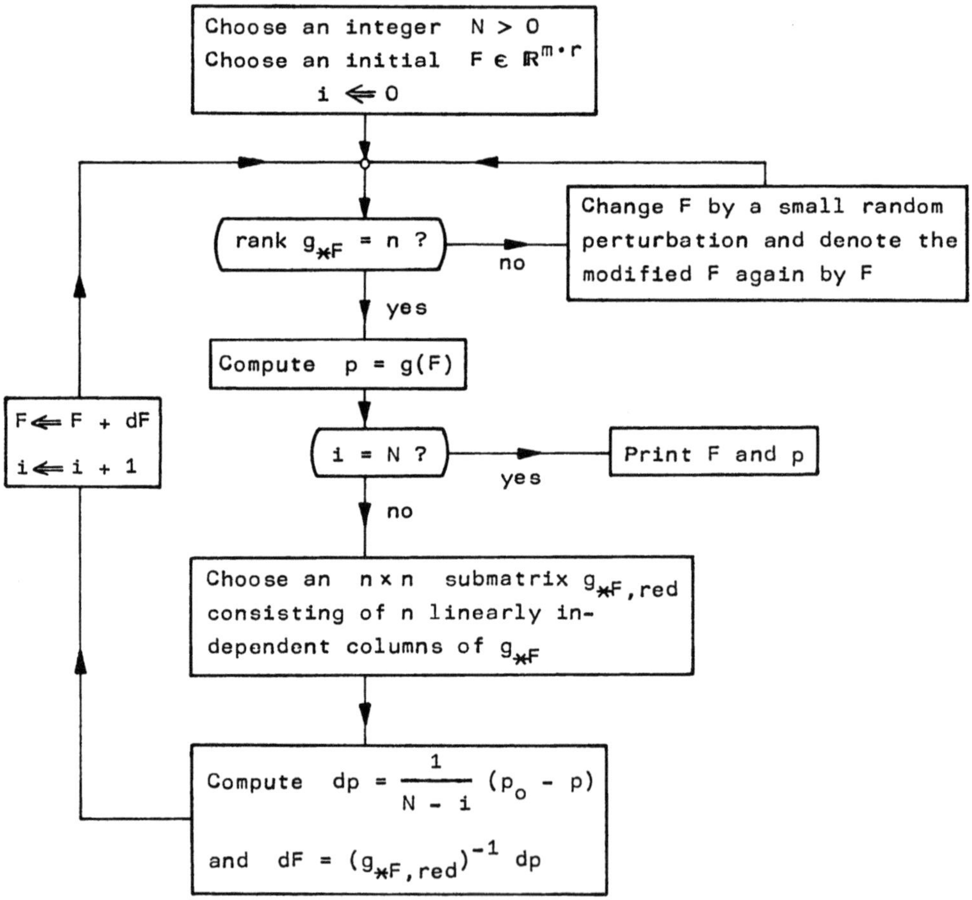

Fig. 33.12

This algorithm yields an $F \in \mathbb{R}^{m\cdot r}$ and a $p \in \mathbb{R}^n$ as output results. If the vector p obtained as output is approximately equal to the given p_o (with sufficient numerical accuracy) then the point $F \in \mathbb{R}^{m\cdot r}$

obtained as output may be regarded as an element of $M = M(p_o)$. Else, refinements of the algorithm, similar to the numerous ramifications of gradient methods well-known in numerical mathematics (see, e.g., Ortega and Rheinboldt 1970), should be applied. Here, the algorithm outlined in the flow-chart is assumed to be sufficient for the numerical computation of an $F \in M(p_o)$ to the given regular value p_o.

In case case of $m \cdot r = n$ the manifold $M(p_o)$ degenerates to the point F and nothing remains to do. Otherwise one can proceed in the following manner: Choose n linearly independent columns of g_{*F}. Let the indices of the remaining columns be $j_1 < j_2 < \dots < j_{m\cdot r-n}$.

Consider the corresponding entries $f_{j_1}, f_{j_2}, \dots, f_{j_{m\cdot r-n}}$ of the computed matrix $F \in \mathbb{R}^{m\cdot r}$.

Fix these F-entries at $f_{j_1}+d_1, f_{j_2}+d_2, \dots, f_{j_{m\cdot r-n}}+d_{m\cdot r-n}$ and repeat the computational procedure sketched in Fig. 33.12. allowing of variation of the other n entries of F only. Generally, another $F \in M(p_o)$ will result. Changing the real differences $d_1, d_2, \dots, d_{m\cdot r-n}$ systematically we shall obtain all points of a grid lying on the smooth $(m\cdot r - n)$-dimensional manifold $M(p_o)$.

Example 33.8: Consider the pole placement in detail for the example system introduced as Example 33.1.

Suppose we would like to place the closed-loop system poles at $z_1 = -2$, $z_2 = -1 + j$, $z_3 = -1 - j$. The corresponding characteristic polynomial is

$$(s + 2)(s + 1 - j)(s + 1 + j) = s^3 + 4s^2 + 6s + 4,$$

and the desired p-vector becomes

$$p_o = (4 \quad 6 \quad 4)' \tag{33.81}$$

It is easily checked that this vector does not belong to N_1, i.e., it is a regular value in the sense of Definition 33.4.

There exists an $m\cdot r-n = 2\cdot 2 - 3 = 1$-dimensional smooth submanifold $M(p_o) \in \mathbb{R}^4$ such that $g(F) = p_o$ for each $F \in M(p_o)$.

Let us try to find $M(p_o)$ according to the procedure outlined in Fig. 33.12.

The attempt to start with $F = 0$ cannot be successful because the corresponding matrix g_{*F} does not have full rank.

The choice

$$F = \begin{pmatrix} 0 & 1 \\ 0 & 1 \end{pmatrix} \tag{33.82}$$

is possible. It corresponds to a regular image,

$$p = g(F) = (1 \quad -2 \quad 0)' \in \mathbb{R}^3 \tag{33.83}$$

and to the Jacobian matrix

$$g_{xF} = \begin{pmatrix} -1 & 0 & 0 & 0 \\ 0 & -1 & -1 & -1 \\ 1 & 0 & -1 & 0 \end{pmatrix} \qquad (33.84)$$

The columns 1, 2, and 3 of the 3×4 matrix (33.84) form a non-singular square matrix that can be used as matrix $g_{*F,red}$.

Following the computational steps of the procedure of Fig.33.12, new regular points $F \in \mathbb{R}^{m \cdot r}$ and their images $p \in \mathbb{R}^n$ may be found with the aid of the inverse matrix $(g_{*F,red})^{-1}$.

For N = 10, one obtains the results summarized in Table 33.1.

i	f_1	f_2	f_3	f_4	p_1	p_2	p_3	z_1	z_2, z_3
0	0	1	0	1	1	-2	0	-2	0 , 1
1	-0.333	0.888	-0.777	1	1.333	-1.111	0.358	-1.984	0.3255±j0.2730
2	-0.666	0.888	-1.665	1	1.666	-0.222	0.812	-1.985	0.1591±j0.6197
3	-1	0.838	-2.554	1	2	0.666	1.268	-1.986	0.0071±j0.7989
4	-1.333	0.888	-3.443	1	2.333	1.555	1.723	-1.987	-0.1731±j0.9150
5	-1.666	0.887	-4.332	1	2.666	2.444	2.178	-1.988	-0.3392±j0.9903
6	-2	0.887	-5.221	1	3	3.333	2.638	-1.990	-0.5050±j1.0336
7	-2.333	0.887	-6.109	1	3.333	4.222	3.089	-1.992	-0.6705±j1.0493
8	-2.666	0.887	-6.999	1	3.666	5.111	3.544	-1.995	-0.8357±j1.0382
9	-3	0.887	-7.887	1	4	6	4	-2	-1 ± j

Table 33.1

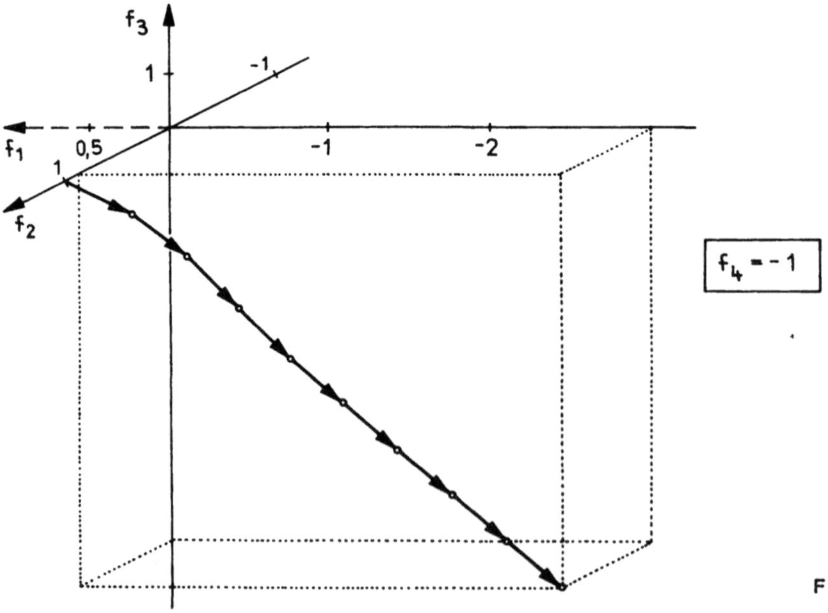

Fig. 33.13.a

Fig. 33.13.a, b, and c illustrate, respectively, the movements of $F \in \mathbb{R}^4$, $p \in \mathbb{R}^3$, and of the corresponding poles in the complex plane.

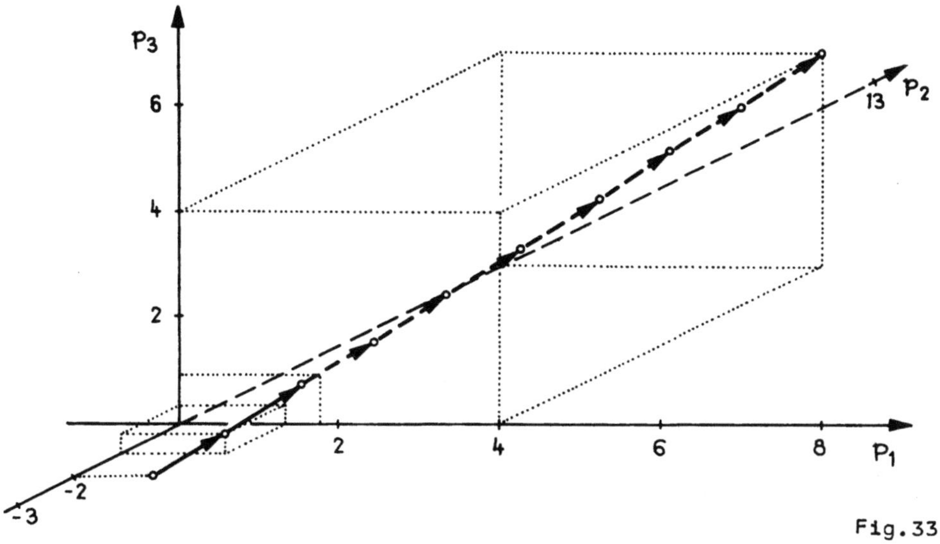

Fig.33.13.b

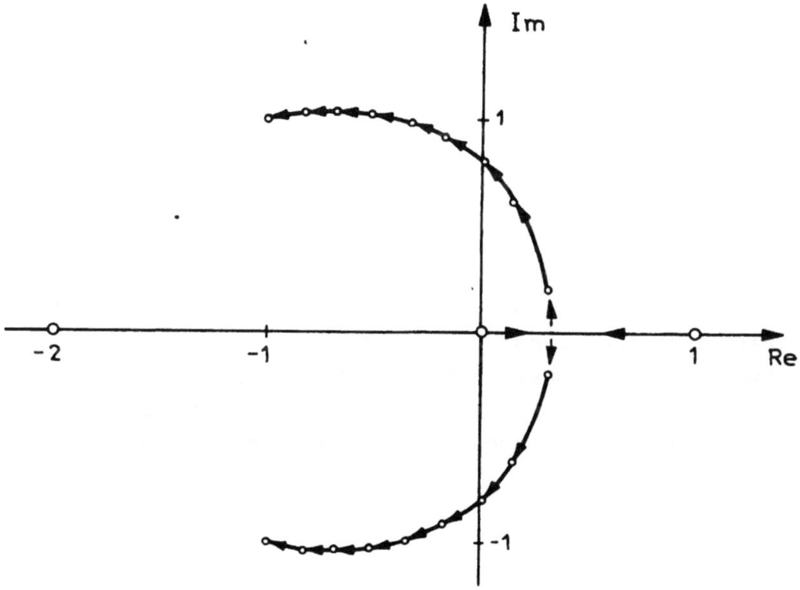

Fig.33.13.c

Finally, we have obtained the feedback matrix

$$F = \begin{pmatrix} -3 & 0.8879 \\ -7.8875 & 1 \end{pmatrix} \tag{33.85}$$

as one element of the one-dimensional smooth manifold $M(p_o)$ for the given p_o, see (33.81).

To obtain a grid of points on $M(p_o)$ the algorithm of Fig. 33.12 was executed for different fixed f_4-values. The results have been written down in Table 33.2.

f_1	f_2	f_3	f_4
-3	1.899	-17.899	10
-3	1.841	-16.841	9
-3	1.770	-15.770	8
-3	1.700	-14.700	7
-3	1.616	-13.616	6
-3	1.518	-12.518	5
-3	1.403	-11.403	4
-3	1.266	-10.266	3
-3	1.099	-9.099	2
-3	0.887	-7.887	1
-3	0.606	-6.606	0
-3	0.193	-5.193	-1
-3	0.082	-4.882	-1.2
-3	-0.044	-4.556	-1.4
-3	-0.190	-4.210	-1.6
-3	-0.365	-3.835	-1.8
-3	-0.586	-3.414	-2
-3	-0.895	-2.905	-2.2
-3	-1.6	-2	-2.4

Table 33.2

Fig.33.14 shows a part of the curve $M(p_o)$.
The picture clearly demonstrates that the front part of this curve becomes almost parallel to the plane $f_4 = -2.4$. This indicates that, for the computation of a prolongation of the curve $M(p_o)$, one should fix the feedback gain f_2 at constant values. Following this proposal, one has to take the columns 1, 3 and 4 of g_{*F} as submatrix $g_{*F,red}$.

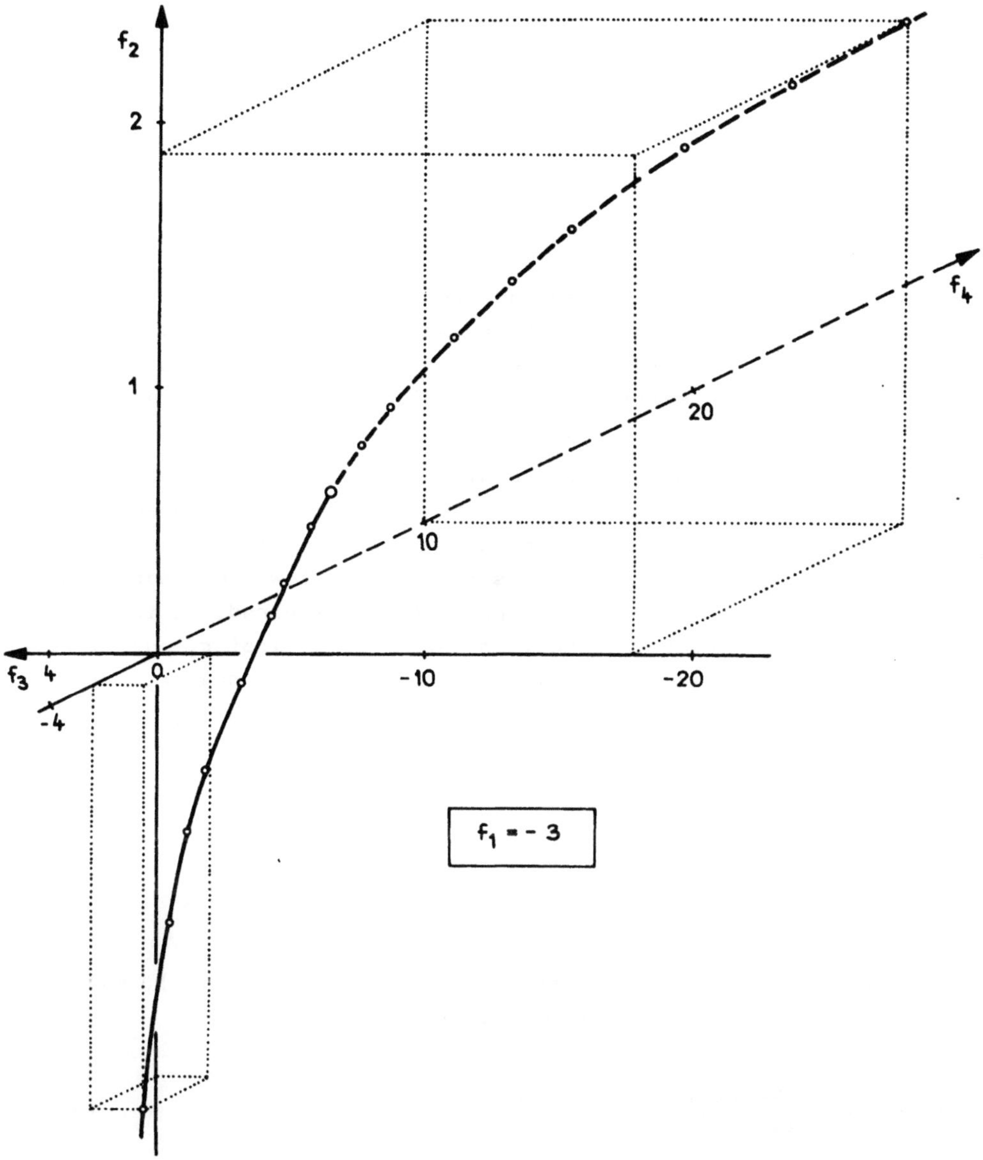

Fig. 33.14

Chapter 4. An outline for further exploitation of the graph-theoretic approach controller synthesis

41 Static output feedback under structural constraints

Let us consider again a control law of the form

$$u = F\,y$$

If the controller may connect any input-output pair of the plant (33.1), (33.2), then F appears as a full $m \times r$ matrix. In many practical applications, however, the control engineer must cope with structural feedback constraints. Such constraints are reflected by fixed zero entries in F. These zero entries characterize the admissible feedback pattern.

So-called "decentralized control" involves special kinds of restrictions on the output-input pairs which the controller may connect.
In terms of the digraph $G(Q_4)$ introduced in Section 11, see Eq.(11.16), the structural constraints may be described as follows. By reordering the output variables and the input variables if required, it is assumed that there exists a feedback edge from each of the first r_1 output vertices to each of the first m_1 input vertices, from the next r_2 output vertices to the next m_2 input vertices,..., from the last r_k output vertices to the last m_k input vertices.
With input and output vectors u and y partitioned as

$$u = \begin{pmatrix} u^1 \\ \vdots \\ u^k \end{pmatrix}, \quad y = \begin{pmatrix} y^1 \\ \vdots \\ y^k \end{pmatrix}$$

the dimensions of u^i and y^i are m_i and r_i $(i = 1,\dots,k)$, and the Eqs. (33.1), (33.2), (33.3) may be written as

$$\dot{x} = A\,x + \sum_{i=1}^{k} B^i u^i \tag{41.1}$$

$$y^i = C^i x \tag{41.2}$$

$$(i = 1,\dots,k)$$

$$u^i = F^i y^i \tag{41.3}$$

That means the feedback matrix is a quasidiagonal matrix in the case of decentralized control,

$$F = \begin{pmatrix} F^1 & 0 & \dots & 0 \\ 0 & F^2 & \dots & 0 \\ \vdots & \vdots & \vdots\vdots & \vdots \\ 0 & 0 & \dots & F^k \end{pmatrix} \tag{41.4}$$

In this Section 41, we do not intend to go into the details of decentralized control or other special kinds of structurally constrained feedback laws. Rather, we shall consider important consequences of an arbitrarily specified structurally constrained feedback pattern characterized by the a priori known zero entries of F.

Owing to the fact that we have managed all entries of F symbolically in this monograph, the consequences of any structural feedback constraints may be seen easily. Provided that a certain feedback gain f_{ij} vanishes identically then all the expressions containing this feedback gain as factor must be omitted. Thus we need not investigate different feedback patterns separately.

41.1 Controllability and observability of systems under structurally constrained output feedback

Usually, controllability and observability are regarded as properties of a given plant described mathematically by the matrices (A, B, C). Similarly, structural controllability and structural observability have been introduced as generic properties of a class of systems defined by the structure matrices [A, B, C], comp. Section 14.
At first glance, the feedback matrix F seems to play no role in this context. However, there are criteria of structural completeness that involve the feedback matrix.
In Theorem 14.5 it has been shown that a class of systems [A, B, C] is both s-controllable and s-observable if and only if the digraph $G([Q_3])$ meets two conditions:
(a) All the state vertices are both input and output connectable.
(b) There is a cycle family of width n in $G([Q_3])$.

In the case of structural feedback constraints it makes no sense to consider the digraph $G([Q_3])$. We have only the digraph $G([Q_4])$ at our disposal. The question is does the digraph $G([Q_4])$ meet both the conditions (a) and (b) or not?

Evidently, the necessary condition (a) cannot be influenced by the feedback pattern, merely condition (b) can be influenced. This condition ensures that the coefficient p_n of the closed-loop characteristic polynomial does not vanish generically. Deletion of feedback edges due to fixed zero entries of F may be considered to be "undangerous" as long as condition (b) remains true. If no cycle family of width n remains in $G([Q_4])$ then the consequences of structural completeness cannot be exploited any longer.

Example 41.1: Consider a class of systems defined by the structure matrices

$$[A] = \begin{pmatrix} 0 & 0 & 0 & 0 \\ L & 0 & 0 & 0 \\ 0 & L & L & L \\ 0 & L & 0 & 0 \end{pmatrix}, \quad [B] = \begin{pmatrix} L & 0 \\ 0 & L \\ 0 & 0 \\ L & 0 \end{pmatrix} \quad [C] = \begin{pmatrix} 0 & L & 0 & 0 \\ 0 & 0 & L & 0 \end{pmatrix} \tag{41.5}$$

In Fig. 41.1, the digraph $G([Q_4])$ for a full 2×2 feedback matrix

$$[F] = \begin{pmatrix} L & L \\ L & L \end{pmatrix} \tag{41.6}$$

has been drawn.

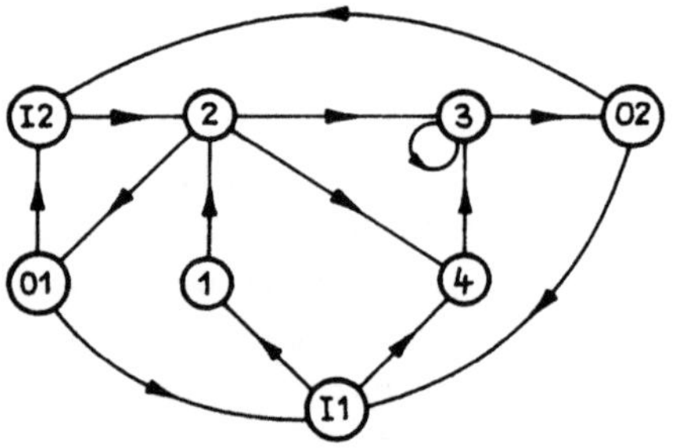

Fig. 41.1

The cycle families associated with the closed-loop characteristic-polynomial coefficients p_1, p_2, p_3, p_4 have been composed in Fig. 41.2.

Now, let us assume the control law to be subjected to structural constraints. Suppose a decentralized feedback pattern defined by the structure matrix

$$[F] = \begin{pmatrix} L & 0 \\ 0 & L \end{pmatrix} \tag{41.7}$$

That is, the feedback edges associated to the entries f_{12} and f_{21} must be deleted in Fig. 41.1. Thus one obtains the digraph $G([Q_4])$ drawn in Fig. 41.3.

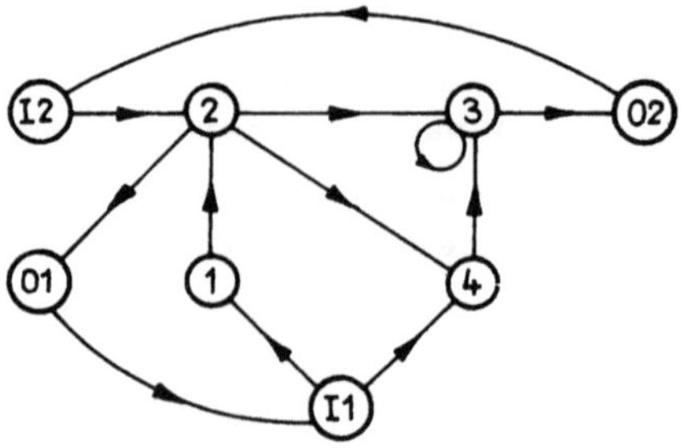

Fig. 41.3

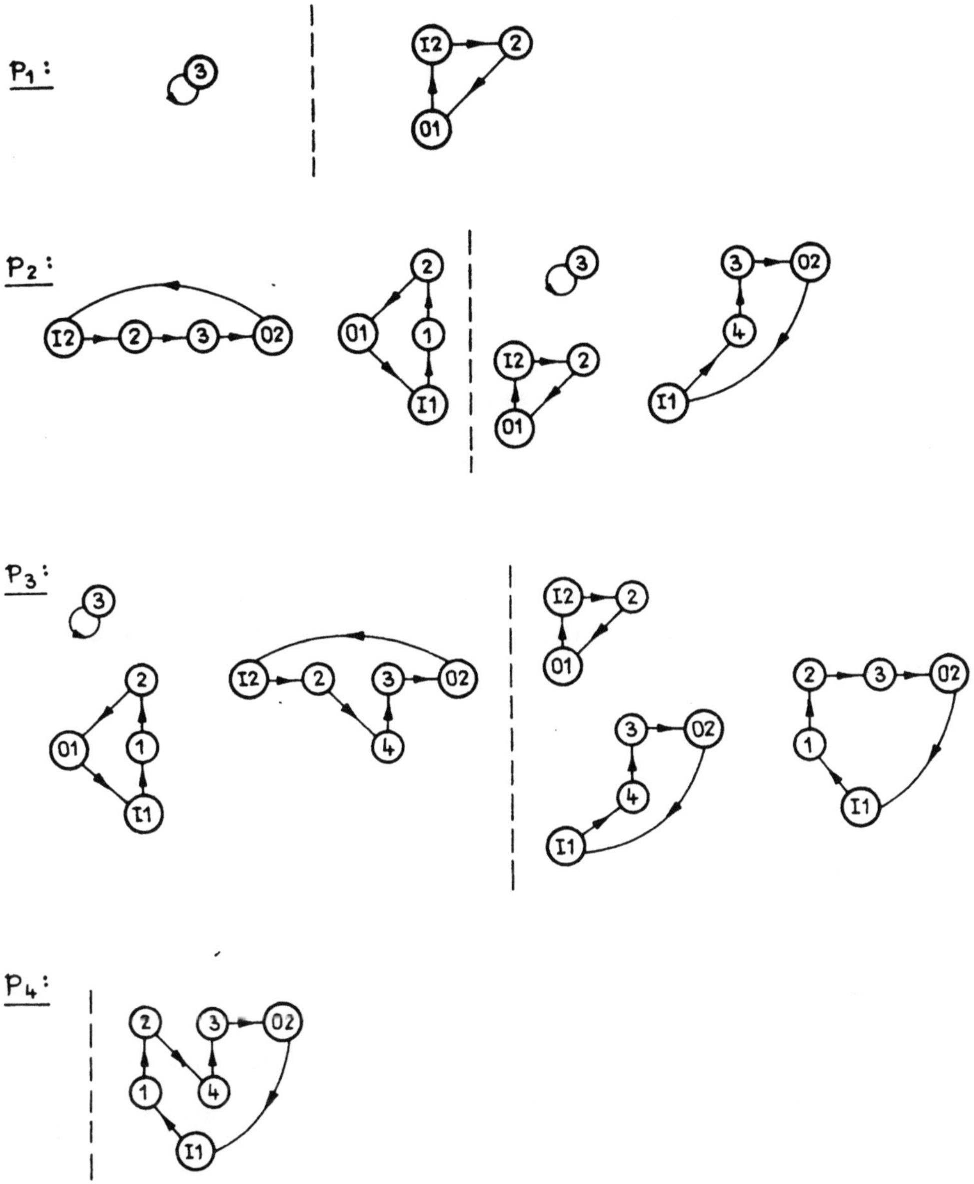

Fig. 41.2

All the cycle families exhibited on the right-hand side seen from the dashed lines in Fig. 41.2 do not exist in the digraph $G([Q_4])$ of Fig. 41.3. Especially, there remains no cycle family of width $n = 4$. Hence, the characteristic-polynomial coefficient p_4 vanishes identically for each admissible realization of the closed-loop system (41.5), (41.7).

41.2 Fixed modes and structurally fixed modes

Investigating the stabilization of decentralized control systems, the notion of fixed modes was introduced by Wang and Davison in 1973. Later on, it has been recognized that this concept can be applied for any kind of static output feedback.

Definition 41.1

If, for a complex value s,

$$\det(sI_n - A - B F C) = 0 \quad \text{for each admissible } F \tag{41.8}$$

then s is called a fixed mode of the closed-loop system.

In other words, a fixed "mode" is a closed-loop system pole invariant with respect to the choice of F. Therefore, the present author would like to prefer the notation "fixed pole" instead of "fixed mode". Nevertheless, to avoid confusion, the traditional notations will be used in the sequel.

The zeros of $\det(sI_n - A - BFC)$ have been investigated in Section 32. The results obtained there may be immediately used for the case of structurally constrained feedback laws.

Lemma 41.1

Let F be a full feedback matrix.
Then there holds

$$\{\text{fixed modes}\} = \{\text{decoupling zeros}\} \tag{41.9}$$

Proof: The relationship (41.9) is a direct consequence of Definition 41.1, Eq. (32.38), Definition 32.4 and Lemma 32.2.

▲

In the case of structurally constrained output feedback, additional fixed modes are possible, for the following reason:

Consider the representation (32.38) of the closed-loop characteristic polynomial. Due to structural feedback constraints it may happen that some minors

$$F^{i_1 \ldots i_q}_{j_1 \ldots j_q} = 0 \quad \text{for all admissible } F. \tag{41.10}$$

Then the product

$$F^{i_1 \dots i_q}_{j_1 \dots j_q} \; P(s)^{1 \dots n, n+j_1 \dots n+j_q}_{1 \dots n, n+i_1 \dots n+i_q}$$

vanishes, too. Thus the closed-loop characteristic polynomial may vanish for all admissible F despite of

$$P(s)^{1 \dots n, n+j_1 \dots n+j_q}_{1 \dots n, n+i_1 \dots n+i_q} \neq 0 \quad \text{for some} \quad 1 \leq j_1 < \dots < j_q \leq r, \quad \text{and} \quad 1 \leq i_1 < \dots < i_q \leq m.$$

In this way, we have substantiated an algebraic characterization of fixed modes.

<u>Theorem 41.1</u>

A fixed mode at a complex value s may be characterized as follows:

(a) $\det(sI_n - A) = 0$ (41.11)

and

(b) for all integers q with $1 \leq q \leq \min(m, r)$ and

$$F^{i_1 \dots i_q}_{j_1 \dots j_q} \neq 0 \quad \text{for some admissible } F \tag{41.12}$$

there holds

$$P(s)^{1 \dots n, n+j_1 \dots n+j_q}_{1 \dots n, n+i_1 \dots n+i_q} = 0. \tag{41.13}$$

Similarly to Lemma 32.2, the following statement can be proved:

<u>Theorem 41.2</u>

For any fixed value s, assume

$$\text{rank}\,(sI_n - A) = n-d \quad \text{where} \quad 1 \leq d \leq n \tag{41.14}$$

Then s is a fixed mode if and only if, for each set of d C-rows $c'_{j_1}, \dots c'_{j_d}$ and each set of d B-columns $b_{i_1}, \dots, b_{i_d}$, there holds

$$\text{rank}(sI_n - A,\, b_{i_1}, \dots, b_{i_d}) < n \quad \text{or} \quad \text{rank}\begin{pmatrix} sI_n - A \\ c'_{j_1} \\ \vdots \\ c'_{j_d} \end{pmatrix} < n \tag{41.15; 41.16}$$

or

$$F^{i_1 \dots i_d}_{j_1 \dots j_d} = 0 \quad \text{for each admissible } F.$$

Remark: For the special case of decentralized control defined by (41.4), Anderson and Clements published another algebraic characterization of fixed modes in 1981.

Definition 41.2

A complex value z is said to be a fixed mode of multiplicity h if, for each admissible F, in the closed-loop characteristic-polynomial representation

$$\det(sI_n - A - BFC) = \det((s-z)I_n - (A - zI_n) - BFC)$$
$$= (s-z)^n + (s-z)^{n-1}p_1(z) + \dots + (s-z)p_{n-1}(z) + p_n(z) \tag{41.17}$$

the last h coefficients vanish, i.e.

$$p_k(z) = 0 \quad \text{for} \quad k = n-h+1, \dots, n-1, n, \tag{41.18}$$

but $p_{n-h}(z) \neq 0$.

Fixed modes (including their multiplicities) may be interpreted graph-theoretically, see Reinschke 1984. For this purpose, consider the digraph associated with the matrix

$$Q_4(z) = \begin{pmatrix} 0 & C & 0 \\ 0 & A-zI_n & B \\ F & 0 & 0 \end{pmatrix} \tag{41.19}$$

Theorem 41.3

A closed-loop system under output feedback, possibly structurally constrained, has a fixed mode of multiplicity h at $s = z$ if and only if, for every $k = n, n-1, \dots, n-h+1$ (but not for $k = n-h$), one of the following two conditions is fulfilled:

(a) There are no cycle families of width k in $G(Q_4(z))$.

(b) There exist two or more cycle families of width k in $G(Q_4(z))$ which cancel each other numerically for all admissible values of the non-vanishing entries of F.

Proof: Theorem 41.3 is a consequence of Definition 41.2 and Theorem 31.4.

▲

In recent years, the so-called structurally fixed modes have had much attention. This notation was introduced by Sezer and Siljak in 1981. According to Definition 41.1, fixed modes reflect an invariance property of a closed-loop system whose matrices A, B, C are numerically fixed while F runs through all admissible realizations with the same structure matrix [F]. It may occur that a slight perturbation of some entries of A, B, C eliminates a fixed mode. Seen from a "structural point of view", this type of fixed modes is of little interest. The more important case is that all admissible systems $(A,B,C) \in [A,B,C]$ have fixed modes.

<u>Definition 41.3</u>

A class of closed-loop systems described by the structure matrices [A], [B], [C], [F] has <u>structurally fixed modes</u> if every actual system $(A,B,C,F) \in [A,B,C,F]$ has fixed modes.

At first sight, it seems to be appropriate to distinguish between two kinds of structurally fixed modes: on the one hand, modes fixed at the same complex value for each system $(A,B,C) \in [A,B,C]$, on the other hand, modes whose values may depend on the actual numerical realization of (A,B,C). In practice, however, s-fixed modes are fixed at zero:

<u>Lemma 41.2</u>

For a class of systems defined by [A,B,C,F], any non-vanishing complex value s_0 can never be a fixed mode for each admissible plant $(A,B,C) \in [A,B,C]$.

<u>Proof</u>: Assume (41.8) to be valid at $s = s_0 \neq 0$ for a given realization $(A,B,C) \in [A,B,C]$.
Then the value s_0 is necessarily an eigenvalue of A.
Consider the set of matrices $\bar{A} = c \cdot A$ where the scalar c runs through the real axis. Obviously, each $\bar{A} \in [A]$.
The scalar c may be chosen in such a way that s_0 is no eigenvalue of $\bar{A} = c \cdot A$. Hence, the value $s_0 \neq 0$ cannot be a fixed mode for all $(\bar{A}, B, C)$.
This completes the proof.

▲

A more general result is incorporated in the subsequent graphical test for structurally fixed modes.

Theorem 41.4

A class of systems defined by the structure matrix

$$[Q_4] = \begin{pmatrix} 0 & [C] & 0 \\ 0 & [A] & [B] \\ [F] & 0 & 0 \end{pmatrix}$$

has structurally fixed modes if and only if the digraph $G([Q_4])$ meets at least one of the following conditions:

(a) For at least one state vertex ν, there is no feedback edge $(O\varsigma, I\mu)$ in $G([Q_4])$ such that there exist both a path from $I\mu$ to ν and a path from ν to $O\varsigma$.

(b) There is no cycle family of width n in $G([Q_4])$.

Theorem 41.4 can be proved in analogy to Theorem 32.5. The detailed proof is left to the reader.

Comparing Theorem 41.4 and Theorem 32.5 one reaches some conclusions.

Corollary 41.1

There holds

$$\{\text{structural decoupling zeros}\} \subset \{\text{structurally fixed modes}\} \qquad (41.20)$$

The difference between both sets is caused by structural feedback constraints.

Corollary 41.2

Only structurally fixed modes that result from condition (a) of Theorem 41.4 may depend on the actual numerical realization of the matrix A. The other s-fixed modes are located at $s = 0$.

As regards the multiplicity of structurally fixed modes at $s = 0$, it should be realized that the indeterminate entries of structure matrices can also vanish for some numerical realizations. To avoid the discussion of atypical cases we formulate

Definition 41.4

A structurally fixed mode at $s = 0$ is said to have the generic multiplicity h^g if in the closed-loop characteristic polynomial

$$\det(sI_n - A - BFC) = s^n + p_1 s^{n-1} + \ldots + p_{n-1}s + p_n$$

the coefficients $p_n, \ldots, p_{n-h^g+1}$ vanish for all $(A,B,C,F) \in [A,B,C,F]$

while $p_{n-h^g} \neq 0$ for some $(A,B,C,F) \in [A,B,C,F]$.

In order to decide on the generic multiplicity of structurally fixed modes at $s = 0$ we consider the digraph $G([Q_4])$.

Theorem 41.5

The class of closed-loop systems $[A,B,C,F]$ has a structurally fixed mode of the genericity h^g at $s = 0$ if and only if, for every $k = n, n-1,\ldots, n-h^g+1$ (but not for $k = n-h^g$), there are no cycle families of width k in $G([Q_4])$.

The proof is left to the reader.

Example 41.2: Consider the example system introduced as Example 41.1. For the decentralized feedback pattern

$$[F] = \begin{pmatrix} L & 0 \\ 0 & L \end{pmatrix}$$

the structure graph $G([Q_4])$ has been drawn in Fig. 41.3, This digraph contains cycle families of widths 1, 2, and 3, but no cycle family of width $4 = n$, comp. Fig. 41.2. That is, there is a structurally fixed mode of the generic multiplicity $h^g = 1$ at $s = 0$.
Further structurally fixed modes do not exist, because all state-vertices in $G([Q_4])$ are both input-connectable and output-connectable.

It can be seen quite easily that in the case of

$$[F] = \begin{pmatrix} L & L \\ 0 & L \end{pmatrix}$$

the closed-loop system would also have a structural fixed mode at $s = 0$ while in the case of

$$[F] = \begin{pmatrix} 0 & 0 \\ L & 0 \end{pmatrix}$$

the closed-loop system would not have any structural fixed mode.

In Chapter 3, we have studied the influence of static output feedback on the behaviour of a given plant. A dynamic feedback system offers more possibilities.

In Fig. 42.1, the generalized structure graph of a composite system controlled by dynamic feedback has been shown.

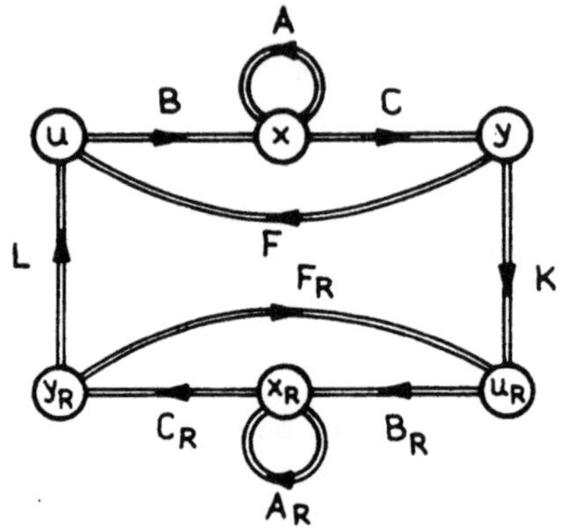

Fig. 42.1

The composite system may be described mathematically by the subsequent equations.

Plant:

$$\dot{x} = A\,x + B\,u$$

$$y = C\,x \qquad \text{where} \quad x \in \mathbb{R}^n, \quad u \in \mathbb{R}^m, \quad y \in \mathbb{R}^r$$

under static output feedback $\quad u = F\,y$

Auxiliary dynamic system (often called "dynamic compensator"):

$$\dot{x}_R = A_R x_R + B_R u_R$$

$$y_R = C_R x_R \qquad \text{where} \quad x_R \in \mathbb{R}^{n_R}, \quad u_R \in \mathbb{R}^{m_R}, \quad y_R \in \mathbb{R}^{r_R}$$

under static output feedback $\quad u_R = F_R\, y_R.$

Couplings:

$$u = L\, y_R, \qquad u_R = K\, y.$$

The composite system may be interpreted as an augmented system of dynamic order $n+n_R$ governed by the equations

$$\dot{x}_{aug} = \begin{pmatrix} \dot{x} \\ \dot{x}_R \end{pmatrix} = \begin{pmatrix} A & 0 \\ 0 & A_R \end{pmatrix} \begin{pmatrix} x \\ x_R \end{pmatrix} + \begin{pmatrix} B & 0 \\ 0 & B_R \end{pmatrix} \begin{pmatrix} u \\ u_R \end{pmatrix} = A_{aug}x_{aug} + B_{aug}u_{aug} \qquad (42.1)$$

$$y_{aug} = \begin{pmatrix} y \\ y_R \end{pmatrix} = \begin{pmatrix} C & 0 \\ 0 & C_R \end{pmatrix} \begin{pmatrix} x \\ x_R \end{pmatrix} = C_{aug}x_{aug} \qquad (42.2)$$

$$u_{aug} = \begin{pmatrix} u \\ u_R \end{pmatrix} = \begin{pmatrix} F & L \\ K & F_R \end{pmatrix} \begin{pmatrix} y \\ y_R \end{pmatrix} = F_{aug}y_{aug} \qquad (42.3)$$

Thus we have traced back the control of a given plant by means of a dynamic compensator to static output feedback of an appropriately augmented system.

All the results derived in Chapter 3 can be applied. For example, instead of the digraph $G(Q_4)$ associated with the $(r+n+m) \times (r+n+m)$ matrix Q_4, see (11.16), we have to consider a digraph $G(Q_{4,aug})$ associated with the $(r+r_R+n+n_R+m+m_R) \times (r+r_R+n+n_R+m+m_R)$ matrix

$$Q_{4,aug} = \begin{array}{c} \\ y \\ y_R \\ x \\ x_R \\ u \\ u_R \end{array} \begin{array}{c} \begin{array}{cccccc} y & y_R & x & x_R & u & u_R \end{array} \\ \left(\begin{array}{cc|cc|cc} \multicolumn{2}{c|}{\multirow{2}{*}{0}} & C & 0 & \multicolumn{2}{c}{\multirow{2}{*}{0}} \\ & & 0 & C_R & & \\ \hline \multicolumn{2}{c|}{\multirow{2}{*}{0}} & A & 0 & B & 0 \\ & & 0 & A_R & 0 & B_R \\ \hline F & L & \multicolumn{2}{c|}{\multirow{2}{*}{0}} & \multicolumn{2}{c}{\multirow{2}{*}{0}} \\ K & F_R & & & & \end{array}\right) \end{array} \qquad (42.4)$$

The submatrix

$$F_{aug} = \begin{pmatrix} F & L \\ K & F_R \end{pmatrix} \qquad (42.5)$$

has $(m+m_R)\cdot(r+r_R)$ entries which can be used to influence the behaviour of the augmented system. Structural feedback constraints are reflected by fixed zero entries of F_{aug}.

Investigating the digraph $G(Q_4,aug)$ the graph-theoretic approach to controller synthesis may easily be extended to the case of dynamic controllers. For lack of space, we are unable to go into the details here. Only one peculiarity should be mentioned: All the cycles in $G(Q_4,aug)$ which contain a K-edge must also contain an L-edge, see Fig. 42.1. Consequently, for most example systems, the number of degrees of freedom that are really at the investigator's disposal for pole assignment or related tasks will be smaller than the number of non-zero entries of F_{aug}.

Two interesting papers concerning the graph-theoretic synthesis of structurally constrained dynamic controllers have been published recently by Ulm and Wend in 1985 and in 1986.

Let us finish this Section 42 with a small example (see Reinschke 1983) illustrating pole placement by dynamic output feedback.

Example 42.2: Consider a single-input single-output plant characterized by

$$[A] = \begin{pmatrix} 0 & 0 & 0 \\ L & 0 & 0 \\ 0 & L & 0 \end{pmatrix}, \quad b = \begin{pmatrix} 1 \\ 0 \\ 0 \end{pmatrix}, \quad c' = (0 \;\; 0 \;\; 1) ,$$

comp. Fig. 42.2a.

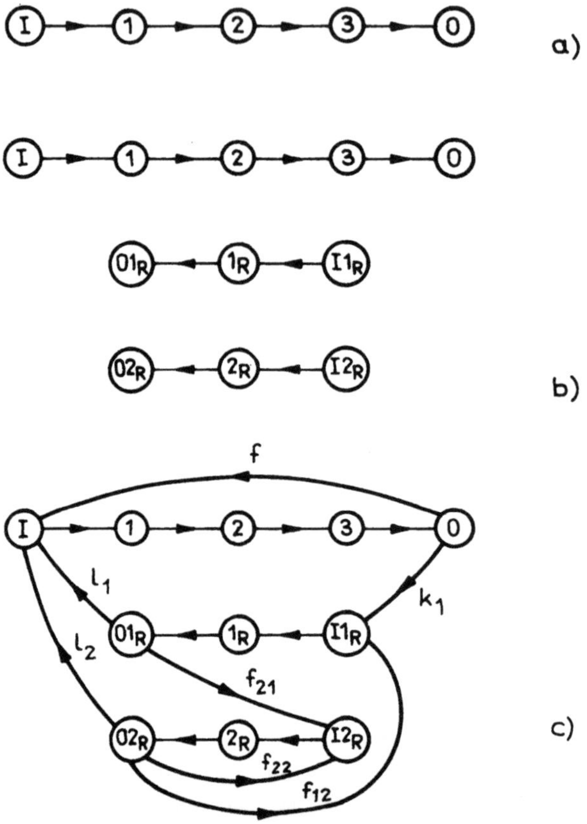

Fig. 42.2

Arbitrary pole placement by static output feedback is impossible, because the coefficients p_1 and p_2 in the closed-loop characteristic polynomial

$$\det(sI_3 - A - bfc') = s^3 + p_1 s^2 + p_2 s + p_3 = s^3 - a_{32}a_{21}f$$

vanish identically.

Now, we connect integrators (as the simplest dynamic feedback systems) to the plant in order to obtain arbitrary pole assignability in the augmented system.

An integrator may be described as

$$\dot{x}_R = u_R \in \mathbb{R}^1, \qquad y_R = x_R.$$

Using one integrator the matrix $Q_{4,aug}$ becomes

$$Q_{4,aug} = \left(\begin{array}{cc|cccc|cc}
0 & 0 & 0 & 0 & 1 & 0 & 0 & 0 \\
0 & 0 & 0 & 0 & 0 & 1 & 0 & 0 \\
\hline
0 & 0 & 0 & 0 & 0 & 0 & 1 & 0 \\
0 & 0 & a_{21} & 0 & 0 & 0 & 0 & 0 \\
0 & 0 & 0 & a_{32} & 0 & 0 & 0 & 0 \\
0 & 0 & 0 & 0 & 0 & 0 & 0 & 1 \\
\hline
f & 1 & 0 & 0 & 0 & 0 & 0 & 0 \\
k & f_R & 0 & 0 & 0 & 0 & 0 & 0
\end{array}\right)$$

The closed-loop augmented system has the dynamical order $n_{aug} = 4$. The feedback matrix F_{aug} has four non-vanishing elements. Nevertheless, pole assignability is impossible, because the coefficient p_2 of the characteristic polynomial of the augmented system vanishes identically. (There is no cycle family of width 2 in $G([Q_4, aug])$.)

Using two integrators, there results a matrix

$$Q_{4,aug} = \left(\begin{array}{ccc|ccccc|ccc}
0 & 0 & 0 & 0 & 0 & 1 & 0 & 0 & 0 & 0 & 0 \\
0 & 0 & 0 & 0 & 0 & 0 & 1 & 0 & 0 & 0 & 0 \\
0 & 0 & 0 & 0 & 0 & 0 & 0 & 1 & 0 & 0 & 0 \\
\hline
0 & 0 & 0 & 0 & 0 & 0 & 0 & 0 & 1 & 0 & 0 \\
0 & 0 & 0 & a_{21} & 0 & 0 & 0 & 0 & 0 & 0 & 0 \\
0 & 0 & 0 & 0 & a_{32} & 0 & 0 & 0 & 0 & 0 & 0 \\
0 & 0 & 0 & 0 & 0 & 0 & 0 & 0 & 0 & 1 & 0 \\
0 & 0 & 0 & 0 & 0 & 0 & 0 & 0 & 0 & 0 & 1 \\
\hline
f & l_1 & l_2 & 0 & 0 & 0 & 0 & 0 & 0 & 0 & 0 \\
k_1 & f_{11} & f_{12} & 0 & 0 & 0 & 0 & 0 & 0 & 0 & 0 \\
k_2 & f_{21} & f_{22} & 0 & 0 & 0 & 0 & 0 & 0 & 0 & 0
\end{array}\right)$$

Now, arbitrary pole assignability of the five poles of the augmented plant (shown in Fig. 42.2b) by "static" output feedback can be achieved. One possibility has been sketched in Fig. 42.2c.
It corresponds to an augmented feedback matrix

$$F_{aug} = \begin{pmatrix} f & l_1 & l_2 \\ k_1 & f_{11} & f_{12} \\ k_2 & f_{21} & f_{22} \end{pmatrix} = \begin{pmatrix} f & l_1 & l_2 \\ k_1 & 0 & f_{12} \\ 0 & f_{21} & f_{22} \end{pmatrix}$$

Fixing k_1 and f_{21} at particular real values ($\neq 0$), say at d_1 and d_2, any given closed-loop characteristic-polynomial coefficients p_1, p_2, p_3, p_4, p_5 determine uniquely the remaining five feedback gains.

Considering the cycle families in the digraph of Fig. 42.2c the following equations may be written down immediately:

$$p_1 = -f_{22},$$
$$p_2 = -f_{12}d_2,$$
$$p_3 = -a_{32}a_{21}f,$$
$$p_4 = -a_{32}a_{21}d_1l_1 + f_{22}a_{32}a_{21}f,$$
$$p_5 = -a_{32}a_{21}d_1d_2l_2 + a_{32}a_{21}d_1l_1f_{22} + a_{32}a_{21}f\cdot d_2f_{12}.$$

The first equation provides f_{22},
the second equation provides f_{12},
the third equation provides f,
the fourth equation provides l_1,
the fifth equation provides l_2.

43 Semi-state system description

In many applications, the system equations do not arise in state-space form, even if we restrict ourselves to linear systems.

Let us consider, for example, linear electrical networks.
They are characterized by the current-voltage relations of each network element and Kirchhoff's rules which restrict the allowable current and voltage distributions taking into account the network topology.
To say it more definitely, we take RLC-networks of arbitrary topology.
There are three types of elements: resistors, capacitors, inductors.
Their current-voltage relations may be written as follows:

$$\text{Resistors:} \qquad u_R = R\, i_R \tag{43.1}$$

$$\text{Capacitors:} \qquad C\, \dot{u}_C = i_C$$

$$\text{Inductors:} \qquad L\, \dot{i}_L = u_L$$

In the sequel, the abbreviation

$$u = \begin{pmatrix} u_R \\ u_C \\ u_L \end{pmatrix}, \qquad i = \begin{pmatrix} i_R \\ i_C \\ i_L \end{pmatrix}$$

will be used.
The network topology may be described by the node-branch incidence matrix K or the mesh-branch incidence matrix M of the network graph.
Then Kirchhoff's rules may be formulated in such a way:

$$\text{"Node law"} \qquad K(i + i^e) = 0 \tag{43.2}$$

$$\text{"Mesh law"} \qquad M(u + u^e) = 0 \tag{43.3}$$

where i^e and u^e symbolize the independent current sources and the independent voltage sources, respectively, that excite the network.

After Laplace transform

$$\mathcal{L}\{i(t)\} = i_*(s)$$

$$\mathcal{L}\{u(t)\} = u_*(s)$$

the Eqs. (43.1 - 3) may be rewritten as

$$\left(\begin{array}{ccc|ccc} R & & & -I_R & & \\ & sL & & & -I_L & \\ & & -I_C & & & sC \\ \hline & K & & & 0 & \\ \hline & 0 & & & M & \end{array}\right) \left(\begin{array}{c} i_* \\ \hline u_* \end{array}\right) = \left(\begin{array}{c} 0 \\ i_L(0) \\ u_C(0) \\ \hline -K\, i_*^e \\ \hline -M\, u_*^e \end{array}\right) \tag{43.4}$$

The system of equations (43.4) may be regarded as a very natural description of RLC-networks. All the physical and topological parameters are reflected in an extremely simple manner. At first glance, the high order of the coefficients matrix on the left-hand side of (43.4) seems to be a disadvantage. Fortunately, this is not the case provided that we have at our disposal appropriate tools for dealing with large sparse matrices.

Of course, Eq. (43.4) can be transformed into a state-space description of the underlying RLC-network. The transformation process, however, is rather cumbersome. In general, the coefficients of the arising state-space system matrix A depend on the physical and topological network parameters in a complicated manner. For details see, e.g., Reinschke and Schwarz 1976.

In the case of other natural or technological systems, a similar reasoning would lead to related conclusions:

A system description of the form

$$\begin{aligned} (sJ - A)\, X(s) &= Jx(0) + BU(s) \\ Y(s) &= CX(s) \end{aligned} \tag{43.5}$$

where J is allowed to be a singular matrix

is more natural and appropriate than the state-space description

$$\begin{aligned} (sI_n - A)\, X(s) &= x(0) + BU(s) \\ Y(s) &= CX(s) \end{aligned} \tag{43.6}$$

Treating large-scale composite systems, the investigator is also confronted with a system description of the form (43.5).

Suppose the composite system to consist of N parts. Then the following relationships must be taken into account:

Description of part i:

$$\begin{aligned} (sJ^i - A^i)X^i(s) &= x^i(0) + B^iU^i(s) \\ Y^i(s) &= C^iX^i(s) \end{aligned} \quad \text{for } i = 1, 2, \dots, N \tag{43.7}$$

Couplings between the parts:

$$\begin{pmatrix} U^1 \\ \vdots \\ U^N \end{pmatrix} = - H \begin{pmatrix} Y^1 \\ \vdots \\ Y^N \end{pmatrix} + K\, U \tag{43.8}$$

where the vector U(s) reflects the system inputs.

System output:

$$Y = L \begin{pmatrix} Y^1 \\ \vdots \\ Y^N \end{pmatrix} \tag{43.9}$$

The Eqs. (43.7 - 9) may be compressed into one compound matrix equation

$$\left(\begin{array}{ccc|ccc|c} sJ^1 - A^1 & & & -B^1 & & & \\ & \ddots & & & \ddots & & 0 \\ & & sJ^N - A^N & & & -B^N & \\ \hline C^1 & & & & & & \\ & \ddots & & & 0 & & -I \\ & & C^N & & & & \\ \hline & 0 & & & I & & H \end{array}\right) \left(\begin{array}{c} X^1 \\ \vdots \\ X^N \\ \hline U^1 \\ \vdots \\ U^N \\ \hline Y^1 \\ \vdots \\ Y^N \end{array}\right) = \left(\begin{array}{c} J^1 x^1(0) \\ \vdots \\ J^N x^N(0) \\ \hline 0 \\ \hline K\,U \end{array}\right) \tag{43.10}$$

Indeed, we have obtained a system description of the form (43.5). Even if all the system parts have a state-space description, i.e., if J^i = unit matrix for all $i = 1,2,\ldots,N$, the characterization (43.10) of the composite system can only classified as a "semi-state description".

In the literature, different notations have been used for a system description like (43.5) and its counterpart in time-domain:

generalized state-space systems,
semi-state description,
descriptor systems,
differential/algebraic systems,
implicit systems with static constraints,
singular systems of differential equations,

see Luenberger 1977, Van Dooren, Verghese and Kailath 1979, Newcomb 1981, Petzold and Gear 1982, Van der Weiden 1983 and many others. Mainly because of its shortness, we use the notation "semi-state system description" here.

Formally spoken, the transition from the state-space description to a semi-state description means merely replacement of the $n \times n$ unit matrix I_n by a possibly singular $k \times k$ matrix J.

Rosenbrock's system matrix

$$P(s) = \begin{pmatrix} sI_n - A & B \\ C & 0 \end{pmatrix} \tag{43.11}$$

which has played a central role in Chapter 3 must be replaced by a system matrix

$$R(s) = \begin{pmatrix} sJ - A & B \\ C & 0 \end{pmatrix} \quad \text{with} \quad \det(sJ-A) \not\equiv 0 \tag{43.12}$$

The entries of the transfer function matrix defined by

$$G(s) = C(sJ - A)^{-1}B \tag{43.13}$$

are rational functions of the complex variable s as in Chapter 3. However, contrary to the entries of T(s) defined by (32.14), the elements

$$G_{ij}(s) = \frac{c_j'(sJ - A)_{adj}\, b_i}{\det(sJ - A)} = \frac{N_{ij}(s)}{D_{ij}(s)}$$

may become non-proper rational functions, because the degree of the numerator polynomial may exceed the degree of the denominator polynomial.

From this result quite a lot of system-theoretic phenomena unknown for systems in state-space description.

There are poles at infinity and decoupling zeros at infinity.

The system order n in state-space splits up into three different structural indices, namely

$$d_1 = \text{rank } J \tag{43.15}$$

$$d_2 = \deg \det(sJ - A) = \text{number of finite poles} \tag{43.16}$$

$$d_3 = \max_{q \geq 0} \max_{\substack{1 \leq j_1 < \ldots < j_q \\ 1 \leq i_1 < \ldots < i_q}} \deg R(s)_{1\ldots k, k+i_1 \ldots k+i_q}^{1\ldots k, k+j_1 \ldots k+j_q} \tag{43.17}$$

$$= d_2 + \text{number of poles at infinity}$$

For details see Van der Weiden 1983 and the references cited there.

As for the graph-theoretic approach, there arise no principal problems. Based on the graph-theoretic interpretation of determinants explained in Appendix A2, graph-theoretic characterizations of the structural properties of systems in semi-state description can be obtained.

This concept will not be developed here.

We shall conclude this Section 43 with two small examples.

Example 43.1: Consider the 4×4 matrix

$$M(s) = A + sB = \begin{pmatrix} a_{11} & 0 & a_{13} & 0 \\ a_{21} & a_{22} & 0 & sb_{24} \\ a_{31} + sb_{31} & 0 & 0 & a_{34} \\ 0 & a_{42} & sb_{43} & sb_{44} \end{pmatrix} \tag{43.18}$$

In Fig. 43.1a, the digraph G(M) has been drawn. The non-vanishing entries of B correspond to bold edges in G(M).

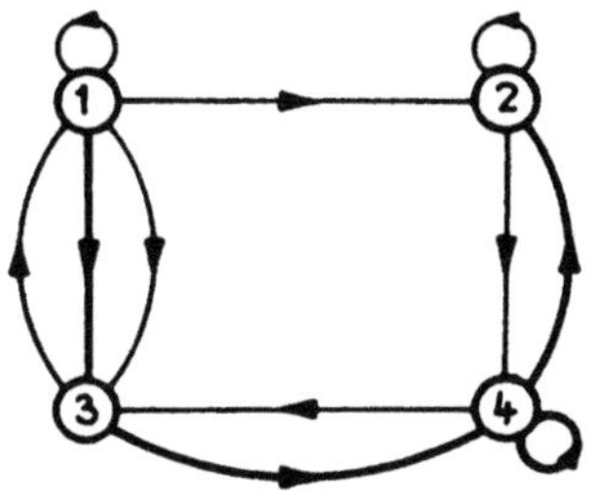

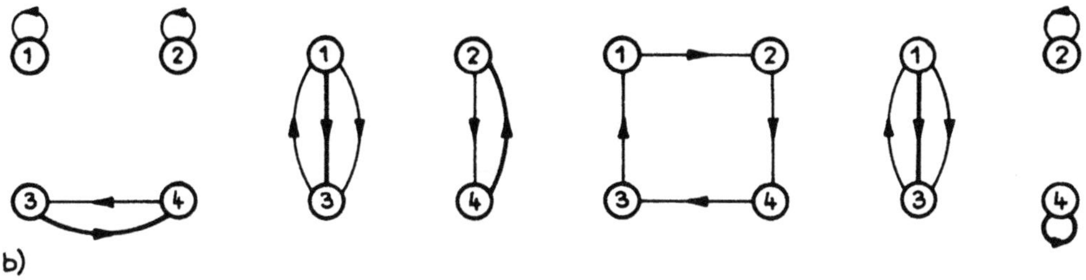

Fig. 43.1

In order to calculate the determinant $\det(A + sB)$ all the cycle families involving the four vertices of $G(M)$ have been composed in Fig. 43.1b. According to Theorem A2.1, one gets

$$\begin{aligned}\det(A+ sB) &= -sb_{43}a_{34}a_{11}a_{22} + (sb_{31}+a_{31})a_{13}(sb_{24}a_{42}- sb_{44}a_{22}) \\ &\quad - a_{21}a_{13}a_{34}a_{42} \\ &= s^2b_{31}(b_{24}a_{42}-b_{44}a_{22}) + s(a_{31}a_{13}(b_{24}a_{42}-b_{44}a_{22})) \\ &\quad - s\, b_{43}a_{34}a_{11}a_{22} \\ &\quad - a_{21}a_{13}a_{34}a_{42}\end{aligned} \tag{43.19}$$

<u>Example 43.2</u>: Let us calculate the transfer function

$$\begin{aligned}G(s) &= (0 \quad 1 \quad 0)(A+sB)^{-1}\begin{pmatrix}1\\0\\0\end{pmatrix} \\ &= [\det(A+sB)]^{-1}(0 \quad 1 \quad 0)(A+sB)_{adj}\begin{pmatrix}1\\0\\0\end{pmatrix}\end{aligned} \tag{43.20}$$

where $(A+sB)$ is given by (43.18).
The denominator polynomial $\det(A+sB)$ is known from (43.19).
To determine the polynomial

$$(0 \; 1 \; 0)(A+sB)_{adj}\begin{pmatrix}1\\0\\0\end{pmatrix} \tag{43.21}$$

we look for all feedback cycle families of width 4 in the digraph G(Q) associated with

$$Q = \begin{pmatrix} 0 & 0 & 1 & 0 & 0 \\ 0 & & & & 1 \\ 0 & & A + sB & & 0 \\ 0 & & & & 0 \\ -1 & 0 & 0 & 0 & 0 \end{pmatrix}$$

see Fig. 43.2a.

There are two such feedback cycle families, see Fig. 43.2b.

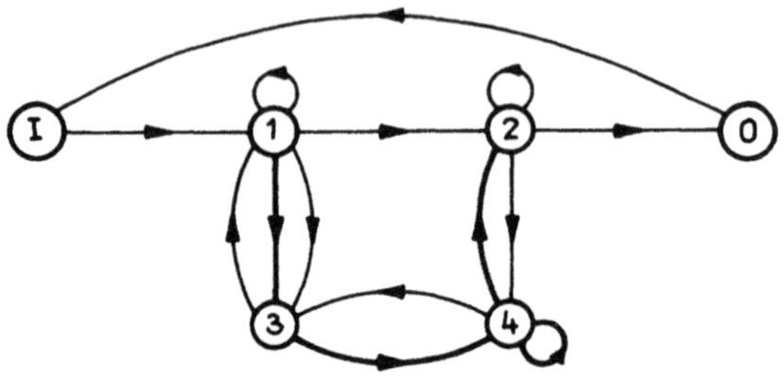

a)

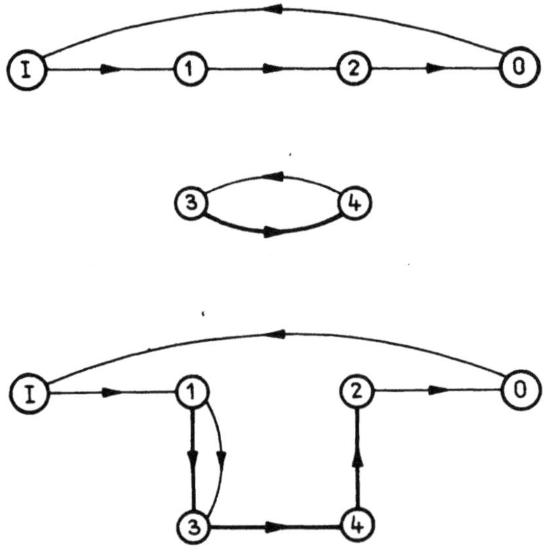

b)

Fig. 43.2

One obtains

$$(0 \quad 1 \quad 0)(A+sB)_{adj}\begin{pmatrix}1\\0\\0\end{pmatrix} = s^3 b_{24}b_{43}b_{31} + s^2 b_{24}b_{43}a_{31} - s\cdot b_{43}a_{34}a_{21}$$

The transfer function (43.20) is non-proper, because the degree 3 of the numerator polynomial exceeds the degree 2 of the denominator polynomial.

Lastly, it should be mentioned that the graph-theoretic approach to control theory is also promising beyond the field of linear continuous plants.

At the very beginning of this monograph, the governing equations of nonlinear smooth plants under feedback have been discussed. The associated digraphs have been derived, too, comp. Fig. 11.1 and Fig. 11.2. These digraphs may be used to characterize important structural properties of nonlinear smooth systems. For instance, Kasinski and Levine (1984) succeeded in taking advantage of this digraph approach to treat the feedback decoupling problem for nonlinear systems.

As for large-scale systems, the partitioning algorithm explained in Section 13 may be applied to nonlinear systems without any alterations: The elements of the structure matrix [A] are given by

$$[a_{ij}] = \begin{cases} L & \text{if } \frac{\partial f_i}{\partial x_j} \not\equiv 0 \\ 0 & \phantom{\text{if } \frac{\partial f_i}{\partial x_j}} \equiv 0 \end{cases} \quad \text{for } i,j = 1,\dots,n$$

where the functions f_i and the state-variables x_j were defined in Section 11.

In the same manner as in Section 13, the n vertices of G([A]) can be subdivided into N equivalence classes of strongly connected vertices, which are 'hierarchically' ordered. Based on such a partitioning, many interesting problems concerning the overall system may be investigated effectively by solving the problem under consideration consecutively for the N disjoint subsystems. This idea has been developed and discussed under different aspects by Wassel and Wend 1979, Vidyasagar 1981, Lunze and Reinschke 1981, Pichai, Sezer and Siljak 1983 and many others.

Denoting by x_{K_i} the state-vector of the i-th subsystem and by u_{K_N}, $u_{K_{N-1}},\dots,u_{K_i}$ the subset of input-vectors acting on the i-th subsystem, the set of Eqs. (11.1) splits up as follows:

$$\begin{aligned}
\dot{x}_{K_N} &= f_{K_N}(x_{K_N};u_{k_N};t) \\
\dot{x}_{K_{N-1}} &= f_{K_{N-1}}(x_{K_N},x_{K_{N-1}};u_{K_N},u_{K_{N-1}};t) \\
&\vdots \\
\dot{x}_{K_2} &= f_{K_2}(x_{K_N},x_{K_{N-1}},\dots,x_{K_2};u_{K_N},u_{K_{N-1}},\dots,u_{K_2};t) \\
\dot{x}_{K_1} &= f_{K_1}(x_{K_N},x_{K_{N-1}},\dots,x_{K_2},x_{K_1};u_{K_N},u_{K_{N-1}},\dots,u_{K_2},u_{K_1};t)
\end{aligned}$$

These blocks of differential equations may be treated one after the other. As a consequence, typical control tasks such as the design of estimators, filters, controllers and observers can be accomplished 'step-by-step', for each subsystem separately.
Obviously, by application of this decomposition approach, the computational expense involved in the design of estimation and control schemes for large-scale systems can be reduced considerably.

In the industrial practice, automated systems consist of technological plants and automated equipments. The control engineer has to cope with combinations of analogue and binary signal processing. Combinatorial as well as sequential logical units are used in automation equipments.

In this monograph, structure matrices and structure graphs have been exploited in order to reveal the structural properties of systems mathematically described by ordinary differential equations. That is, we used a binary model to investigate continuous systems. The other units of automated systems which accomplish logical perations show a priori a binary behaviour. Thus it seems to be useful to form an unified binary model for the whole automated system. Such a model may be represented by a digraph. Its structure will change when different units of the automation equipment are switched on or switched off, possibly with a delay of time due to the effects of sequential logical signal processing. This basic idea was explained and illustrated for an example of power station automation by Reinschke, Röder and Rösel in 1984.

Appendix

A1 Introduction to Graph theory

Graph theory provides a useful tool to tackle problems occurring in completely different fields. Seen from a general point of view, graph theory is a comparatively simple part of algebraic topology. In this appendix the basic vocabulary of graph theory is outlined. The definitions and concepts introduced here serve as a most important foundation for the approach to control developed in this book. Such an introduction seems to be necessary because there is no well-standardized terminology of graph-theoretic notions whose knowledge may be taken for granted within the control engineers' community.

A1.1 Graphs and digraphs

A graph (or line complex) is a mathematical structure which consists of two kind of sets that may be interpreted geometrically as vertices (or nodes or points or 0-simplexes) and as edges (or lines or arcs or branches or 1-simplexes) whose two endpoints are vertices.

We shall denote the vertex-set and the edge-set by $V = \{v_1, v_2, \ldots\}$ and $E = \{e_1, e_2, \ldots\}$, respectively. If both V and E are finite sets, as we always presume to be the case, then the graph is called finite.

Fig. A1.1 shows a graph with $v = 4$ vertices and $e = 6$ edges.

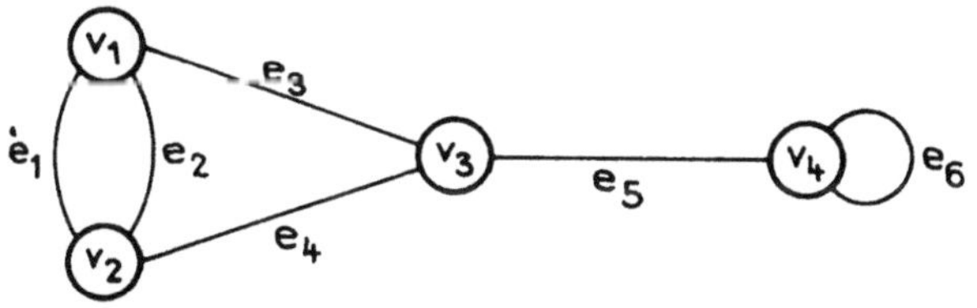

Fig. A1.1

Any edge may be specified by the vertices that it connects. These vertices are called end-vertices of the edge under consideration. A self-loop is an edge whose end-vertices coincide. Edges which have the same end-vertices are said to be parallel.

For the graph of Fig. A1.1, the vertices v_1 and v_3 are the end-vertices of edge e_3. Edge e_6 is a self-loop. Edge e_1 is parallel to edge e_2.

One says that any edge is incident with its end-vertices. Thus, incidence concerns topological relations between edges and vertices and vice versa, whereas <u>adjacency</u> relates edges to edges and vertices to vertices. Two edges are called <u>adjacent</u> if they have a common end-vertex. Two vertices are called <u>adjacent</u> if there is at least one edge whose end-vertices are both the vertices under consideration.

A diagram that visualizes a graph can be drawn in different ways in most cases. For example, Fig. A1.2 shows two diagrams for one graph with $v = 6$ vertices and $e = 13$ edges. It is not difficult to verify that the incidence relations as well as the adjacency relations are the same for both diagrams.

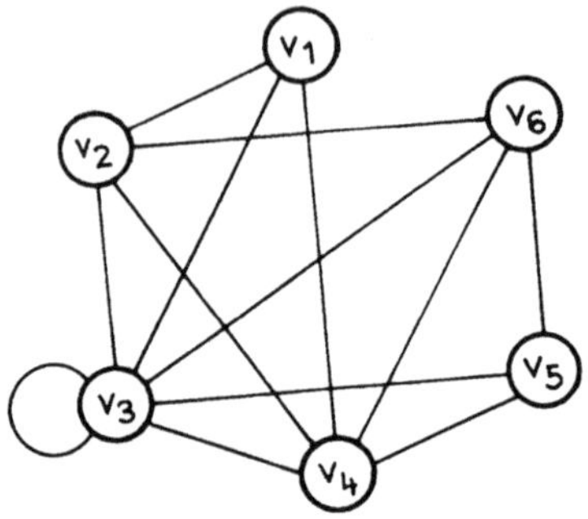

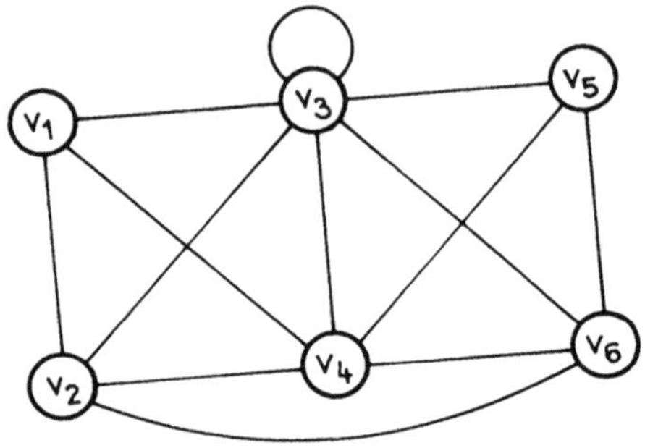

Fig. A1.2

In practice, it is often a question of experience and skillness to find a well-suited representation of a graph.

If the vertex-set V of a graph G may be partitioned into two disjoint subsets, V_1 and V_2, such taht every edge of G lies between a vertex in V_1 and a vertex in V_2 then G is called <u>bipartite</u>.

Fig. A1.3 shows an example where $V_1 = \{v_1, v_3, v_6\}$ and $V_2 = \{v_2, v_4, v_5, v_7\}$.

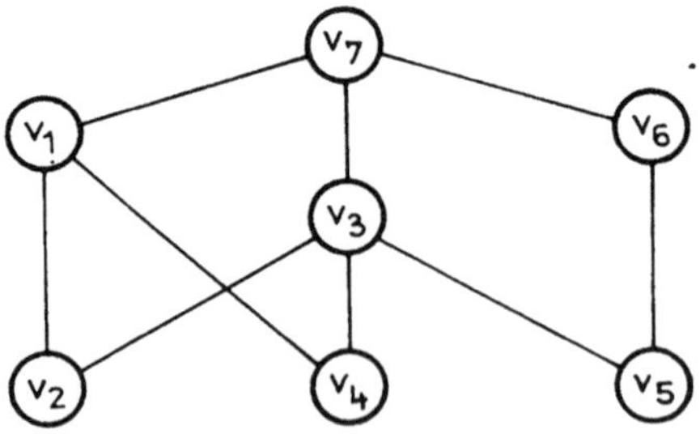

Fig. A1.3

Removing some edges and/or vertices of a given graph G one obtains a <u>subgraph</u> G' of G, for short $G' \subset G$. The <u>removal of a vertex</u> implies the removal of every edge incident with it. The <u>removal of edges</u>, however, does not imply the removal of vertices.

A graph G is said to be connected if, for every pair of vertices, say v_j and v_k, there exists a sequence of adjacent edges e_1, e_2, ..., e_i such that v_j and v_k are end-vertices of e_1 and e_i, respectively.

All the graphs depicted in Fig. A1.1-3 are connected graphs. The removal of edge e_5 in Fig. A1.1 would result in two separated subgraphs, see Fig. A1.4.

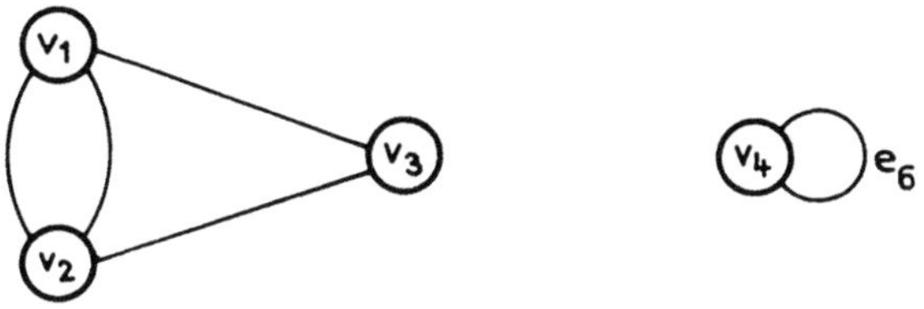

Fig. A1.4

A cut-set of a connected graph is defined as a minimal set of edges whose removal would disconnect the graph G. (The attribute "minimal" means that no proper subset of a cut-set is able to cause disconnection.) In the graph of Fig. A1.1, for example, the edge sets $\{e_3, e_4\}$, $\{e_5\}$, $\{e_1, e_2, e_3\}$ are cut-sets, but the sets $\{e_3, e_5\}$, $\{e_1, e_3\}$, $\{e_3, e_4, e_6\}$ are no cut-sets.

A spanning subgraph of a connected graph G results from G by removing edges where the set of removed edges must not contain a cut-set.

Studying dynamical systems it is appropriate in most cases to assign a direction (or orientation) to each edge of a graph. To visualize this in the diagram of the graph, each edge is equipped with an arrow. A graph modified in this manner is called a directed graph (or oriented graph), for short, digraph. Then each edge e_i is directed from one of its end-vertices to the other one, say, from v_j to v_k. We write $e_i = (v_j, v_k)$ and call v_j the initial vertex and v_k the final vertex of e_i.

An example of a digraph is shown in Fig. A1.5.

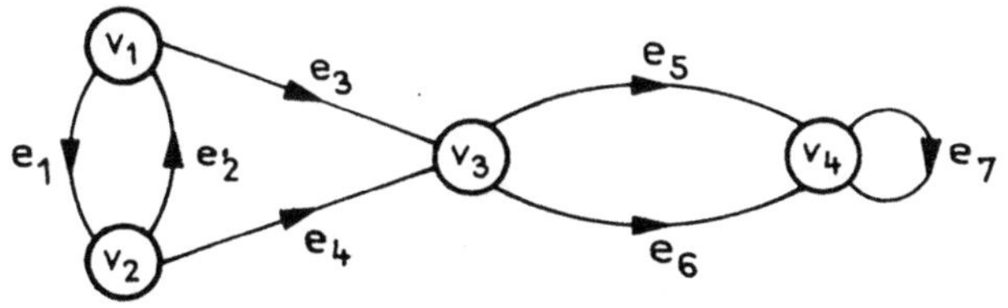

Fig. A1.5

Obviously, if a digraph contains an edge (v_j, v_k) then it may or may not contain an edge (v_k, v_j). In digraphs, parallel edges have both the same initial vertex and the same final vertex. The edges e_5 and e_6 in Fig. A1.5 are parallel edges, whereas e_1 and e_2 are not. Parallel edges in digraphs cannot be uniquely characterized by identifying its initial and final vertices only. As a rule, in this book we are concerned with digraphs containing no parallel edges.

Finally, we have to explain the concept of a weighted digraph. It is often useful to assign a number to each edge of a digraph and to call it edge weight. Then the digraph augmented in this way is called a weighted digraph. Sometimes it has proved to be appropriate to introduce the weight of special subgraphs as follows: The weight of a subdigraph is equal to the product of the weights of all its edges. As for applications of this idea, see Section A1.3.

A1.2 Paths, cycles, and trees

Paths, cycles, and trees are important classes of (sub-)graphs. In graph theory, they are separately defined for the case of graphs as well as for the case of digraphs. We need these concepts for digraphs only. Therefore, paths, cycles, and trees are here introduced as digraphs of special structure. Of course, these digraphs consist of a vertex-set as well as an edge-set. In a few of the following definitions, however, only the edge-sets are mentioned explicitly. It will then be tacitly assumed that those vertices which are incident with at least one edge of the edge-set under consideration form the vertex-set of the digraph.

A (directed) path is a sequence of edges $\{e_i, e_j, \ldots\}$ such that the initial vertex of the succeeding edge is the final vertex of the preceding edge. The initial vertex of the first edge and the final vertex of the last edge are called initial vertex of the path and final vertex of the path, respectively. The edges occurring in the sequence $\{e_i, e_j, \ldots\}$ are not necessarily distinct. The number of edges contained in the sequence $\{e_i, e_j, \ldots\}$ is called the length of the path.

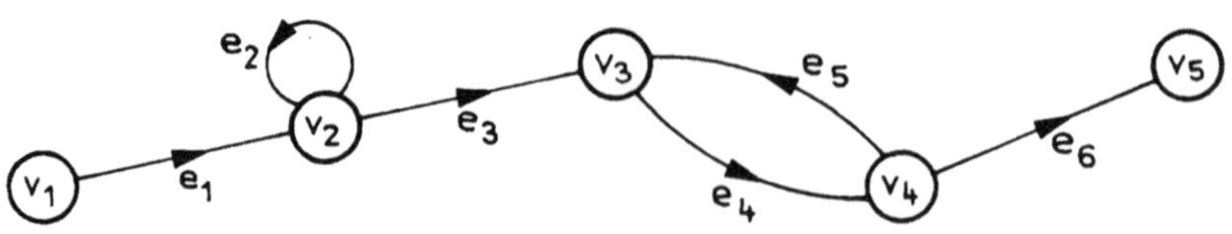

Fig. A1.6

Fig. A1.6 shows an example: the edge sequence $\{e_1, e_2, e_3, e_4, e_5, e_4, e_6\}$ is a path of length 7 leading from initial vertex v_1 to final vertex v_5.

It is often convenient to describe the edges of a path by their initial and final vertices. For example, the path in Fig. A1.6 just mentioned may equivalently be written as $\{(v_1,v_2), (v_2,v_2), (v_2,v_3), (v_3,v_4), (v_4,v_3), (v_3,v_4), (v_4,v_5)\}$.

A path is called simple if one reaches going along the path from its initial to its final vertex no vertex more than once. Fig. A1.7 represents a simple path of length 4.

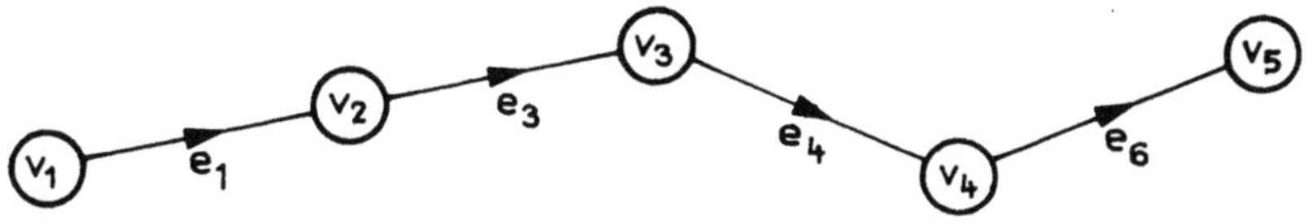

Fig. A1.7

By means of paths, we define an important type of connectedness in digraphs. Two vertices, v_i and v_k, are said to be strongly connected if there is both a path from v_i to v_k and a path from v_k to v_i.
A digraph is said to be strongly connected if for every vertex v_i and every vertex v_k there is a path from v_i to v_k.

For example, the digraph in Fig. A1.6 is not strongly connected, the digraph in Fig. A1.8 is strongly connected.

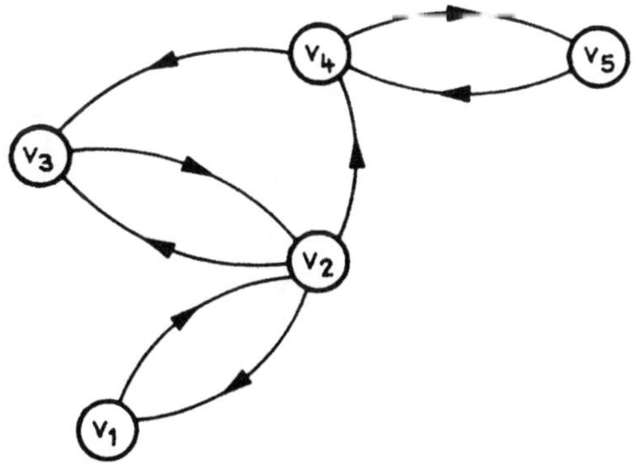

Fig. A1.8

A closed path is a path whose initial and final vertices are the same. A closed path is said to be a cycle if one reaches going along the path no vertex, other than the initial-final vertex, more than once.

Fig. A1.9 shows a closed path obtained from the path of Fig. A1.6 by adding the edge e_7.

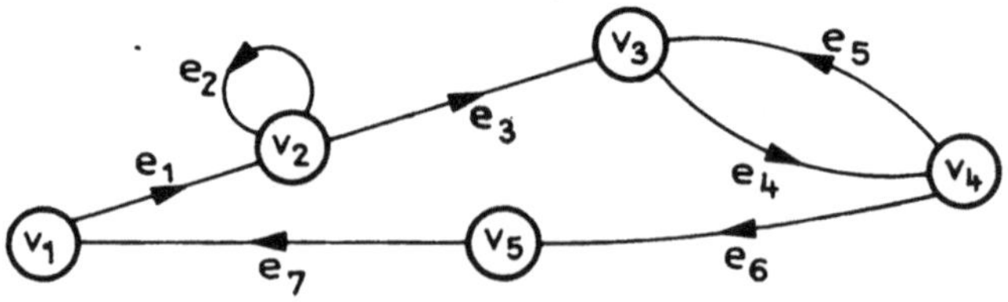

Fig. A1.9

The graph of Fig. A1.9 contains three cycles shown in Fig. A1.10.

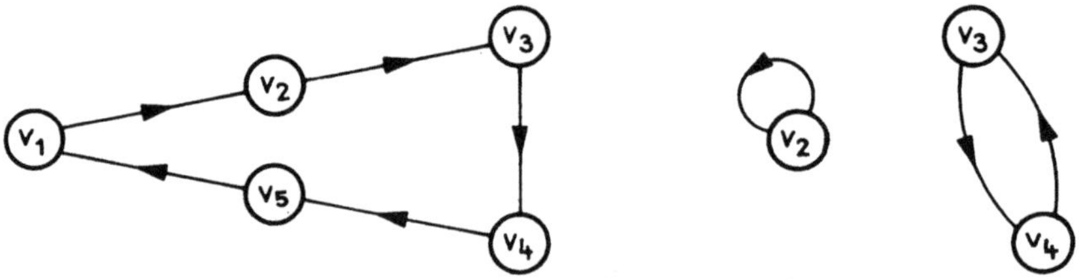

Fig. A1.10

The number of edges contained in a cycle defines the length of this cycle.
The cycles in Fig. A1.10 are of lengths 5, 1, and 2, respectively.
Cycles of length 1 are called <u>self-cycles</u>.

A set of vertex disjoint cycles is said to be a <u>cycle family</u>.
Fig. A1.11 shows a cycle family consisting of three individual cycles.

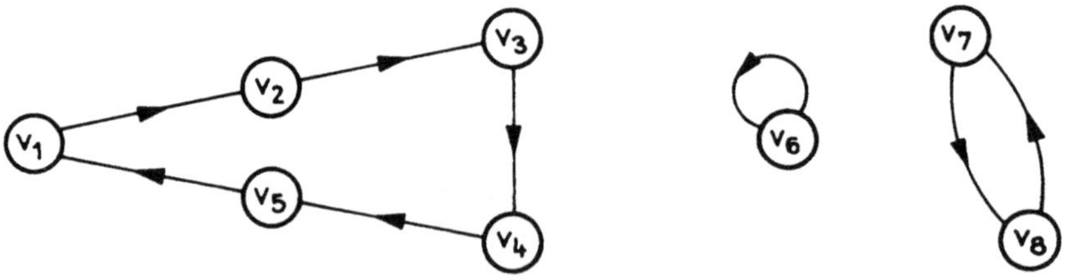

Fig. A1.11

Note the difference between Fig. A1.11 and Fig. A1.10. The graph of Fig. A1.9 does not contain a cycle family with three individual cycles.

Fig. A1.12 shows another closed path, $\{(v_1,v_2), (v_2,v_3), (v_3,v_4), (v_4,v_3), (v_3,v_1)\}$, that is neither a cycle nor a cycle family.

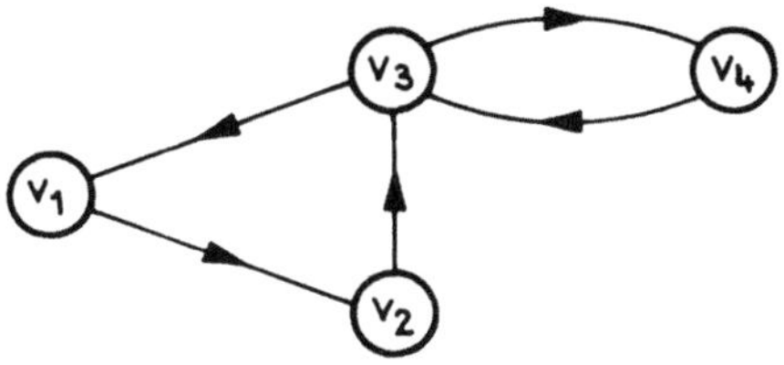

Fig. A1.12

A <u>spanning cycle family</u> of a digraph G is a cycle family which touches all the vertices of G.

A digraph is said to be <u>acyclic</u> if it contains no cycle.

So-called trees are another frequently occurring type of graphs.
The name was introduced by A. Cayley in 1857 for the case of undirected graphs:
A <u>tree</u> is a connected graph whose number of edges is one less than the number of vertices.
A <u>spanning tree</u> (or complete tree) of a connected graph G is a subgraph which is a tree that involves all the vertices of G.
For control purposes, we need primarily trees in digraphs in which one vertex, denoted as root or ground, is distinguished. They appear, for example, in describing 'genealogical trees' (in German: Stammbäume).

A digraph is said to have a <u>root</u> r if r is a vertex and, for every other vertex v, there is a path which starts in r and ends in v.
A digraph G is called a <u>rooted tree</u> if G has a root from which there is a unique path to every other vertex.
Fig. A1.13 shows two rooted trees contained in the digraph of Fig.A1.8 where $r = v_1$.

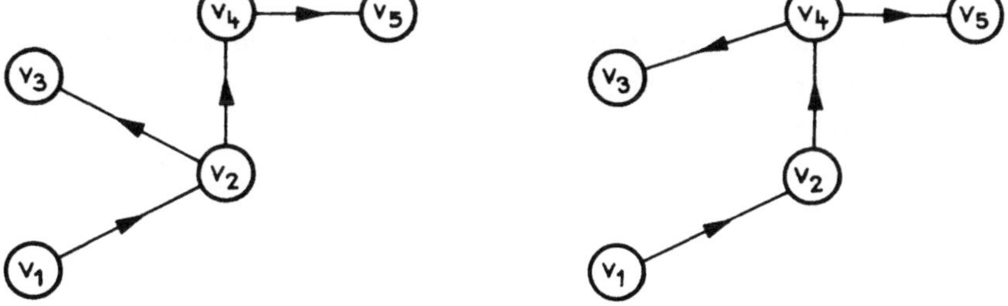

Fig. A1.13

It should be realized that every vertex of the digraph in Fig. A1.8 may be regarded as a root.

By assigning the opposite orientation to each edge of a rooted tree one obtains a grounded tree.
A digraph is said to have a ground g if g is a vertex and, for every other vertex v, there is a path which starts in v and ends in g.
A digraph G is called a grounded tree if G has a ground to which there is a unique path from every other vertex.

A1.3 Graph-theoretic characterizations of square matrices

Let A be a square matrix of order n,

$$A = (a_{ij}) \quad \text{for} \quad i,j = 1,2,\dots,n, \tag{A1.1}$$

where the matrix elements a_{ij} are real numbers.

For example, consider a matrix A of order n = 3,

$$A = \begin{pmatrix} a_{11} & 0 & a_{13} \\ a_{21} & a_{22} & a_{23} \\ 0 & a_{32} & 0 \end{pmatrix} \tag{A1.2}$$

There are several possibilities of constructing weighted graphs that have a one-to-one correspondence with a given square matrix A.

In his classical text on graph theory published in 1936, D. König uses the following first graph-theoretic characterization of square matrices

I. To the given matrix A one may assign a bipartite graph such that
- each row of A corresponds to one of the vertices $u_1, u_2, \dots, u_n$;
- each column of A corresponds to one of the vertices $v_1, v_2, \dots, v_n$; and
- there is an edge from v_j to u_i with the weight a_{ij} if and only if $a_{ij} \neq 0$. (Other edges are not introduced.)

In this way the example matrix (A1.2) may be characterized by the weighted bipartite digraph
shown in Fig. A1.14.

Fig. A1.14

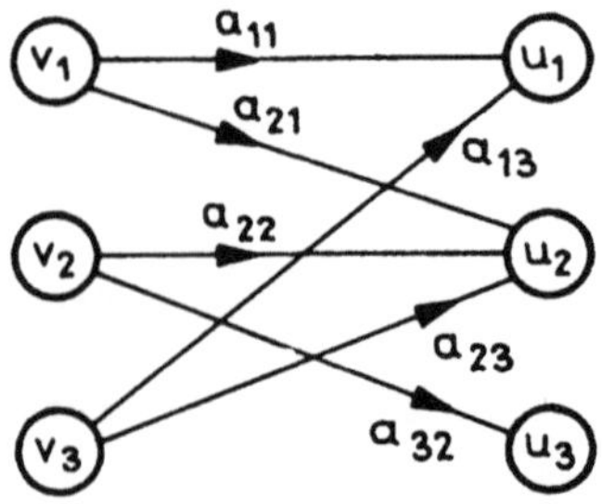

If one unites, for all $i = 1,2,\ldots,n$, each vertex u_i with the corresponding vertex v_i into one coalesced vertex again denoted by v_i, then a second graph-theoretic characterization of a square matrix A results:

II. There is a one-to-one correspondence between the matrix (A1.1) and a weighted digraph G(A) which has n vertices $v_1,v_2,\ldots,v_n$ and a directed edge (v_j,v_i) from the initial vertex v_j to the final vertex v_i if the matrix element a_{ij} does not vanish $(i,j = 1,2,\ldots,n)$. The edge weight is given by the value of a_{ij}.

As for the example matrix (A1.2), one obtains a graph-theoretic characterization of the given matrix depicted in Fig. A1.15.

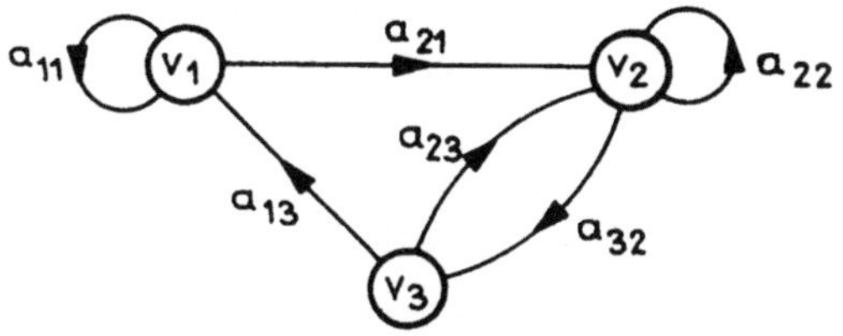

Fig. A1.15

The characterization II was used by C.L. Coates in the context of his "flow-graph" solutions of linear algebraic equations published in 1959. S.J. Mason's "signal-flow graphs" also introduced in the fifties are closely related to Coates' flow-graphs.

Applying Mason's approach, any square matrix (A1.1) has the following third graph-theoretic characterization:

III. There is a one-to-one correspondence between the matrix (A1.1) and a weighted digraph which has n vertices $v_1,v_2,\ldots,v_n$ and, for $i \neq j$, a directed edge from v_j to v_i with weight a_{ij} if $a_{ij} \neq 0$, and, for $i = j$, a self-cycle at v_i with weight $a_{ii}+1$ if $a_{ii} \neq -1$. (Edges with vanishing weights are not introduced.)

In this way, the example matrix (A1.2) is graph-theoretically characterized as shown in Fig. A1.16. In case of $a_{11} = -1$ $(a_{22} = -1)$ the self-cycle at v_1 (at v_2) is omitted.

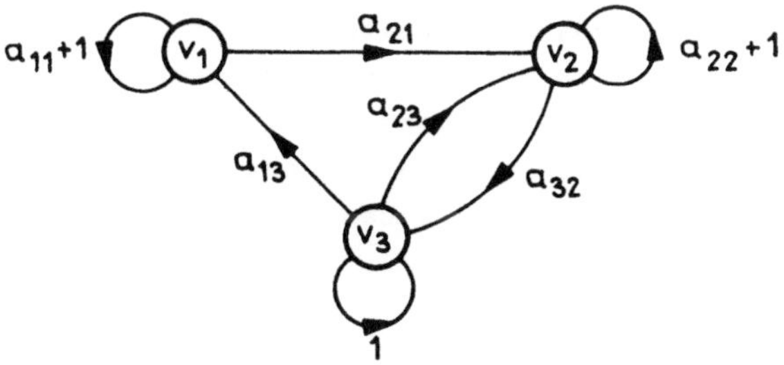

Fig. A1.16

Comparing Fig. A1.16 with Fig. A1.15 the impression might arise that the characterization III is unnatural because there are introduced more edges than necessary. In fact, this is true in many cases. But it should be realized that Mason started from systems of linear equations with a special structure such that all main diagonal elements of A were equal to (-1). Then the graph-theoretic representation III does not contain self-cycles and is more concise than the representation II.

Finally, let us consider a <u>fourth graph-theoretic characterization</u> of square matrices which entirely avoids self-cycles.

IV. There is a one-to-one correspondence between the matrix (A1.1) and a weighted digraph G'(A) which has $n+1$ vertices $v_1, v_2, \ldots, v_n, g$ and, for $i \neq j = 1,2,\ldots,n$, and edge from v_j to v_i with weight a_{ij} if $a_{ij} \neq 0$, and, for $j = 1,2,\ldots,n$, an edge from v_j to the 'ground vertex' g with weight

$$(-\sum_{i=1}^{n} a_{ij}) \quad \text{provided that} \quad \sum_{i=1}^{n} a_{ij} \neq 0.$$

(Again, edges with vanishing weights are omitted.)

For the example matrix (A1.2), Fig. A1.17 shows its digraph representation IV.

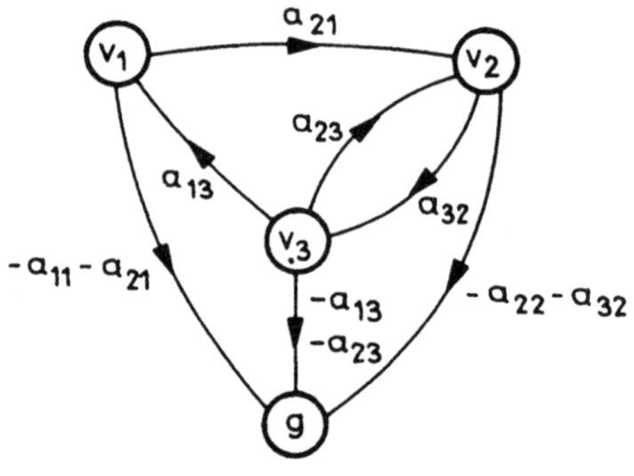

Fig. A1.17

At first glance, the characterization IV seems to be more complicated than the characterizations I or II. For important classes of matrices occurring in context of Markov chains, of electrical networks and elsewhere, however, there are typical dependencies between the main diagonal elements and the elements in the same row and/or column. Then the characterization IV can be used profitably.

A2 Digraphs and determinants

Almost all texts on linear algebra introduce determinants as follows:

A determinant det A of an $n \times n$ matrix A, see (A1.1), is defined by

$$\det A = \sum_{\substack{\text{even} \\ \text{permutations}}} \prod_{i=1}^{n} a_{it_i} - \sum_{\substack{\text{odd} \\ \text{permutations}}} \prod_{i=1}^{n} a_{it_i} \qquad \text{(A2.1)}$$

where $\{t_1, t_2, \ldots, t_n\}$ is a permutation of $\{1, 2, \ldots, n\}$.

For the example matrix (A1.2) one obtains

$$\det A = a_{13}a_{21}a_{32} - a_{11}a_{23}a_{32} \qquad \text{(A2.2)}$$

A2.1 Computation of determinants with the aid of weighted digraphs

In Section A1.3, four different graph-theoretic characterizations of square matrices have been introduced. In the sequel it will be shown that each of those characterizations may be used as a starting point for the graph-theoretic characterization of determinants.

Based on the graph-theoretic characterization I of square matrices, treated in Section A1.3, each non-vanishing expression

$$a_{1t_1}a_{2t_2}\ldots a_{nt_n}$$

of the determinant (A2.1) corresponds to a set of n edges leading from vertex v_{t_i} to vertex u_i $(i = 1,2,\ldots,n)$ in the bipartite digraph. Such a subset of n edges being incident with all the 2n vertices of the bipartite graph was called a factor of first degree by D. König.

The bipartite digraph of Fig. A1.14 contains two factors of first degree which are exhibited in Fig. A2.1.

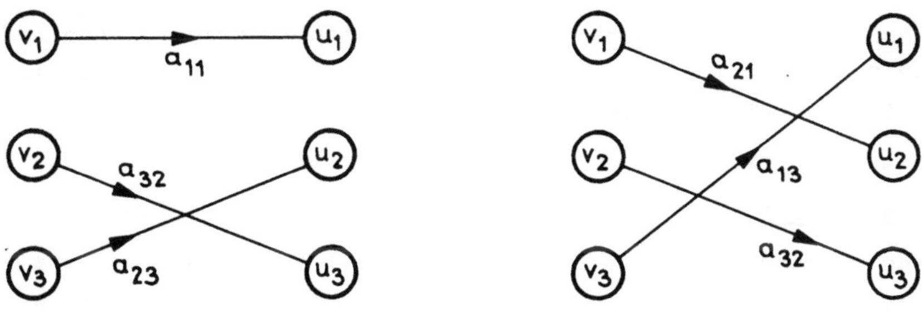

Fig. A2.1

During the process of coalescing the vertices u_1 and v_1, u_2 and $v_2, \ldots,$ u_n and v_n, respectively, by which the bipartite graph representation I of a matrix A is transformed into the digraph representation II, each factor of first degree is converted into a cycle family that touches all the n vertices.

As for the example (A2.2), there are two cycle families that touch all the $n = 3$ vertices (Fig. A2.2).

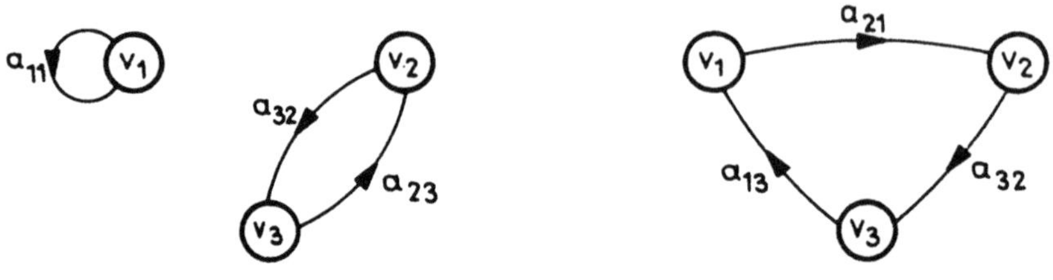

Fig. A2.2

Obviously, the weight of the cycle family, i.e. the product of the weights of all edges involved, is equal to the numerical value of the corresponding expression

$$a_{1t_1} a_{2t_2} \ldots a_{nt_n} \tag{A2.3}$$

of the determinant. It remains to recognize graph-theoretically if the permutation $\{t_1, t_2, \ldots, t_n\}$ is even or odd.

The following <u>sign rule</u> was found by A.L. Cauchy as early as in 1812 and published in 1815, reformulated in Latin by C.G.J. Jacobi in 1841, and reinvented by C.L. Coates in 1959:

> If the cycle family under consideration splits up into d individual cycles, the sign factor of its corresponding expression
>
> $a_{1t_1} a_{2t_2} \ldots a_{nt_n}$ is $(-1)^{n-d}$.

For example, the sign factor of the cycle family drawn on the left of Fig. A2.2 is $(-1)^{3-2} = -1$, whereas the cycle family on the right has a sign factor $(-1)^{3-1} = (-1)^2 = 1$.

What has been said until now can be summarized in the following way:

<u>Theorem A2.1</u>

Each summand (A2.3) of det A (see (A2.1)) corresponds to a spanning cycle family in the weighted digraph G(A) introduced above as digraph representation II of A. The value of (A2.3) is given by the product of the weights of the n edges involved. If this cycle family consists of d disjoint cycles, then the sign factor of the summand under consideration is $(-1)^{n-d}$.

Based on the digraph characterization III of an arbitrary square matrix A, see Section A1.3, its determinant det A may be calculated as follows:

$$\det A = (-1)^n + \sum_{k=1}^{n} (-1)^{n-k} s^{(k)} \tag{A2.4}$$

where $s^{(k)}$ symbolizes the sum of weights of all the subgraphs consisting of k vertex disjoint cycles.

This is Mason's rule for evaluating the determinant of an arbitrary square matrix A.

Formula (A2.4) may be derived from Theorem A2.1 by simple but space consuming transformations which will not be repeated here. (See, for example, Reinschke and Schwarz 1976.)

For the example matrix (A1.2), the digraph representation III (comp. Fig. A1.16) contains five individual cycles shown in Fig. A2.3.

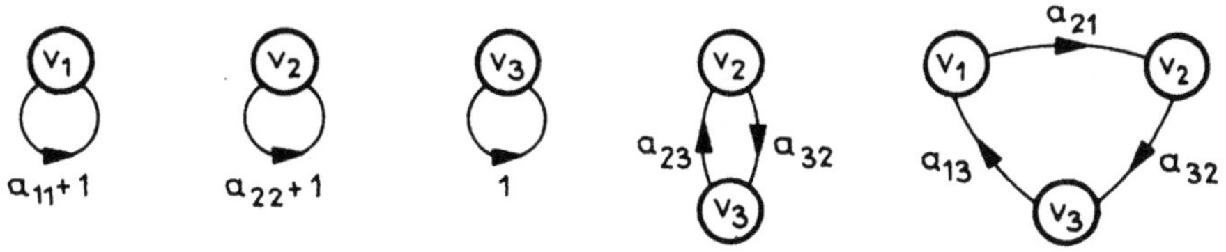

Fig. A2.3

From the Figure it is immediately seen that

$$s^{(1)} = a_{11} + 1 + a_{22} + 1 + 1 + a_{23}a_{32} + a_{13}a_{21}a_{32}$$

$$s^{(2)} = (a_{11} + 1)(a_{22} + 1 + 1 + a_{23}a_{32}) + (a_{22} + 1)\cdot 1$$

$$s^{(3)} = (a_{11} + 1)(a_{22} + 1)\cdot 1$$

whence

$$\det A = (-1)^3 + (-1)^2 s^{(1)} + (-1)^1 s^{(2)} + (-1)^0 s^{(3)}$$

$$= -1 + s^{(1)} - s^{(2)} + s^{(3)}$$

$$= a_{13}a_{21}a_{32} - a_{11}a_{23}a_{32}$$

Starting from the digraph characterization IV of an arbitrary square matrix A, see Section A1.3, its determinant may be calculated in the following manner:

Theorem A2.2

The determinant det A can be determined as the sum of the weights of all spanning grounded trees contained in the weighted digraph G'(A) introduced as digraph representation IV of square matrices in Section A1.3.

Proof: It seems to be difficult to derive Theorem A2.2 as a consequence of Theorem A2.1. Therefore, an independent proof of Theorem A2.2 will be sketched here.

To a given $n \times n$ matrix A, the weighted digraph G'(A) is constructed in accordance with representation IV:

G'(A) has $n+1$ vertices $v_1, \ldots, v_n, g$ and,
for $i \neq j$ and $a_{ij} \neq 0$, directed edges from v_j to v_i with weights a_{ij},
for $j = 1,2,\ldots,n$ and $\sum_{i=1}^{n} a_{ij} \neq 0$, edges from v_j to g with weight $(-\sum_{i=1}^{n} a_{ij})$.

Assume the graph G'(A) to have e edges, and the edges to be enumerated from 1 to e. According to the chosen order of the edges, we write down their negative weights as main diagonal elements of an $e \times e$ diagonal matrix D.
Furthermore, two $n \times e$ matrices $K = (k_{ij})$ and $L = (l_{ij})$, where $i = 1,2,\ldots,n$, $j = 1,2,\ldots,e$, are defined:

$$k_{ij} = \begin{Bmatrix} 1 \\ -1 \\ 0 \end{Bmatrix} \text{if edge } j \begin{Bmatrix} \text{starts from} \\ \text{ends in} \\ \text{is not incident with} \end{Bmatrix} \text{vertex } v_i$$

$$l_{ij} = \begin{cases} 1 & \text{if edge } j \text{ starts from vertex } v_i \\ 0 & \text{else} \end{cases}$$

There holds the fundamental identity

$$A = K\,D\,L' \tag{A2.5}$$

In the sequel, we denote the $p \times p$ minors of A by

$$A^{i_1 i_2 \ldots i_p}_{j_1 j_2 \ldots j_p} = \det \begin{pmatrix} a_{i_1 j_1} & a_{i_1 j_2} & \cdots & a_{i_1 j_p} \\ a_{i_2 j_1} & a_{i_2 j_2} & \cdots & a_{i_2 j_p} \\ \vdots & \vdots & \ddots & \vdots \\ a_{i_p j_1} & a_{i_p j_2} & \cdots & a_{i_p j_p} \end{pmatrix} \tag{A2.6}$$

Applying the Binet-Cauchy formula (see Cauchy 1815 or, for example, Gantmacher 1966, Ch.1) to (A2.4) we may write

$$\det A = \sum_{\{e_1 \ldots e_n\}} K^{1\,2\,\ldots n}_{e_1 e_2 \ldots e_n} (D\,L')^{e_1 e_2 \ldots e_n}_{1\;\,2 \ldots\; n}$$

$$= \sum_{\{e_1 \ldots e_n\}} K^{1\,2\,\ldots\,n}_{e_1 e_2 \ldots e_n} L^{1\,2\,\ldots\,n}_{e_1 e_2 \ldots e_n} D^{e_1 e_2 \ldots e_n}_{e_1 e_2 \ldots e_n} \tag{A2.7}$$

where $\{e_1 \ldots e_n\}$ symbolizes a set of n edges with $e_1 < e_2 < \ldots < e_n$.

It can be shown that

$$K^{1\,2\,\ldots n}_{e_1 e_2 \ldots e_n} \; L^{1\,2\,\ldots n}_{e_1 e_2 \ldots e_n} = \begin{cases} 1 & \text{if the edges } \{e_1,\ldots,e_n\} \text{ form a grounded tree in } G'(-A) \\ 0 & \text{otherwise} \end{cases} \qquad (A2.8)$$

Hence,

det A = sum of the weights of all spanning grounded trees in G'(-A)

or, equivalently, (A2.9)

det(-A) = sum of the weights of all spanning grounded trees in G'(-A),

q.e.d. ▲

Illustration of the foregoing proof:
Consider the 3×3 matrix (A1.2) as example. Fig. A1.17 shows the digraph G'(A). If the $e = 7$ edges of G'(A) are enumerated according to Fig. A2.4, then the matrices D, K, and L are uniquely determined:

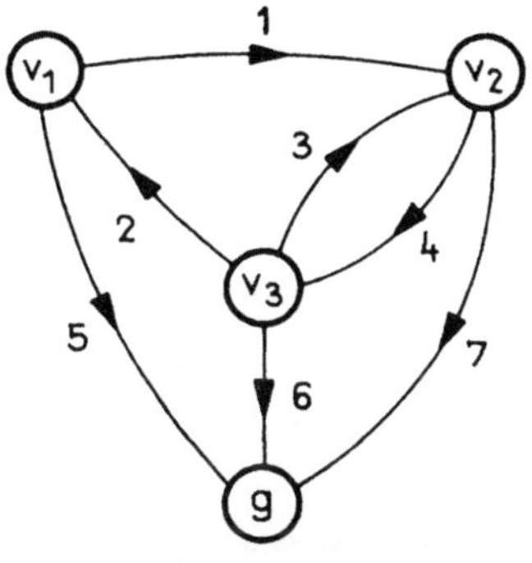

Fig. A2.4

$$D = \langle -a_{21}, -a_{13}, -a_{23}, -a_{32}, a_{11}+a_{21}, a_{13}+a_{23}, a_{22}+a_{32} \rangle$$

$$K = \begin{pmatrix} 1 & -1 & 0 & 0 & 1 & 0 & 0 \\ -1 & 0 & -1 & 1 & 0 & 0 & 1 \\ 0 & 1 & 1 & -1 & 0 & 1 & 0 \end{pmatrix}, \quad L = \begin{pmatrix} 1 & 0 & 0 & 0 & 1 & 0 & 0 \\ 0 & 0 & 0 & 1 & 0 & 0 & 1 \\ 0 & 1 & 1 & 0 & 0 & 1 & 0 \end{pmatrix}$$

The validity of the identity (A2.5) can be easily checked.

There exist nine spanning grounded trees shown in Fig. A2.5.
Formula (A2.9) yields

$$\begin{aligned} \det(-A) = {} & a_{13}a_{32}(-a_{11}-a_{21}) + a_{13}a_{21}(-a_{22}-a_{32}) + a_{21}a_{32}(-a_{13}-a_{23}) \\ & + a_{21}(-a_{22}-a_{32})(-a_{13}-a_{23}) + a_{21}a_{23}(-a_{22}-a_{32}) \\ & + a_{13}(-a_{11}-a_{21})(-a_{22}-a_{32}) + a_{23}(-a_{11}-a_{21})(-a_{22}-a_{32}) \\ & + a_{32}(-a_{11}-a_{21})(-a_{13}-a_{23}) + (-a_{11}-a_{21})(-a_{13}-a_{23})(-a_{22}-a_{32}) \end{aligned}$$

$= -a_{32}a_{21}a_{13} + a_{11}a_{23}a_{32}$

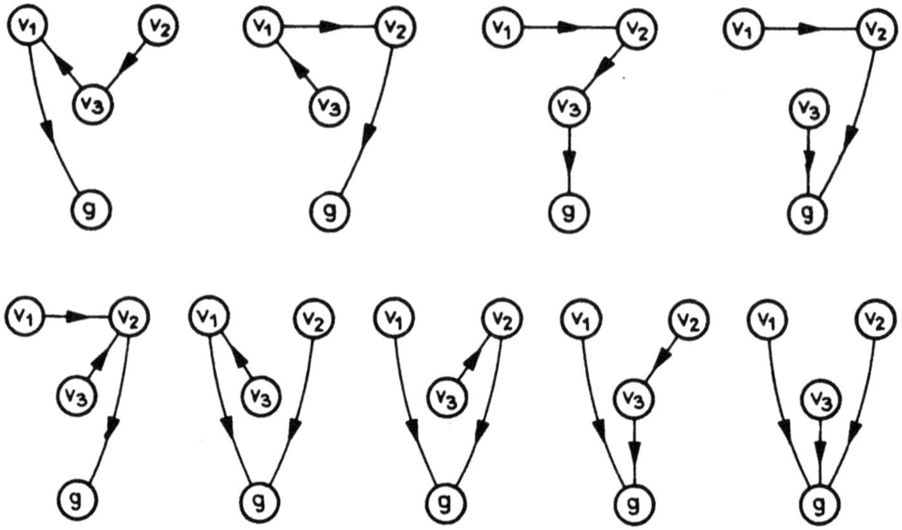

Fig. A2.5

Corollary A2.1

The number of spanning grounded trees in $G'(A)$ is equal to the determinant of an $n \times n$ integer matrix M whose entries are

$$m_{ij} = \begin{cases} -1 \\ 0 \end{cases} \text{ if } a_{ij} \begin{cases} \neq 0 \\ = 0 \end{cases} \text{ for } i \neq j = 1,2,\dots,n \qquad \text{(A2.10)}$$

$$m_{jj} = 1 + \text{number of non-diagonal elements } a_{ij} \neq 0 \text{ in column } j$$

Proof: The graph $G'(M)$ has the same structure as the graph $G'(A)$. The edges may be enumerated in the same way. Then the incidence matrices K and L are also the same for $G'(A)$ and $G'(M)$. Each entry of the diagonal matrix that corresponds to D becomes 1. This means the weight of each spanning grounded tree of $G'(M)$ is equal to 1. In analogy to the identity (A2.9), the determinant det M is equal to the number of all spanning grounded trees in $G'(A)$. This completes the proof.

▲

As for the example matrix (A1.2), the corresponding 3×3 integer matrix M may be obtained from (A2.10) as follows:

$$M = \begin{pmatrix} 2 & 0 & -1 \\ -1 & 2 & -1 \\ 0 & -1 & 3 \end{pmatrix}.$$

It is easily checked that
det M = 9
in accordance with Fig. A2.5

A2.2 Computation of characteristic polynomials in several variables with the aid of weighted digraphs

Let be

A any $n \times n$ matrix, see (A1.1),

and

$$\Lambda = \langle \lambda_1, \lambda_2, \ldots, \lambda_n \rangle \qquad \text{(A2.11)}$$

an $n \times n$ diagonal matrix.

Consider the generalized characteristic polynomial

$$\det(- A) = \prod_{i=1}^{n} \lambda_i + \sum_{j=1}^{n} \Big(\prod_{\substack{i=1 \\ i \neq j}}^{n} \lambda_i \Big) c_j + \sum_{j<k} \sum \Big(\prod_{\substack{i=1 \\ i \neq j \\ i \neq k}}^{n} \lambda_i \Big) c_{jk} \qquad \text{(A2.12)}$$

$$+ \ldots + \sum_{i=1}^{n} \lambda_i c_{1 \ldots i-1, i+1 \ldots n} + c_{1\, 2 \ldots i \ldots n}$$

In the sequel, the coefficients $c_j, c_{jk}, \ldots, c_{1\,2 \ldots n}$ occurring in (A2.12) will be interpreted graph-theoretically.

Again, each of the four graph-theoretic characterizations of square matrices introduced in Section A1.3 may be used as basis for different interpretations of those coefficients. In order to save space we shall confine ourselves to the graph-theoretic matrix characterizations II and IV.

Theorem A2.3

Suppose the given $n \times n$ matrix A to be represented by a weighted digraph G(A) according to the characterization II of Section A1.3. Then the c-coefficients occurring in (A2.12) may be computed with the aid of the cycle families in G(A) as follows:

c_j is equal to the weight of all cycles which touch only the vertex v_j, multiplied by (-1),

c_{jk} is equal to the sign-weighted sum of the weights of those cycle families each of which touches the vertices v_j and v_k and no other vertices,

c_{ijk} is equal to the sign-weighted sum of the weights of those cycle families each of which touches exactly the vertices v_i, v_j, and v_k,

$\vdots$

$c_{1 \ldots i-1, i+1 \ldots n}$ is equal to the sign-weighted sum of the weights of those cycle families each of which touches all but the ith vertices,

$c_{1\,2\,\dots\,n}$ is equal to the sign-weithed sum of the weights of those cycle families each of which touches all n vertices.

The sign-factor of any summand is +1 or -1 if the associated cycle family contains an even or an odd number of disjoint individual cycles, respectively.

Proof: We apply Theorem A2.1 taking into account the special structure of the matrix

$$Q = \Lambda - A.$$

The digraph $G(Q)$ results from $G(A)$ by changing the sign of all edge weights and then supplementing additional self-cycles with weight λ_i at each vertex v_i for $i = 1,2,\dots,n$.

Each summand on the right-hand side of (A2.12) corresponds to a set of cycle families in G(Q). For example, let us consider the summand

$$\Big(\sum_{\substack{i=1 \\ i \neq j,k}}^{n} \lambda_i\Big)\, c_{jk}$$

It is associated with a set of cycle families each of which consists of n-2 self-cycles with weight λ_i, for all i different from j and k, and all the cycle families of G(A) touching exactly the vertices v_j and v_k. Each cycle family of the set under consideration consists of n or n-1 individual cycles because the involved cycle families of G(A) may consist of one or two individual cycles. In the former case the sign-factor of the corresponding summand of c_{jk} will be

$$(-1)^{n-(n-2)-1}\,(-1)^2 = -1,$$

and in the latter case,

$$(-1)^{n-(n-2)-2}\,(-1)^2 = +1$$

The factor $(-1)^2$ reflects the sign changing of the weights of the involved edges of G(A).

So far, as for the coefficients c_{jk}, the statement of Theorem A2.3 has been verified.

The statements concerning the other coefficients can be proved similarly. The details may be left to the reader interested in a complete proof.

▲

In order to illustrate Theorem A2.3 by an example, we choose a 3×3 matrix A as above, see (A1.2). The graph G(A) has been shown in Fig. A1.15. From this Figure, the coefficients c_1, c_2, c_3, c_{12}, c_{13}, c_{23}, and c_{123} can immediately read as follows:

$$c_1 = -a_{11},\ c_2 = -a_{22},\ c_3 = 0,\ c_{12} = a_{11}a_{22},\ c_{13} = 0,\ c_{23} = -a_{23}a_{32},$$

$$c_{123} = a_{11}a_{23}a_{32} - a_{13}a_{32}a_{21}.$$

Let us now turn to another interpretation of the c-coefficients based on the graph-theoretic characterization IV of a square matrix A, denoted by G'(A) as above.
For this purpose, we introduce reduced graphs resulting from G'(A) by means of "short-circuiting" of vertices. Here the phrase "short-circuiting of vertex v_i" means successive execution of the steps
1. coalescing of v_i with the ground vertex g,
2. deleting of all the edges starting in g,
3. uniting of parallel edges.

Fig. A2.6 gives an example: short circuiting of vertex v_2 in the digraph G'(A) drawn in Fig. A1.17.

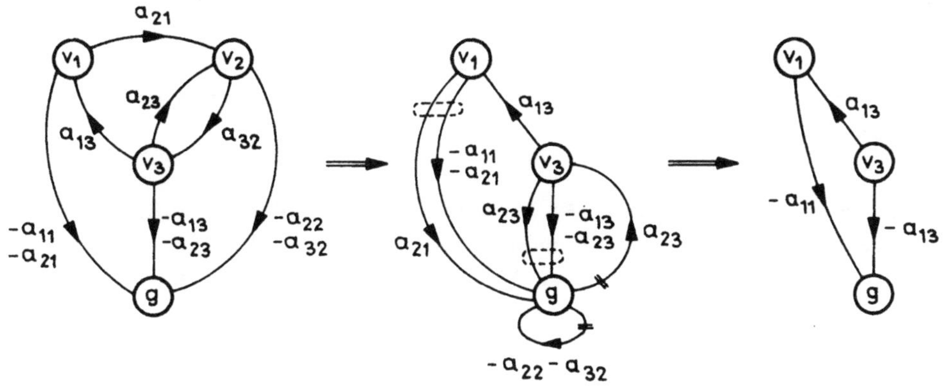

Fig. A2.6

Theorem A2.4

Suppose the given nxn matrix A to be represented by a weighted digraph G'(A) according to the characterization IV.
Then the c-coefficients occurring in (A2.12) can be determined as the sum of the weights of all spanning grounded trees contained in reduced digraphs which result from G'(A) by short-circuiting of vertices as follows:

c_j short-circuiting of all vertices but vertex v_j,

c_{jk} short-circuiting of all vertices but vertices v_j and v_k,

c_{ijk} short-circuiting of all vertices but vertices v_i, v_j, and v_k,

⋮

$c_{1 \ldots i-1, i+1 \ldots n}$ short-circuiting of vertex v_i,

$c_{1\,2\,\ldots\,n}$ short-circuiting of no vertex.

Proof: We apply Theorem A2.2 taking into account the special structure of the matrix $Q = \Lambda - A$.
The digraph G'(-Q) results from G'(A) by increasing the weight of each

edge (v_i,g) by λ_i for $i = 1,2,\ldots,n$.

Each summand on the right-hand side of (A2.12) is associated with a set of spanning grounded trees in G'(Q). For example, let us consider the summand

$$(\prod_{\substack{i=1 \\ i\neq j,k}}^{n} \lambda_i)\, c_{jk}$$

It corresponds to a set of spanning grounded trees each of which consists of the $n-2$ edges (v_i,g) with weight λ_i for $i = 1,2,\ldots,n$, $i \neq j$, $i \neq k$, and a spanning grounded tree in the reduced graph resulting from G'(A) by short-circuiting of the $n-2$ vertices v_i just considered. Thus, as for the coefficient c_{jk}, the statement has been verified.

The statements concerning the other coefficients can be proved similarly. The details may be left to the reader interested in a complete proof of Theorem A2.4.

▲

Now, we consider again the 3×3 matrix A defined in (A1.2) as illustrative example. Fig. A1.17 shows the digraph G'(A). Its spanning grounded trees depicted in Fig. A2.5 provide the coefficient $c_{123} = \det(-A)$.

The reduced graphs and their spanning grounded trees, which supply us with the coefficients c_{12}, c_{13}, c_{23}, c_1, c_2, and c_3, are depicted in Fig. A2.7.

From the drawings of Fig. A2.7, the desired coefficients can be determined,

$c_{12} = -a_{22}a_{21} - a_{22}(-a_{11}-a_{21}) = a_{11}a_{22}$, $c_{13} = -a_{11}(-a_{13}) + a_{13}(-a_{11}) = 0$,
$c_{23} = -a_{23}(-a_{22}-a_{32}) + a_{23}(-a_{22}-a_{32}) - a_{23}a_{32} = -a_{23}a_{32}$, $c = -a_{11}$,
$c_2 = -a_{22}$, $c_3 = 0$.

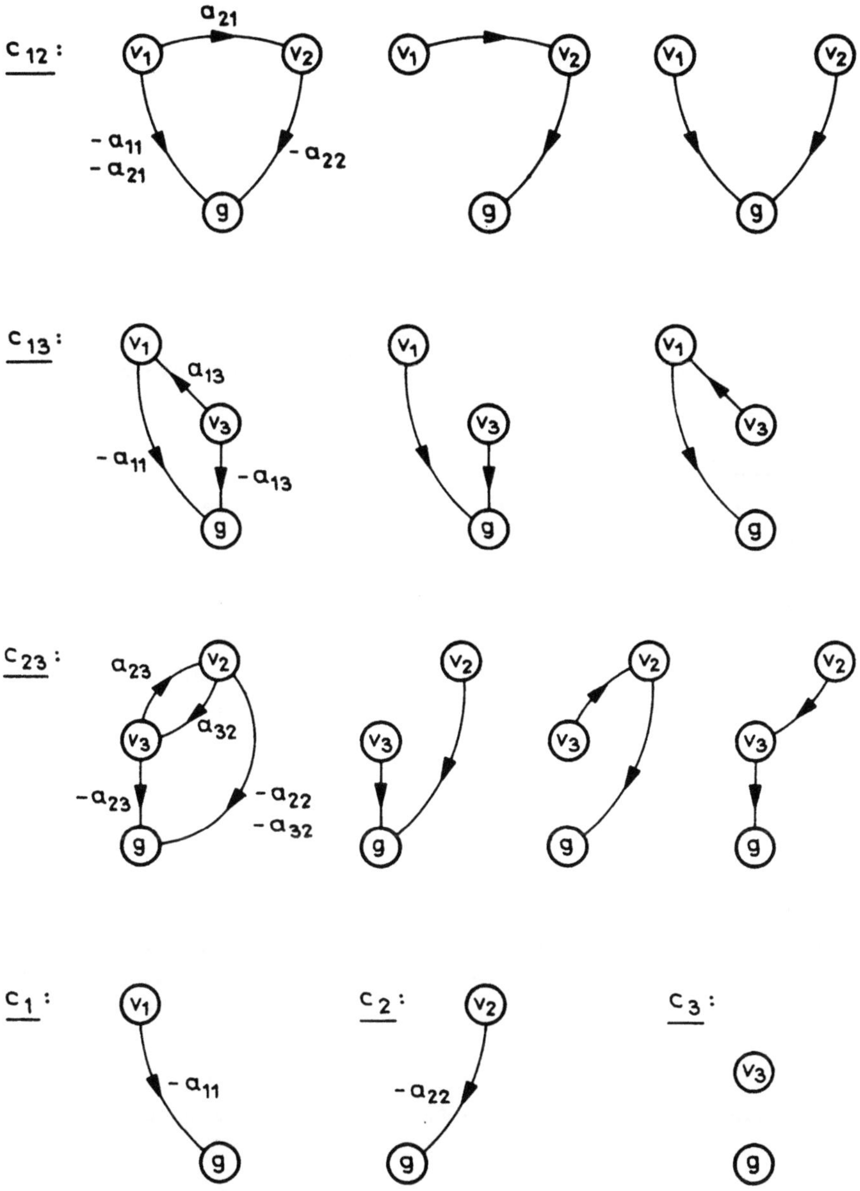
c12:
a21
-a11
-a21
-a22
c13:
a13
-a11
-a13
c23:
a23
a32
-a23
-a22
-a32
c1:
-a11
c2:
-a22
c3:
v1
v2
v3
g

Fig. A2.7

A2.3 Graph-theoretic characterization of characteristic-polynomial coefficients and their sensitivities

In this Section, we shall restrict ourselves to the digraph representation II of square matrices introduced in Section A1.3.

The characteristic polynomial of any $n \times n$ matrix A can be written as

$$\det(sI_n - A) = s^n + \sum_{i=1}^{n} p_i \, s^{n-i} \tag{A2.13}$$

Based on Theorem A2.3, the coefficients p_i $(1 \le i \le n)$ can immediately be interpreted with the aid of cycle families in the digraph G(A) introduced in Section A1.3.

Theorem A2.5

The coefficients p_i $(1 \le i \le n)$ of the characteristic polynomial (A2.13) are determined by the cycle families of length i within the digraph G(A).

Each cycle family of length i corresponds to one summand of p_i. The numerical value of the summand results from the weight of the corresponding cycle family. This value must be multiplied by a sign factor $(-1)^d$ if the cycle family under consideration consists of d disjoint cycles.

In particular,

p_1 results from all self-cycles, and the common sign-factor is (-1);

p_2 results from all cycles of length 2, each with sign-factor (-1), and all disjoint pairs of self-cycles, each pair with sign-factor +1;

p_3 results from all cycles of length 3, each with sign-factor (-1), all disjoint pairs with one cycle of length w and one self-cycle, each pair with sign-factor +1, and all disjoint triples of self-cycles, each triple with sign-factor (-1);

etc.

To give an example of how Theorem A2.5 can be applied consider a 4×4 matrix

$$A = \begin{pmatrix} a_{11} & a_{12} & a_{13} & 0 \\ 0 & a_{22} & 0 & a_{24} \\ a_{31} & 0 & a_{33} & a_{34} \\ 0 & 0 & a_{43} & 0 \end{pmatrix}$$

and its associated digraph (Fig. A2.8).

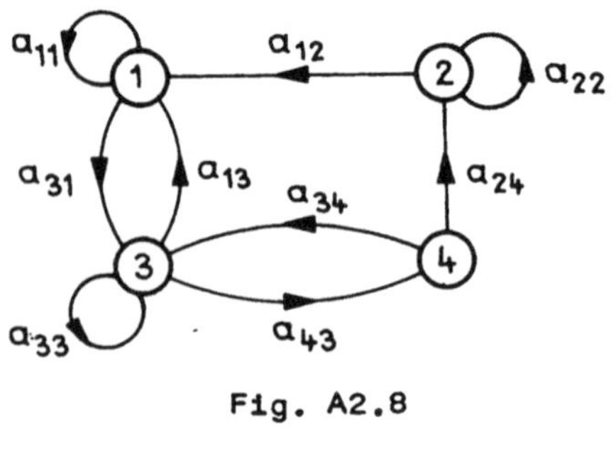

Fig. A2.8

For this example, the algebraic expressions for p_i ($1 \leq i \leq n$) and the corresponding cycle families have been shown in Fig. A2.9.

Coefficient p_i	Symbolic expression for p_i	Corresponding cycle families
p_1	$-a_{11}$ $-a_{22}$ $-a_{33}$	
p_2	$a_{11}a_{22}$ $+ a_{11}a_{33}$ $+ a_{22}a_{33}$ $- a_{13}a_{31}$ $- a_{34}a_{43}$	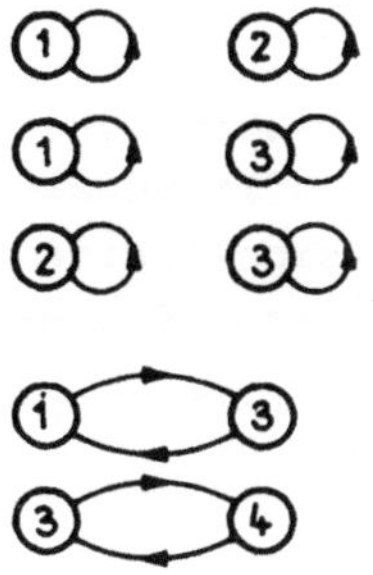
p_3	$- a_{11}a_{22}a_{33}$ $+ a_{13}a_{31}a_{22}$ $+ a_{11}a_{34}a_{43}$	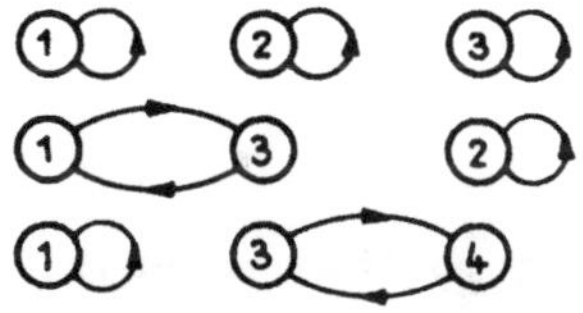
p_4	$- a_{31}a_{43}a_{24}a_{12}$ $- a_{11}a_{22}a_{34}a_{43}$	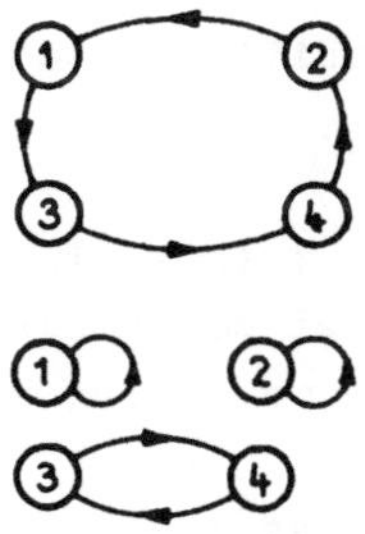

Fig. A2.9

The sensitivities of the characteristic-polynomial coefficients with respect to a varying matrix element as parameter can also be determined.

Theorem A2.6

The sensitivity function

$$\frac{\partial p_k}{\partial a_{ij}} \quad (1 \le i,j,k \le n) \qquad (A2.14)$$

for the coefficient p_k of the characteristic polynomial (A2.13) is uniquely determined by the set of cycle families of length k containing the edge from vertex j to vertex i in the digraph G(A).

The symbolic expression for (A2.14) can be found as follows:

1. Look for all cycle families of length k which contain the edge from vertex j to vertex i.
2. Replace the edge weight a_{ij} by +1 within all cycle families just found.
3. The sign-weighted sum of the weights of the cycle families obtained is equal to the sensitivity function (A2.14).
   The sign-factor is $(-1)^d$ if the cycle family under consideration consists of d disjoint cycles.

Proof: It is well-known that p_k depends linearly on each matrix element a_{ij} (see, for example, Theorem A2.5), more exactly,

$$p_k = x_k + a_{ij}y_k$$

where both the expressions x_k and y_k do not depend on a_{ij}.
This means

$$\frac{\partial p_k}{\partial a_{ij}} = 1 \cdot y_k = y_k$$

According to Theorem A2.5, the expression $a_{ij}y_k$ represents the sign-weighted sum of those cycle families of length k that contain the edge from vertex j to vertex i. After replacing the edge weight a_{ij} by +1 the desired sensitivity function (A2.14) results. This completes the proof.

▲

For the example matrix of Fig. A2.8, the sensitivities of the characteristic-polynomial coefficients p_1, p_2, p_3, and p_4 with respect to all not identically vanishing matrix elements a_{ij} are put together in Fig. A2.10.
It should be noted that Theorem A2.6 can be applied even in the case of vanishing nominal values of the variable matrix elements.
As for the example under discussion, presume the nominal parameter values of the not identically vanishing matrix elements to be

$\check{a}_{11} = 0$, $\check{a}_{12} = 1$, $\check{a}_{13} = 2$, $\check{a}_{22} = -1$, $\check{a}_{24} = -2$,
$\check{a}_{31} = 3$, $\check{a}_{33} = -3$, $\check{a}_{34} = 0$, $\check{a}_{43} = 0.5$,

	$\frac{\partial p_4}{\partial a_{ij}}$ expression	cycle families	$\frac{\partial p_3}{\partial a_{ij}}$ expr.	cycle families	$\frac{\partial p_2}{\partial a_{ij}}$ expr.	cycle families	$\frac{\partial p_1}{\partial a_{ij}}$ expr.	c.f.
a_{11}	$-a_{22}a_{34}a_{43}$	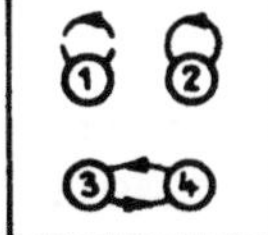	$-a_{22}a_{33}$ $+a_{34}a_{43}$	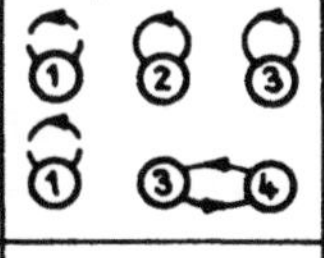	a_{22} $+a_{33}$	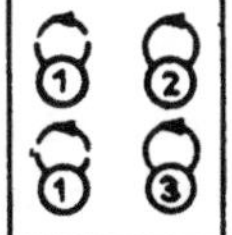	-1	
a_{12}	$-a_{31}a_{43}a_{24}$	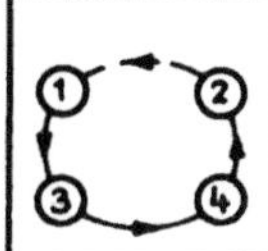	—	—	—	—	—	—
a_{13}	—	—	$a_{31}a_{22}$	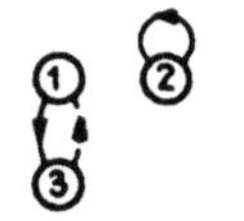	$-a_{31}$		—	—
a_{22}	$-a_{11}a_{34}a_{43}$	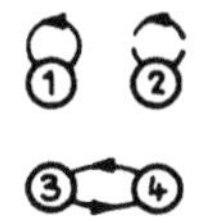	$-a_{11}a_{33}$ $+a_{13}a_{31}$	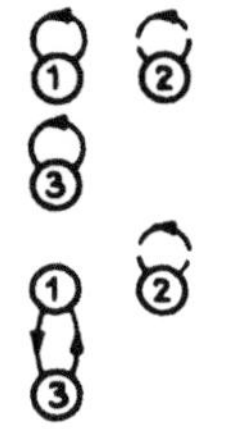	a_{11} $+a_{33}$	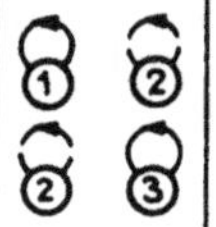	-1	
a_{24}	$-a_{43}a_{31}a_{12}$	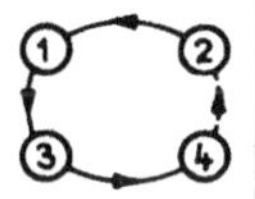	—	—	—	—	—	—
a_{31}	$-a_{43}a_{24}a_{12}$	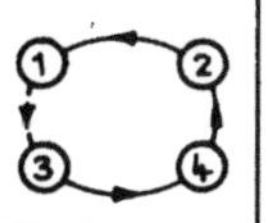	$a_{13}a_{22}$	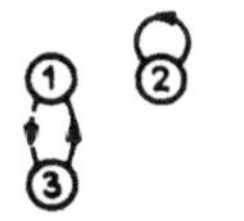	$-a_{13}$		—	—
a_{33}	—	—	$-a_{11}a_{22}$		a_{11} a_{22}	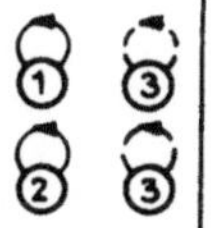	-1	
a_{34}	$-a_{11}a_{22}a_{43}$	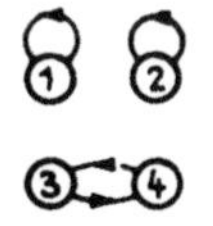	$a_{11}a_{43}$	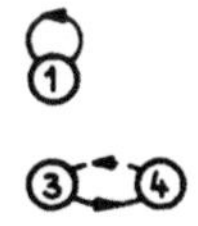	$-a_{43}$		—	—
a_{43}	$-a_{31}a_{24}a_{12}$ $-a_{11}a_{22}a_{34}$	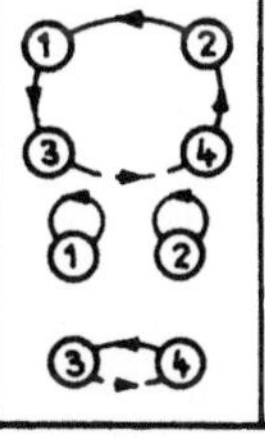	$a_{11}a_{34}$	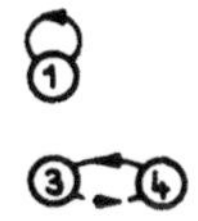	$-a_{34}$		—	—

Fig. A2.10

in other words,

$$\check{A} = \begin{pmatrix} 0 & 1 & 2 & 0 \\ 0 & -1 & 0 & -2 \\ 3 & 0 & -3 & 0 \\ 0 & 0 & .5 & 0 \end{pmatrix}$$

Then the characteristic polynomial coefficients sensitivities at $A = \check{A}$ have the following numerical values (comp. Fig. A2.10):

	$\frac{\partial p_4}{\partial a_{ij}}$	$\frac{\partial p_3}{\partial a_{ij}}$	$\frac{\partial p_2}{\partial a_{ij}}$	$\frac{\partial p_1}{\partial a_{ij}}$
a_{11}	0	-3	-4	-1
a_{12}	3	0	0	0
a_{13}	0	-6	-3	0
a_{22}	0	6	-3	-1
a_{24}	-1.5	0	0	0
a_{31}	1	-2	-2	0
a_{33}	0	0	-1	-1
a_{34}	0	0	-0.5	0
a_{43}	6	0	0	0

References

[1] Ackermann, J.: Abtastregelung. Band I: Analyse und Synthese. Band II: Entwurf robuster Systeme. Springer-Verlag, Berlin 1983

[2] Anderson, B.D.O. and H.-M. Hong: Structural controllability and matrix nets. Int. J. Control 35 (1982), pp.397-416

[3] Andrei, N.: Sparse systems. Digraph approach of large-scale linear systems theory. Verlag TÜV Rheinland, Köln 1985

[4] Benedetto, M.D. Di and A. Isidori: The matching of nonlinear models via dynamic state feedback. Rap. 04.84. University of Rome "La Sapienza", 1984

[5] Biggs, N.L., E.K. Lloyd and R.J. Wilson: Graph Theory 1736-1936. Clarendon Press: Oxford 1976

[6] Boksenbom, A.S., R. Hood: General algebraic method applied to control analysis of complex engine types. NACA, Techn. Report 980, 1950

[7] Brasch, Jr.,F.M. and J.B. Pearson: Pole placement using dynamic compensators. IEEE Trans. AC-15 (1970), pp.34-43

[8] Bröcker, Th. and K. Jänicke: Einführung in die Differentialtopologie. Springer-Verlag: Heidelberg 1973

[9] Cauchy, A.L.: Memoire sur les fonctions qui ne peuvent obtenir que deux valeurs... J.l'Ecole Polytechnique 10 (1815), pp.29-97. Reprinted in Oeuvres Completes, Serie II, Tome I, pp.91-169

[10] Cayley, A.: On the theory of the analytical forms called trees. Philosophical Magazine (4) 13 (1857), pp.172-176

[11] Chen, C.F. and M.B.Ahmad: Evaluation the gain of a flow graph by the Grassmann algebra. Int.J.Control 39(1984), pp. 1329-1337

[12] Coates, C.L.: Flow-graph solutions of algebraic equations. IEEE Trans. CT-6 (1959), pp.170-187

[13] Commault, C. and J.M. Dion: Smith-McMillan factorization at infinity of rational matrix functions and their control interpretation. Systems&Control Letters 1 (1982), pp. 312-320

[14] Commault, C., J.M. Dion and S. Perez: Transfer matrix approach to the disturbance decoupling problem. Proc. 9th IFAC World Congress, vol. 8, pp.130-133, Budapest 1984

[15] Corfmat, J.P. and A.S. Morse: Structurally controllable and structurally canonical systems. IEEE Trans. AC-21 (1976), pp.129-131

[16] Cremer, M.: A precompensator of minimal order for decoupling a linear multivariable system. Int. J. Control 14 (1971), pp.1089-1103

[17] Cvetkovic, D.M., M. Doob and H. Sachs: Spectra of graphs. VEB Deutscher Verlag der Wissenschaften, Berlin 1980

[18] Davison, E.J.: On pole assignment in linear systems with incomplete state feedback. IEEE Trans. AC-15 (1970), pp. 348-351

[19] Davison, E.J.: Connectability and structural controllability of composite systems. Automatica 13 (1977), pp.109-123

[20] Davison, E.J. and S.H. Wang: Properties and calculation of transmission zeros of linear multivariable systems. Automatica 10 (1974), pp.643-658

[21] Davison, E.J. and S.H. Wang: On pole assignment in linear multivariable systems using output feedback. IEEE Trans. AC-20 (1975), pp.516-518

[22] Denham, M.J.: A necessary and sufficient condition for decoupling by output feedback. IEEE Trans.AC-18(1973),pp.535-537

[23] Descusse, J., J.F. Lafay and M. Malabre: On the structure at infinity of linear block-decouplable systems: the general case. IEEE Trans. AC-28 (1983), pp.1115-1118

[24] Desoer, C.A. and J.D. Schulman: Zeros and poles of matrix transfer functions and their dynamical interpretation. IEEE Trans. CAS-21 (1974), pp.3-8

[25] Dion, J.M.: Feedback block decoupling and infinite structure of linear systems. Int. J. Control 37 (1983),pp.521-533

[26] Evans, F.J. and P.M. Larsen: Structural design of control systems System Research Reports A7, University of Technology, Helsinki 1984

[27] Evans, F.J., Schizas, C. and J. Chan: Control system design using graphical decomposition techniques. IEE Proc., part D, 128 (1981), pp.77-84

[28] Falb, P.L. and W.A. Wolovich: Decoupling in the design and synthesis of multivariable control systems. IEEE Trans. AC-12 (1967), pp.651-659

[29] Ferreira, P.G. and S.P. Bhattacharyya: On blocking zeros. IEEE Trans. AC-22 (1977), pp.258-259

[30] Föllinger, O.: Entwurf konstanter Ausgangsrückführungen im Zustandsraum. Automatisierungstechnik 34 (1986),pp.5-15

[31] Ford. L.R. and D.R. Fulkerson: Flows in networks. Princeton University Press, Princeton, N.J. 1962

[32] Franksen, O.I., P. Falster and F.J. Evans: Qualitative aspects of large-scale systems. Springer-Verlag, New York 1979

[33] Gantmacher, F.R.: Theory of matrices. Sec.ed. (in Russian). Nauka: Moscow 1966

[34] Gilbert, E.G.: The decoupling of multivariable systems by state feedback. SIAM J.Control 7 (1969), pp.50-63

[35] Glover, K. and L.M. Silverman: Characterization of structural controllability. IEEE Trans. AC-21(1976),pp.534-537

[36] Grassmann, H.: Die lineale Ausdehnungslehre - ein neuer Zweig der Mathematik. Verlag von Otto Wigand, Leipzig 1844. Reprinted in 'Hermann Grassmann's gesammelte mathematische und physikalische Werke'. Band I.1. B.G. Teubner: Leipzig 1894

[37] Hadamard, J.: Sur les transformations ponctuelles. Bull.Soc.Math.France 34 (1906), pp.71-84

[38] Hartshorne, R.: Algebraic Geometry. Springer-Verlag, New York 1977

[39] Hautus, M.L.J.: Controllability and observability of linear autonomous systems. Proc.Kon.Akad.Wetensch.Ser.A, 72 (1969), pp.443-448

[40] Hautus, M.L.J.: The formal Laplace transform for smooth linear systems. Pp.29-47 in G. Marchesi and S.K. Mitter (eds.): Mathematical Systems Theory. LNEMS 131. Springer-Verlag, New York 1976

[41] Hermann, R. and C.F. Martin: Applications of algebraic geometry to systems theory. Part I. IEEE Trans. AC-22 (1977), pp.19-25

[42] Hosoe, S.: Determination of generic dimensions of controllable subspaces and its application. IEEE Trans. AC-25 (1980), pp.1192-1196

[43] Hosoe, S., Y.Hayakawa and T. Aoki: Structural controllability for linear systems in linearly parametrized descriptor form. Proc. 9th IFAC World Congress, vol.8,pp.115-120, Budapest 1984

[44] Jacobi, C.G.J.: De formatione et proprietatibus determinantium. Crelle's J.f.reine & angew.Mathem.22 (1841), pp.285-318 Reprinted in 'C.G.J. Jacobi's gesammelte Werke', Band 3, pp.355-392 (1894)

[45] Johnston, R.D. and G.W. Barton and M.L. Brisk: Determination of the generic rank of structural matrices. Int. J. Control 40 (1984), pp.257-264

[46] Kailath, T.: Linear systems. Prentice Hall,Inc., Englewood Cliffs 1980

[47] Kamiyama, S. and K. Furuta: Decoupling by restricted state feedback. IEEE Trans. AC-21 (1976), pp.413-415

[48] Kabamba, P.T. and R.W. Longman: Exact pole assignment using direct or dynamic output feedback. IEEE Trans. AC-27 (1982), pp.1244-1246

[49] Kasinski, A. and J. Levine: A fast graph-theoretic algorithm for the feedback decoupling problem of nonlinear systems. Pp.550-562 in P.A. Fuhrmann (ed.): Mathematical Theory of Networks and Systems. LNCIS 58. Springer-Verlag, Berlin 1984

[50] Kaufmann, A.: Introduction à la combinatorique en vue des applications. Dunod, Paris 1968

[50'] Kaufman, I.: On poles and zeros of linear systems. IEEE Trans. CT-20 (1973), pp.93-101

[51] Kemeny, J.G. and J.L. Snell: Finite Markov chains. D.van Nostrand Co., Princeton 1960

[52] Kevorkian, A.K.: Structural aspects of large dynamic systems. Proc.6th IFAC World Congress, Part IIIA, Paper 19.3. Boston 1975

[53] Kimura, H.: Pole assignment by gain output feedback. IEEE Trans. AC-20 (1975),pp.509-516

[54] Kimura, H.: A further result on the problem of pole assignment by output feedback. IEEE Trans. AC-22 (1977),pp.458-463

[55] König, D.: Theorie der endlichen und unendlichen Graphen. Akademische Verlagsgesellschaft, Leipzig 1936

[56] Kono, M. and I. Sugiura: Generalization of decoupling control. IEEE Trans. AC-19 (1974), pp.281-282

[57] Kouvaritakis, B. and A.G.J. MacFarlane: Geometric approach to analysis and synthesis of system zeros. Int. J. Control 23 (1976), pp.149-181

[58] Lin, C.T.: Structural controllability. IEEE Trans. AC-19 (1974), pp.201-208

[59] Linnemann, A.: Fixed modes in parametrized systems. Report No.77. Forschungsschwerpunkt Dynamische Systeme, Universität Bremen 1982

[60] Luenberger, D.G.: Dynamic equations in descriptor form. IEEE Trans. AC-22 (1977), pp.312-321

[61] MacFarlane, A.G.J. and N. Karcanias: Poles and zeros of linear multivariable systems: a survey of the algebraic, geometric and complex-variable theory. Int. J. Control 24 (1976), pp.33-74

[62] Malabre, M.: Structure à l'infini des triplets invariants. Application à la poursuite parfaite de modèle. in A. Bensoussan & J.L. Lions (eds): Analysis and optimization of systems. LNCIS 44, Springer-Verlag, New York 1982

[63] Malabre, M. and V. Kucera: Infinite structure and exact model matching problem: a geometric approach. IEEE Trans. AC-29 (1984), pp.266-268

[64] Mason, S.J.: Feedback theory, some properties of signal-flow graphs. Proc. IRE 41 (1953), pp.1144-1156

[65] Mason, S.J.: Feedback theory, further properties of signal-flow graphs. Proc. IRE 44 (1956), pp.920-926

[66] Mayeda, H. and T. Yamada: Strong structural controllability. SIAM J. Control & Optimiz. 17 (1979), pp. 123-138

[67] Miscenko, A.S. and A.T. Fomenko: Introduction to differential geometry and topology, in Russian. Univ. Publ. House, Moscow 1980

[68] Momen, S.: Structural controllability and stability of decentralised systems. Ph.D.thesis, Queen Mary College, University of London 1982

[69] Moore, B.C.: On the flexibility offered by state feedback in multivariable systems beyond closed-loop eigenvalue assignment. IEEE Trans.AC-21(1976), pp.689-692

[70] Morgan, Jr., B.S.: The synthesis of linear multivariable systems by state variable feedback. IEEE Trans. AC-9 (1964), pp.405-411

[71] Morse, A.S.: Structural invariants of linear multivariable systems. SIAM J. Control 11 (1973), pp.446-465

[72] Morse, A.S. and W.M. Wonham: Status of noninteracting control. IEEE Trans. AC-16 (1971), pp.568-580

[73] Munro, N. and A. Vardulakis: Pole-shifting using output feedback. Int. J. Control 18 (1973), pp.1267-1273

[74] Nakamizo, T. and N. Kobayashi: On decoupling of a linear multivariable system. Trans.Soc.Intrum.&Control Engrs. 16 (1980), pp. 615-622

[75] Newcomb, R.W.: The semistate description of nonlinear and timevariable circuits. IEEE Trans. CAS-28(1981),pp-62-71

[76] Nijmeijer, H. and J.M. Schumacher: Zeros at infinity for affine nonlinear control systems. Memorandum No.441, Twente University of Technology (1983). IEEE Trans. AC-30 (1985), pp.566-573

[77] Ortega, J.M. and W.C. Rheinboldt: Iterative solution of nonlinear equations in several variables. Academic Press, New York 1970

[78] Patel, R.V.: On output feedback pole assignability. Int. J. Control 20 (1974), pp.955-960

[79] Patel, R.V.: On zeros of multivariable systems. Int. J. Control 21 (1975), pp. 599-608

[80] Patel, R.V.: On blocking zeros in linear multivariable systems. IEEE Trans. AC-31 (1986), pp.239-241

[81] Petzold, L.R. and C.W. Gear: ODE methods for the solution of differential/algebraic systems. Sandia Report SAND 82-8051, 1982

[82] Pugh, A.C. and V. Krishnaswamy: Algebraic and dynamic characterization of poles and zeros at infinity. Int. J. Control 42 (1985), pp. 1145-1153

[83] Pugh, A.C. and P.A. Ratcliffe: On the zeros and poles of a rational matrix. Int. J. Control 30 (1979), pp.213-226

[84] Raske, F.: Ein Beitrag zur Dekomposition von linearen zeitinvarianten Groß-Systemen. Dr.-Ing. Diss., Universität Hannover 1981

[85] Reinisch, K.: Analyse und Synthese kontinuierlicher Steuerungssysteme. VEB Verlag Technik, Berlin 1979

[86] Reinschke, K.: Struktureller Zugang zum Reglerentwurf durch Polvorgabe. Z.msr, Berlin 26 (1983), pp.313-318

[87] Reinschke, K.: Graph-theoretic characterization of fixed modes in centralized and decentralized control. Int. J. Control 39 (1984), pp. 715-729

[88] Reinschke, K., H.-W. Röder and G.-S. Rösel: Strukturmodell für komplexe Automatisierungsanlagen und seine Anwendung in der Kraftwerksautomatisierung. Z.msr, Berlin 27 (1984), pp. 29-33

[89] Reinschke, K. and P. Schwarz: Verfahren zur rechnergestützten Analyse linearer Netzwerke. Akademie-Verlag, Berlin 1976

[90] Rekasius, Z.V.: Decoupling of multivariable systems by means of state variable feedback. Proc. 3rd Allerton Conference Circuit & System Theory, Univ.Illinois, 1965,pp.439-447

[91] Roppenecker, G.: Polvorgabe durch Zustandsrückführung. Regelungstechnik 29 (1981), pp.228-233

[92] Roppenecker, G.: Vollständige modale Synthese linearer Systeme und ihre Anwendung zum Entwurf strukturbeschränkter Zustandsrückführungen. VDI-Fortschrittsberichte, Reihe 8, Nr.59, VDI-Verlag, Düsseldorf 1983

[93] Rosenbrock, H.H.: State space and multivariable theory. William Clowes & Sons, London 1970

[94] Rosenbrock, H.H.: Computer-aided control system design. Academic Press, London 1974

[95] Rosenbrock, H.H.: Structural properties of linear dynamical systems. Int. J. Control 20 (1974), pp.191-202

[96] Sard, A.: The measure of the critical values of differentiable maps. Bull.Amer.Math.Soc. 48 (1942), pp.883-890

[97] Schwarz, H.: Mehrfachregelungen. Grundlagen einer Systemtheorie. Band 1. Springer-Verlag, Berlin 1967

[98] Sevaston, G.E.: A new approach to the problem of pole assignment through output feedback. Proc.Amer.Control Conf., San Diego,CA, 1984, pp.378-380

[99] Sezer, M.E. and D.D. Siljak: Structurally fixed modes. Systems & Control Letters 1 (1981), pp.60-64

[100] Shields, R.W. and J.B. Pearson: Structural controllability of multi-input linear systems. IEEE Trans. AC-21 (1976), pp.203-212

[101] Siljak, D.D., V. Pichai and M.E. Sezer: Graph-theoretic analysis of dynamic systems. Report DE-AC 037 ET 29138-35, University of Santa Clara, Calif., 1982

[102] Siljak, D.D.: Large-scale dynamic systems: stability and structure. North-Holland, New York 1978

[103] Silverman, L.M.: Inversion of multivariable linear systems. IEEE Trans. AC-14 (1969), pp.270-276

[104] Silverman, L.M., H.J. Payne: Input-output structure of linear systems with application to the decoupling problem. SIAM J. Control 9 (1971), pp.199-233

[105] Söte, W.: Eine grafische Methode zur Ermittlung der Nullstellen in Mehrgrößensystemen. Regelungstechnik 28 (1980), pp.346-348

[106] Suda, N. and E. Mutsuyoshi: Invariant zeros and input-output structure of linear, time-invariant systems. Int. J. Control 28 (1978), pp.525-535

[107] Suda, N. and K. Umahashi: Decoupling of nonsquare systems. A necessary and sufficient condition in terms of infinite zeros. Proc. 9th IFAC World Congress, vol.8, pp. 88-93, Budapest 1984

[108] Svaricek, F.: Nullstellen im Unendlichen von linearen Mehrgrößensystemen. Forschungsbericht Nr.6/85, MSR-Technik, Universität - GH - Duisburg, 1985

[109] Tchoń, K.: On stable and typical properties of control systems. Monograph Nr.12, Technical University of Wroclaw, 1986

[110] Tsien, H.S.: Engineering Cybernetics. McGraw-Hill, New York 1954

[111] Ulm, M. and H.D. Wend: A graph-theoretic method for the characterization of decentrally stabilizable control systems. Proc.2nd Int.Symp."Systems Anal. & Simul.", Berlin 1985, pp. 272-275

[112] Ulm, M. and H.D. Wend: Parameter independent evaluation of decentralized stabilizability. Proc.IFAC Symp. "Large-scale systems", Zürich 1986

[113] Van der Weiden, A.J.J.: The use of structural properties in linear multivariable control system design. Ph.D.thesis, Technical University Delft, 1983

[114] Van Dooren, P., G.C. Verghese and T. Kailath: Properties of a system matrix of a generalized state-space system. Int. J. Control 30 (1979), pp.235-243

[115] Vardulakis, A.I.G.: On infinite zeros. Int. J. Control 32 (1980), pp.849-866

[116] Vardulakis, A.I.G. and N. Karcanias: Relations between strict equivalence invariants and structure at infinity of matrix pencils. IEEE Trans. AC-28 (1983), pp.514-518

[117] Verghese, G.C.: Infinite frquency behaviour in generalized dynamical systems. Ph.D.thesis, Dept.Electr.Eng., Stanford University, 1978

[118] Verghese, G.C., B. Levy and T. Kailath: A generalized state-space for singular systems. IEEE Trans. AC-26 (1981), pp.811-831

[119] Wang, S.H. and E.J. Davison: A note on the simple modes of multivariable systems. IEEE Trans. AC-17 (1972), p.548

[120] Willems, J.C. and W.H. Hesselink: General properties of the pole placement problem. Proc. 7th IFAC World Congress, vol. 3, pp.1725-1729, Helsinki 1978

[121] Wolovich, W.A.: Multivariable system zeros. Pp.226-236 in F. Fallside (ed.): Control system design by pole-zero assignment. Academic Press, London 1977

[122] Wonham, W.M.: On pole assignment in multi-input controllable linear systems. IEEE Trans.AC-12(1967), pp.660-665

[123] Wonham, W.M.: Linear multivariable control. A geometric approach. Springer-Verlag, Berlin 1974

[124] Wonham, W.M. and A.S. Morse: Decoupling and pole assignment in linear multivariable systems: a geometric approach. SIAM J. Control 8 (1970), pp.1-18

[125] Anderson, B.D.O. and D.J. Clements: Algebraic characterization of fixed modes in decentralized control. Automatica 17 (1981), pp.703-712

[126] Bunch, J.R. and D.J. Rose: Sparse matrix computations. Academic Press Inc., New York 1976

[127] Lin, C.T.: System structure and minimal structure controllability. IEEE Trans. AC-22 (1977), pp.855-862

[128] Lunze, J. and K. Reinschke: Analyse unvollständig bekannter Regelungssysteme. ZKI-Informationen Nr. 2/81 and 3/81 (214 p.) Akademie der Wissenschaften der DDR, Berlin 1981

[129] Pichai, V., M.E. Sezer and D.D. Siljak: A graph-theoretic algorithm for hierarchical decomposition of dynamic systems with applications to estimation and control. IEEE Trans. SMC-13 (1983), pp.197-207

[130] Vidyasagar, M.: Input-output analysis of large-scale interconnected systems. Springer-Verlag, New York 1981

[131] Wang, S.H. and E.J. Davison: On the stabilization of decentralized control systems. IEEE Trans. AC-18 (1973), pp.473-478

[132] Wassel, M. and D. Wend: On the implementation of hierarchical optimization algorithms. Pp. 96-109 in M.G. Singh and A. Titli (eds.): Handbook of large scale systems. North-Holland Publ.Comp., Amsterdam 1979

[133] Yip, E.L. and R.F. Sincovec: Solvability, controllability, and observability of continuous descriptor systems. IEEE Trans. AC-26 (1981), pp.702-707

Subject Index

Band 1

Inverse and Improperly Posed Problems in Differential Equations

Proceedings of the Conference on Mathematical and Numerical Methods, held in Halle/Saale (GDR) from May 29 to June 2, 1979

Edited by Gottfried Anger

296 Seiten, 1979 (vergriffen)

Band 2

Fundamentals of Computation Theory FCT '79

Proceedings of the Conference on Algebraic, Arithmetic and Categorial Methods in Computation Theory, held in Berlin/Wendisch-Rietz (GDR) September 17–21, 1979

Edited by Lothar Budach

576 Seiten, 1979

Band 3

Decidability and Generalized Quantifiers

By Andreas Baudisch, Detlef Seese, Hans-Peter Tuschik, and Martin Weese

XII, 235 Seiten, 1980

Band 4

Zur Elastizitäts- und Thermoelastizitätstheorie

Potentialtheoretische Methoden in Anwendung auf verschiedene Modelle

Von Lothar Jentsch und Johannes Maul

157 Seiten, 1980

Band 5

Systems Analysis and Simulation 1980

Proceedings of the International Symposium held at Berlin (GDR) September 1–5, 1980

Edited by Achim Sydow

496 Seiten, 1980 (vergriffen)

Band 6

Nichtlineare diskrete Optimierung

von Eberhard Girlich und Michail M. Kowaljow

218 Seiten, 1981

Band 7

Completeness Problems in the Structural Theory of Automata

By Jürgen Dassow

148 Seiten, 1981

Band 8

Darstellung von Lösungen linearer elliptischer Differentialgleichungen

Von Günther Wildenhain

92 Seiten, 1981

Band 9

Introduction to the Spectral Theory of Operators in Spaces with an Indefinite Metric

By I. S. Iohvidov, M. G. Krein, and H. Langer

120 Seiten, 1982 (vergriffen)

Band 10

Nonlinear Time-discrete Systems

A General Approach by Nonlinear Superposition

By Michael Gössel

115 Seiten, 1982 (vergriffen)

Band 11

Preprimal Algebras

By Klaus Denecke

162 Seiten, 1982

Band 12

Ergodic Theory and Related Topics

Proceedings of the Conference held in Vitte/Hiddensee (GDR) October 19–23, 1981

Edited by Horst Michel

232 Seiten, 1982

Band 13

Singularly Perturbed Differential Equations

By H. Goering, A. Felgenhauer, G. Lube, H.-G. Roos, and L. Tobiska

176 Seiten, 1983

Band 14

Leben wir in einer Volterra-Welt?

Ein ökologischer Zugang zur angewandten Systemanalyse

Von Manfred Peschel und Werner Mende

233 Seiten, 1983

Band 15

Markov Decision Problems with Countable State Spaces

Optimality Criteria – Algorithms – Clustering

By Hans Michael Dietz and Volker Nollau

174 Seiten, 1983

Band 16

Stochastische Verfahren zur Suche nach dem Minimum einer Funktion

Von Ryszard Zieliński und Peter Neumann

133 Seiten, 1983

Band 17

Invertibility and Asymptotics of Toeplitz Matrices

By Albrecht Böttcher and Bernd Silbermann

200 Seiten, 1983

Band 18

Structural Induction on Partial Algebras

Introduction to Theory and Application of Partial Algebras – Part II

By Horst Reichel

205 pages, 1984

Band 19

Algebraic Methods for Toeplitz-like Matrices and Operators

By Georg Heinig and Karla Rost

212 pages, 1984

Band 20

Frege Conference 1984

Proceedings of the International Conference held at Schwerin (GDR), September 10–14, 1984
Edited by Gerd Wechsung
408 pages, 1984

Band 21

Numerical Analysis of Selected Semilinear Differential Equations

Edited by Christian Großmann, Thomas Riedrich, and Hartmut Schönheinz
203 pages, 1984

Band 22

Singular Ordinary Differential Operators and Pseudodifferential Equations

By Johannes Elschner
200 pages, 1985

Band 23

Lotka-Volterra-Approach to Cooperation and Competition in Dynamic Systems

Proceedings of the 5th Meeting of UNESCO's Working Group on Systems Theory held on the Wartburg, Eisenach (GDR) at March 5–9, 1984
Edited by Werner Ebeling and Manfred Peschel
271 pages, 1985

Band 24

Convergence Structures 1984

Proceedings of the Conference on Convergence held in Bechyně (Czechoslovakia), September 24–28, 1984
Edited by Josef Novák, Werner Gähler, Horst Herrlich, and Jan Mikusiński
254 pages, 1985

Band 25

Parcella '84

Proceedings of the II. International Workshop on Parallel Processing by Cellular Automata and Arrays held in Berlin (GDR), September 25–27, 1984
Edited by Wolfgang Händler, Tamás Legendi, and Gottfried Wolf
239 pages, 1985

Band 26

Multiobjective and Stochastic Optimization Based on Parametric Optimization

By Jürgen Guddat, Francisco Guerra Vasquez, Klaus Tammer, and Klaus Wendler
175 pages, 1985

Band 27/28

Systems Analysis and Simulation 1985
I: Theory and Foundations
II: Applications

Proceedings of the International Symposium held in Berlin (GDR), August 26–31, 1985
Edited by Achim Sydow, Manfred Thoma, and Robert Vichnevetsky
per volume 495 pages, 1985

Band 29

Parcella '86

Proceedings of the III. International Workshop on Parallel Processing by Cellular Automata and Arrays held in Berlin (GDR), September 9–11, 1986
Edited by Tamás Legendi, Dennis Parkinson, Roland Vollmar, and Gottfried Wolf
272 pages, 1986

Band 30

Fuzzy Sets Applications, Methodological Approaches, and Results

Proceedings of the International Workshop on Fuzzy Sets Applications cosponsored by IIASA and convened in cooperation with IAEA held on the Wartburg, Eisenach (GDR), at March 3–8, 1985
Edited by Steffen Bocklisch, Sergei Orlovski, Manfred Peschel, and Yasushi Nishiwaki
233 pages, 1986

Band 31

Mathematical Methods of Specification and Synthesis of Software Systems '85

Proceedings of the International Spring School, held in Wendisch-Rietz (GDR), April 22–26, 1985
Edited by Wolfgang Bibel and Klaus P. Jantke
245 pages, 1986

Band 32

A Model Theoretic Oriented Approach to Partial Algebras

Introduction to Theory and Application of Partial Algebras – Part I
By Peter Burmeister
319 pages, 1986

Band 33

Finite-difference Methods on Irregular Networks

A Generalized Approach to Elliptic Problems of Second Order
By Bernd Heinrich
206 pages, 1986

Band 34

Modern Algorithms for Large Sparse Eigenvalue Problems

By Arnd Meyer
125 pages, 1987

Band 35

Parametric Optimization and Related Topics

Edited by Jürgen Guddat, Hubertus Th. Jongen, Bernd Kummer, and František Nožička
407 pages, 1987

Band 36

Discretization in Differential Equations and Enclosures

Edited by Ernst Adams, Rainer Ansorge, Christian Großmann, and Hans-Görg Roos
250 pages, 1987

Band 37

Factorization of Measurable Matrix Functions

By Georgi Semenovich Litvinchuk and
Ilya Matveyevich Spitkovskiĭ

372 pages, 1987

Band 38

Parallel Algorithms and Architectures

Proceedings of the International Workshop
on Parallel Algorithms and Architectures
held in Suhl (GDR), May 25–30, 1987

Edited by Andreas Albrecht, Hermann Jung,
and Kurt Mehlhorn

205 pages, 1987

Band 39

Parametric Integer Optimization

By Bernd Bank and Reinhard Mandel

138 pages, 1988

Band 40

Computer Analysis of Images and Patterns 1987

Proceedings of the II. International Conference
CAIP '87 on Automatic Image Processing,
held in Wismar (GDR), September 2–4, 1987

Edited by Leonid P. Jaroslavskiĭ,
Azriel Rosenfeld, and Wolfgang Wilhelmi

282 pages, 1987

Band 41

Multivariable Control

A Graph-theoretic Approach

By Kurt Reinschke

274 pages, 1988

Band 42

Equilibrium Distributions of Branching Processes

By Alfred Liemant, Klaus Matthes, and Anton
Wakolbinger

240 pages, 1988

Band 43

Andreotti-Grauert Theory by Integral Formulas

By Gennadi Markovich Henkin and
Jürgen Leiterer

270 pages, 1988

Band 44

Continual Means and Boundary Value Problems in Function Spaces

By Efim Mikhailovich Polishchuk

160 pages, 1988

Band 45

Advances in Mathematical Optimization

Invited papers dedicated to Prof. Dr. Dr. h. c.
F. Nožička on occasion of his 70th birthday

Edited by Jürgen Guddat et al.

232 pages, 1988

This monograph reports a new approach to control theory, especially suited for multivariable systems with uncertain parameter values. Although it introduces all the needed concepts of applied graph theory (finite graphs and digraphs, paths, cycles, trees, graph-theoretic interpretation of determinants and characteristic polynomials), the emphasis is on problems of controller synthesis: appropriate state enumeration in case of large-scale systems, structural controllability and observability, pole assignment under state feedback, disturbance rejection, noninteracting control, pole assignability by static output feedback, decentralized control and output feedback under structural constraints, dynamic compensators.
The book provides both original research contributions and a comprehensive survey of the subject under consideration. Numerous examples are includeded to illustrate the material of the text.

Die vorliegende Monographie behandelt einen neuen Zugang zur Regelungstheorie, der für mehrvariable Systeme mit unbestimmten Parameterwerten besonders geeignet ist. Obwohl alle benötigten Begriffe der angewandten Graphentheorie (endliche Graphen und gerichtete Graphen, Pfade, Zyklen, Bäume, graphentheoretische Interpretation von Determinanten und charakteristischen Polynomen) eingeführt werden, liegt die Betonung auf Problemen der Reglersynthese: zweckmäßige Indizierung der Zustände bei großen Systemen, strukturelle Steuerbarkeit und Beobachtbarkeit, Polzuweisung bei Zustandsrückführung, Störgrößenbekämpfung, Entkopplung der Regelgrößen, Polzuweisbarkeit bei statischen Ausgangsrückführungen, dezentrale Regelung und Ausgangsrückführungen unter strukturellen Beschränkungen, dynamische Kompensatoren.
Das Buch liefert sowohl neue Forschungsbeiträge als auch eine zusammenfassende Übersicht über das Gesamtgebiet. Zur Illustration der Ergebnisse wurden zahlreiche Beispiele in den Text eingestreut.

ISBN 3-05-500452-3
ISSN 0138-3019

Bei Fragen zur Produktsicherheit wenden Sie sich bitte an:
If you have any questions regarding product safety,
please contact:

Walter de Gruyter GmbH
Genthiner Straße 13
10785 Berlin
productsafety@degruyterbrill.com